World Sustainability Series

Series Editor

Walter Leal Filho, European School of Sustainability Science and Research, Research and Transfer Centre "Sustainable Development and Climate Change Management", Hamburg University of Applied Sciences, Hamburg, Germany

Due to its scope and nature, sustainable development is a matter which is very interdisciplinary, and draws from knowledge and inputs from the social sciences and environmental sciences on the one hand, but also from physical sciences and arts on the other. As such, there is a perceived need to foster integrative approaches, whereby the combination of inputs from various fields may contribute to a better understanding of what sustainability is, and means to people. But despite the need for and the relevance of integrative approaches towards sustainable development, there is a paucity of literature which address matters related to sustainability in an integrated way.

Notes on the quality assurance and peer review of this publication

Prior to publication, the works published in this book are initially assessed and reviewed by an in-house editor. If suitable for publication, manuscripts are sent for further review, which includes a combined effort by the editorial board and appointed subject experts, who provide independent peer-review. The feedback obtained in this way was communicated to authors, and with manuscripts checked upon return before finally accepted. The peer-reviewed nature of the books in the "World Sustainability Series" means that contributions to them have, over many years, been officially accepted for tenure and promotion purposes.

Tanupriya Choudhury · Amit Awasthi ·
Jung-Sup UM
Editors

Open Access as a Knowledge Platform for Implementing the United Nations Sustainable Development Goals

Open Access and Sustainable Development Goals

Springer

Editors
Tanupriya Choudhury
University of Petroleum and Energy Studies (UPES)
Dehradun, Uttarakhand, India

Amit Awasthi
Department of Applied Science
University of Petroleum and Energy Studies (UPES)
Dehradun, Uttarakhand, India

Jung-Sup UM
Department of Geography
College of Social Sciences
Kyungpook National University
Daegu, Korea (Republic of)

ISSN 2199-7373 ISSN 2199-7381 (electronic)
World Sustainability Series
ISBN 978-3-031-99604-7 ISBN 978-3-031-99605-4 (eBook)
https://doi.org/10.1007/978-3-031-99605-4

This Springer imprint is published by the registered company Springer Nature Switzerland AG
The registered company address is: Gewerbestrasse 11, 6330 Cham, Switzerland

Contents

Introduction to Sustainable Development Goals: A Historical Review of Their Development

Kailash Pandey, Mamta Bansal, and Amit Awasthi

Abstract

To address the challenges like climate change, poverty, environmental degradation and inequality, etc. Sustainable Development Goals (SDGs) were introduced which serve as a universal blueprint for achieving sustainable, equitable, and inclusive development by 2030. By understanding the importance of SDG, in this introductory chapter, Journey of the SDG is discussed that start from the Stockholm Conference (1972), the Brundtland Report (1987), and the Earth Summit (1992). This review chapter examines the conceptual framing, historical evolution, and policy milestones that led to the adoption of the 2030 Agenda. Assessment on the efforts of the implementation on a global scale, persistent gaps and monitoring frameworks in achieving SDG targets has been done. It also highlights that; the importance of these SDG is not merely for academic and government understanding but must be understood by all individual and consider these as their collective not imposed responsibility. For better future and amendment of SDG, there is need for stronger global cooperation, technological innovation, and flexible governance. Overall, this chapter provides an understanding of the SDGs' journey and their vital role in shaping a strong and sustainable world.

Keywords

Introduction • Journey • Overview • Sustainable development goal • 2030 agenda

K. Pandey (✉) · A. Awasthi
Department of Physics, UPES, Dehradun, Uttarakhand, India
e-mail: kpandey@ddn.upes.ac.in

M. Bansal
School of Basic and Applied Sciences, Maharaja Agarasen University, Solan, Himachal Pradesh, India

T. Choudhury et al. (eds.), *Open Access as a Knowledge Platform for Implementing the United Nations Sustainable Development Goals*, World Sustainability Series,
https://doi.org/10.1007/978-3-031-99605-4_1

1 Introduction

Many sectors, both industrial as well as non-industrial shows significant advancement towards an impressive economic growth. Although this advancement has come with many serious results like deforestation, social inequality, overuse of natural resources, which leads to environmental degradations (Dijoo & Khurshid, 2022). Many challenges like climate change, air & water quality, disappearance of biological diversity, hunger, unequal availability of healthcare and education for everyone are the complex and substantial issues facing the world today (Awasthi et al., 2025; Fletcher et al., 2024; Sharma et al., 2024). Hence, developmental which are needs of a civilization had seriously impacted the environment in form of natural calamities. As a result, concerned societies across the globe started envisioning developmental models in a reformed way, which could be in synchronization with environment as well as the needs of future generations (Kumar et al., 2024; Pathak & Deshkar, 2023). By observing and understanding different environmental challenges and issues, there is need of some steps and regulation which must be accepted and adopted throughout the word for the safe, clean and sustainable environment (Mohanty et al., 2024). The ideas for framing policies, such that the societal development along with environmental and cultural sustainability is preserved and forms what are called as development goals. To materialize the ideas of a sustainable development plan, the main point of concern remains the adaptability of the policies to both the local and the global community. Formation of the United Nations proved to be a major milestone in this direction, as it became the major body for putting up the ideas for sustainable development along with taking care of the present needs and the future concerns (Ester, 2024).

To address the world's complex and interconnected challenges, including poverty, climate change, inequality, and environmental degradation Sustainable Development Goals (SDGs) were developed. They provide a universal framework for all countries to work toward inclusive and sustainable growth, social justice, and environmental protection. Lot of research throughout the world has been done and country wise number of publication details are shown in the Fig. 1 and Table 1.

A co-authorship network visualization using VOS viewer was performed to scrutinize patterns of research collaborations in the topic area of SDG's between 1990 and 2025 and shown in Table 1 and Fig. 1. Bibliometric information extracted showed 72 nations which divided into 6 clusters. Every node within the visualization is a country, with node size determining the number of publications and colored links showcasing the intensity of co-authorship relationships. Most prominent countries are United Kingdom, United States, China, Canada, Australia, India (Table 1).

Lot of discussion and implementation start after the development of SDG that clearly infer from the Table 1. After the development of SDG, different organizations start working on the implementation of SDG, but the COVID-19 pandemic disturbed the progress toward achieving the SDGs significantly, increasing the existing inequalities and reducing improvements in health, education, and poverty reduction (Ameli et al., 2023). Lot of studies has been done to study the effect

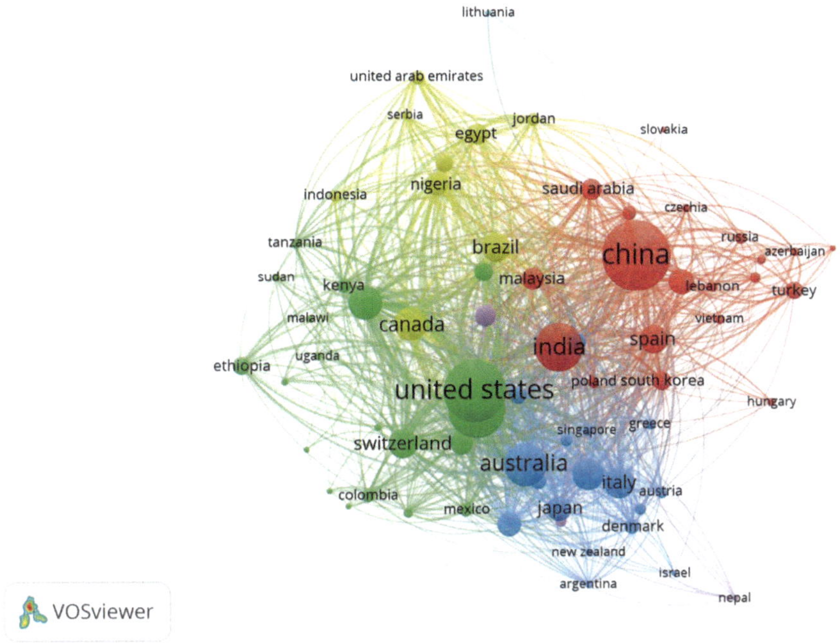

Fig. 1 Country-level co-authorship network on SDG's (1990–2025)

of COVID19 on the different areas Shukla and Choudhury (2022), Awasthi et al., (2021). Studies highlighted the urgent need for resilient healthcare systems, inclusive social protection, and sustainable development, reinforcing the importance of the SDGs as a guiding framework for recovery and long-term global resilience (Ameli et al., 2023). The pandemic gave a deep shock to the peoples health in the whole world and decreased the SDGs goals, the pandemic also includes climate-related disasters, widening inequalities, and going geopolitical conflicts have added layers of the complexity lot of research has been done during the period of COVID19 (Ameli et al., 2023).

In this chapter, attempt has been made to offers a comprehensive overview and journey of the development of the SDGs that begins by introducing the concept of the SDGs in the context of global sustainability. The chapter discusses the major historical global events start from the Stockholm Conference (1972), the Brundtland Report (1987), and the Earth Summit (1992), leading up to the Millennium Development Goals (2000–2015). Discussion on the arrival of the SDGs across critical milestones such as the Rio + 20 Conference (2012), global consultations, and the adoption of the 2030 Agenda in 2015 is also presented. Further analyses on the structure, implementation and ongoing progress of the SDGs along with the key challenges are discussed. Finally, future perspectives and the collective global responsibility toward achieving a sustainable and equitable world beyond 2030 are presented.

Table 1 Top 20 most prominent countries in a research co-authorship network of SDG

Rank	Country	Total link strength	Citations	Documents
1	United Kingdom	903	6068	352
2	United States	894	5210	409
3	China	647	7020	481
4	Canada	505	2438	153
5	Australia	455	2419	218
6	India	403	3668	264
7	South Africa	402	1381	149
8	Germany	382	2952	135
9	Switzerland	365	2752	115
10	Brazil	317	908	117
11	Nigeria	307	707	77
12	Italy	283	2882	134
13	Netherlands	277	1515	86
14	Saudi Arabia	221	656	71
15	Egypt	219	942	68
16	Norway	218	668	51
17	Spain	211	1538	112
18	Pakistan	202	1399	94
19	Sweden	194	1028	74
20	Portugal	181	541	55

2 Foundation of or Idea Behind the SDG

Development and idea of SDGs did not arise overnight. Their development was shaped in over four decades of international cooperation, evolving ideas, and growing awareness of global challenges (Silva, 2021). Originated from earlier global efforts such as the Stockholm Conference of 1972, the Brundtland Commission's 1987 report, and MDGs adopted in 2000, the SDGs signify a transformative plan that builds on historical initiatives while addressing the multidimensional and interconnected nature of today's global challenges. These timeframe for conceptualization of the ideas of SDGs can be categorised as detailed in the next sections.

2.1 The Stockholm Conference (1972)

The concept of sustainable development had been proposed since ancient times by different civilizations throughout the globe. Since the formation of United nations in 1945 after world war-II, there has been documented progress in the evolution of development goals. It was in year 1972, at Stockholm when the UN held a Human Environment Conference, which proved to be a seminal moment, and it introduced environmental concerns to the global community. It laid the groundwork for integrating ecological issues with human welfare. The conference was the first gathering of representatives of about 113 states. The conference concluded with the formation of United Nations Environment Programme (UNEP). So, the world now has a dedicated organization for issues of environmental concerns. The conference although being first ever documented record of the human concern for environment proved to be a milestone in framing directions for the future (Jacob, 2024).

The conference provided a set of around 109 recommendations for a sustainable growth, which is well coordinated with social, cultural and environmental factors. The conference resulted in adoption of major resolutions pertaining to sustainable environment growth. It also recommended to celebrate world environment day on 5th of June, so that activities related to environment becomes a regular practice, throughout the globe (Jacob, 2024). Hence, conference which was held on the SDGs in 1972 at Stockholm considered as the roots of the SDGs discovered on Human Environment, which is the first-time concern on environment on the global agenda. That conference sent a big message to the world that economic growth and environmental protection must go hand in hand. One of the major outcomes of the conference was the creation of the United Nations Environment Programme (UNEP), which shaped international awareness with a strong environmental focus.

2.2 The Brundtland Report (1987)

The formation of World Commission on Environment and Development (WCED), in year 1983, laid down the guiding principles for sustainable development, in the form of what is now known as the Brundtland report, after the name of its chairwomen Gro Harlem Brundtland who presented it in year 1987 under the title "**Our Common Future**". The term "sustainable development" got introduced, highlighting the need to balance economic growth, environment conservation, and social equity. The report emphasised on exploring the possibilities for new era of economic growth, through the policies which considers the expansion of environmental resources in a sustained manner. The report mainly focused on three key objectives viz common concerns, challenges and endeavours. The common concerns which got addressed were, the future threat, sustainable development and global economy. The definition of sustainability was coined out as "***Development that meets the needs of the present without compromising the ability of future generations to meet their own needs.***" Hence, the year 1987 marked a major turning point with the release of the Brundtland Commission's report that defined sustainable development with the consideration of both present and future scenariao. The report displayed the linkage of social, environmental, and economic dimensions, shaping the principles behind the SDGs (Hajian & Kashani, 2021).

2.3 Earth Summit and Agenda 21 in the Year 1992

A step towards sustainable development was in the form of a comprehensive non-binding action plans at the 1992 United Nations Conference on Environment and Development (UNCED) at the capital city of Brazil -Rio de Janeiro which produced "Agenda 21". It stressed role of global partnerships in addressing issues such as deforestation, biodiversity loss, and poverty (Yang & Xiu, 2023). The landmark outcomes can be expressed as:

- Declaration on Environment and Development.
- Agenda 21.
- Forest Principles.
- Legally binding convections on biodiversity and the UN framework convection on climatic change.

Agenda 21 refers to the futuristic action plans for the twenty-first century, with main goal to promote sustainable development, well aligned with environmental preservation, economic growth and social equity at global, national and local levels. It is primarily divided into four main parts with 40 chapters. The first section pertains to social and economic dimensions and focuses on poverty handling, changing consumption pattern, promoting good health, sustaining population growth and synchronizing development with environment in any policy decision making. Second section addresses the preservation and sensible utilization of resources used for development. It focuses on issues of atmospheric protection, deforestation, conservation of biodiversity, pollution control and maintenance of water, energy and land resources. The third section deals with strengthening the roles of major stakeholders and involvement of local groups in implementing sustainable development. The fourth section describes the methods of implementation through technology transfer, education, developing institutions and financial mechanisms necessary to achieve plan's objective (Mhatre & Shukla, 2024). After Brundtland Commission's report, the momentum continued to build at the 1992 Earth Summit in Rio de Janeiro, wherein an Agenda 21 was introduced. This summit expanded global discussions which include poverty, public participation, and equity, and introduced key idea of "common but differentiated responsibilities," which continues to guide environmental negotiations till today (Bryner, 2023).

2.4 Millennium Development Goals (MDGs) from 2000 to 2015

An international information campaign was started in 1998 for 2 years, which preceded with the Millenium summit in year 2000. The summit was a three-day event wherein world leaders across the globe assembled in New York city to formulate an innovative developmental strategy which was adopted in 2000, in the form of the MDGs (UN reports). These focused on eradicating extreme poverty, improving health and education, and ensuring gender equality. Though limited in scope and geography, the MDGs achieved notable successes and highlighted the importance of measurable targets in development agendas.

This was the largest gathering of heads of the states and government, of all times which concluded with the adoption of millennium declaration by 189 member states (De Jong & Vijge, 2021). The summit aimed to address the utmost challenges of development by 2015 through the adoption of eight MDG which is summarised in the Table 2.

These MDGs although regarded as landmark in international development for mobilizing global efforts and resources to eradicate poverty and improving wellbeing, they lacked in certain aspects such as, broader scope and inclusivity, universal applicability, socioeconomic and environment integration etc. To account for the limitations of the MDG's, the United Nations adopted a set of 17 global goals, to address global challenges and promote sustainable development by 2030.

Table 2 Eight millenium declarations adopted in 2000

S.N.	Development goal	Focus point
1	Extreme poverty elimination and hunger management	Generating innovative employment opportunities and decent work culture for all
2	Achieving Universal primary education	Ensuring primary education to all
3	Gender equality and women empowerment	Eliminating gender disparity in primary and secondary education
4	Minimizing child mortality rate	Minimizing death rate among children below five years of age by two third
5	Improved maternal health	Decreasing the mother mortality by three quarter and achieving universal access to reproduction health
6	Combating HIV/AIDS, malaria and other types of lethal diseases	Attaining negative growth rate of spread of HIV/AIDS, malaria and other diseases
7	Environment sustainability	Merging sustainable development with state policy for reducing biodiversity loss, safe drinkable water to all and enhancing life standard of slum dwellers
8	Developing international partnership	Addressing needs of least developed countries by developing open trading and finance systems, so as to provide employment and access to essential medicines and technologies

3 Birth of the SDGs

The MDGs mainly focused on reducing poverty, improving education and health, and promoting gender equality, particularly in the developing countries. The MDGs made progress in many areas, they were also targeted for being too narrow in scope, lacking environmental depth, and excluding developed countries from shared responsibility. These gaps highlighted that the MDGs need for a more expansive and globally interactive framework. MDGs were adopted for a time frame of 15 years from 2000 till 2015, and to build up the progress of the MDG's and address the limitations along with providing a roadmap for sustainable future by 2030, the sustainable development goals got accepted by the United Nations member states in 2015 (De Jong & Vijge, 2021). The development of these SDG's involves decades of dedicated efforts starting from the 1992 earth summit when agenda 21 was adopted by all the member and the United Nations, which also comprised, the department of Economic and Social Welfare. The SDG's became the successors of the MDG's, soon after the necessity of the MDG's to transform from specific to more global participation.

3.1 Rio + 20 and "The Future We Want"

Idea for the sustainable development goals was first proposed by member state Columbia in year 2011, at a preparatory event for the Rio + 20 summit to be held next year. It gained major thrust at the 64th UN Department of Public Information NGO conferences in Bonn, Germany. The birth year of the SDG's which are also known as the global goals is 2012, at the Rio + 20 Conference, wherein all the UN member states agreed to develop a new set of goals to replace the MDGs as agenda point for 2030 sustainable development. General Assembly proposed the vision document, "The Future We Want," by recalling its resolution 64/236 of 24th December 2009 and the resolution 66/197 of 22nd December 2011 (Guerra et al., 2022). The outcome document called for an inclusive and participatory process in shaping these goals. The member state agreed upon to initiate the development of SDGs which could serve as the advancement of the MDG's and establish an advent level of political forum for sustainable development. Outcome of the conference also suggested methods for implementation of sustainable mandates for future strategies on financial development.

The common vision encompasses the following main points.

- Ensuring economically, socially and environmentally sustainable future for earth and its present and future generations.
- Eliminating poverty and hunger at utmost urgency from the human society.
- Mainstreaming sustainable development at all levels by integrating the environmental, social and economic aspects.

- Promoting sustained, inclusive and equitable economic growth, creating greater opportunities for all, reducing inequalities, improving basic living standards and managing natural resources as per the ecosystem conservation requirement.
- Providing impulsive growth for achievement of internationally agreed development goals.
- Ensuring freedom, mental peace, social security and respect.
- Promotion of effective, transparent, accountable and democratic institutions worldwide.
- Strengthening international cooperation to address persistent challenges related to sustainable development for all.
- Addressing the theme for green economy in context to sustainable development and poverty eradication.
- Providing equal opportunities to people from all sections of society in decision making and forming broad alliances of people, governments, civil society and private sector.
- Renewing political commitments.

3.2 Global Consultations and Open Working Group (2013–2014)

The Rio + 20 Conference in 2012 launched discussions on new set of the universal goals. Unlike the MDGs, these goals would be shaped through global consultations which involves not just governments but also civil society, academia and business. The open working group held hundreds of the discussions and public statements to make sure that the arising SDGs reflected the voices, challenges in between 2013 and 2014. The formal beginning for the development of the SDG's was with the formation of a 30 member open working group by the General Assembly in year 2013 (Griggs et al., 2013). The Rio + 20 outcome document set the background for mandating the establishment of open working group for conceptualising and framing the sustainable development goals to be considered and required action to be taken at the 68th session. The open working group had completed eight sessions by early 2014 which were described as the stocktaking phase. It involved broad consultations with the member states, observers, UN agencies and major agencies by addressing thematic issues such as oceans, forests, biodiversity, equality, gender, peace and governance. Mid 2014 was the decision-making phase in which the OWG reviewed the inputs from stakeholders including civil societies, development experts and the UN bodies, to ensure transparency in the process.

Mediation process for the development agenda post 2015, started on January 2015 by the General Assembly and it culminated with the adoption of agenda 2030 pertaining to Sustainable development, and encompassing 17 SDGs as its focus point, at the UN Summit of 2015 (Guerra et al., 2022). Through an unprecedented global consultation involving governments, civil society, and experts, the Open Working Group (OWG) proposed 17 goals and 169 targets, emphasizing universality, integration, and inclusiveness.

3.3 Adoption of the 2030 Agenda in the Year of 2015

The SDGs were formally adopted on September 25, 2015, by 193 UN member states as part of the 2030 Agenda for Sustainable Development. The member states unanimously agreed to the resolution entitled "Transforming our world" the 2030 agenda for sustainable development (A/RES//70/1). Unlike the MDGs, the SDGs apply to all countries and incorporate a broader array of interconnected challenges. The key features of the 2030 agenda were universal plan of action, the 17 SDG's, Global commitment, consultative processes and implementation (Curtó-Pagès et al., 2021).

The SDG's adopted by the UN General Assembly are set of seventeen goals covering a broad range of sustainable development issues, which includes: eradication of poverty, hunger elimination, sound health and wellbeing, inclusive and equitable learning opportunities, no discrimination on gender basis, providing clean water and hygienic sanitation for all, affordable and clean energy, respectable work environment and economic growth, building industry and foster innovation, reducing inequalities, developing sustainable cities and communities, responsible consumption and production, climatic action, water life, life on land, peace justice and strong innovation and partnership for the goals.

An annual progress of the sustainable development goals is monitored by the Secretary General and progress report is generated annually in coordination with UN and based on the feedback at the regional level and the global indicator framework. In addition to the annual report a global sustainable development report is also produced at a frequency of four years, to provide information about the quadrennial SDG review deliberations at the General Assembly.

4 Structure and Objectives of the SDGs

In September 2015 the official adoption of the 2030 agenda for sustainable development marked a crucial turning point in global cooperation. The SDGs cover a big spectrum with the 17 interconnected goals as depicted in Fig. 2, like from ending poverty and ensuring quality education for all the children in the world, clean energy, sustainable cities in the world, and strong institutions. The SDGs focused on all the nations while on the other hand the MDGs which focused on developing countries. The SDGs acknowledge that the sustainable development is a world's challenge not only for some countries it requires contributions from every country, regardless of income level (Cling & Delecourt, 2022).

The SDG are the building blocks for achieving a range of milestones pertaining to a balanced standard of living for every life form. These development goals are seventeen in number and serve as an urgent call of action for both developed and developing countries. These 17 goals pertain to specific targets which together constitutes 169 in number and these targets specifies the benchmarks and the outcomes of the development goals to be achieved by year 2030 (Moyer & Hedden, 2020;

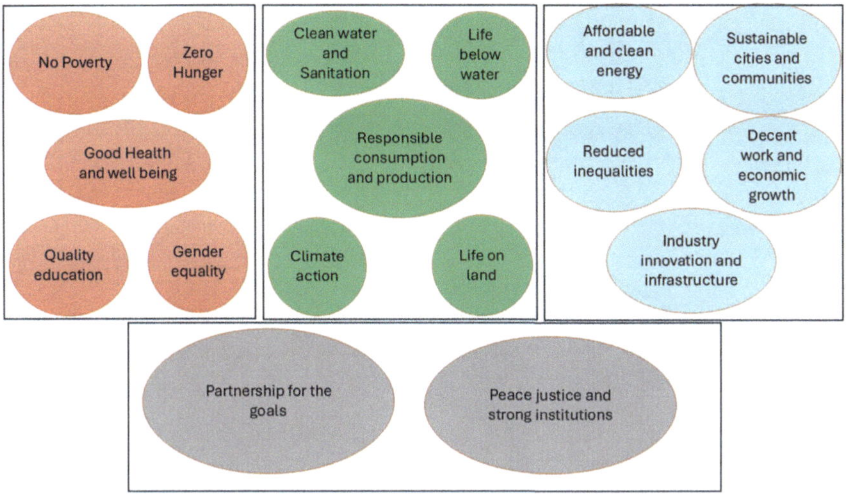

Fig. 2 The seventeen sustainable development goals

Golding et al. 2017). Each of these 169 targets is quantified in terms of the indicators to monitor the progress and ensure accountability, the SDGs are framed in such a way that they are simultaneously independent as well as integrated with the three main core elements of the sustainable development: the economic growth, the social inclusion and environment protection. The structure of these SDGs can be understood from Fig. 3, which gives a framework for development and implementations of these SDGs.

All these SDG's cover five key areas known as the "5 Ps": People, Planet, Prosperity, Peace, and Partnerships. Each goal is linked with specific targets and indicators to enable monitoring and accountability. These serve as the foundation pillars for the 2030 agenda for sustainable development and provide directions for implementation of all the 17 SDGs throughout the globe. These are elaborated as given in Table 3.

These makes up the objectives to be achieved by 2030, through dedicated implementation and periodic monitoring of the sustainable development goals.

5 Implementation and Global Progress

A well-coordinated effort at global, national and local level is the key for ensuring the proper implementation of the sustainable development goals, so that desired outcomes are attained well within the timeframe. It requires an integration of the policy with the socioeconomic, cultural and environmental aspects.

Fig. 3 The flowchart to express the structure of the SDG's

Table 3 Five key areas encompassing all the seventeen SDG's

5Ps	Associate SDGs	
People	1–5	No poverty, zero hunger, good health and well being, quality education and gender equality
Planet	6, 12–15	Clean water and sanitation, responsible consumption and production, climate action, life below water and life on land
Prosperity	7–11	Affordable and clean energy, decent work and economic growth, industry innovation and infrastructure, reduced inequalities and sustainable cities and communities
Peace	16	Peace, justice and strong institutions
Partnership	17	Partnership for the goals

5.1 National Adaptation and Localization

The progress of a nation depends on how well the policies are framed for a sustainable development. Countries worldwide have integrated the SDGs into national policies, with varying degrees of success. Localizing the goals to address context-specific challenges has proven critical. The integration of the development goals as per agenda 2030 can be achieved by aligning the SDGs with the national policy framework in a strategic way, so as to maintain a proper balance between economic, social and environmental dimensions. The strategies and the actions implemented constitutes the mechanism for the development of National Adaptation Plans (NAPs), which aims to reduce vulnerability to climate impacts and integrate climate adaptation coherently into national policies (Cling & Delecourt, 2022).

The alignment of the SDGs with regional and continental agendas ensures coherence among global, national and regional priorities.

5.2 Monitoring and Reporting

The implementation process of the SDGs needs robust monitoring and reporting mechanism, so that any rectification can be smoothly incorporated to the implemented goals. The process is complex and involves multiple levels, with range of actors, data systems and standard indicators. The prime aim of the monitoring and reporting process is to track progress, ensure accountability, guiding policy anr resource allocation to meet the global agenda 2030 (Henderson and Loreau 2023). The main elements for a well frame monitoring and reporting network involves indicators framework, levels of Monitoring and data sources and challenges. The UN, through agencies like the UNDP and DESA, provides annual progress reports. Platforms like the SDG Index, rank countries based on their performance, encouraging transparency and competition.

Main components of reporting mechanism constitute the (1) Annual SDG progress report, (2) SDG dashboards and indices, (3) Open data platforms and (4) Ground practices and peer learning.

The monitoring and reporting of SDGs is a dynamic and evolving process, that requires proper managed data systems, international cooperation and transparent reporting. All these elements of the monitoring and reporting framework can be explained from Table 4.

Table 4 The monitoring and reporting framework for SDG's

Component	Description
Indicators	98–100 indicators, plus national and thematic indicators
Data sources	Official statistics, administrative data
Main actors	National governments, UN agencies, civil societies, private actors
Reporting tools	UN SDG progress report, SDG dashboards, SDG tracker, national portals
Challenges	Data gaps, coordination, capacity

Table 5 The main challenges and the gaps for implementation of SDGs

Challenge area	Description
Slow progress	Global advancement is too slow to meet 2030 targets: setbacks from pandemic, conflict, climate
Data gaps	Missing or incomplete indicators and lack of disaggregated data for many goals and regions
Reporting burden	Expanding indicators increases workload for national statistical systems
Inequality	Persistent social, economic, and geographic disparities hinder progress
Environmental issues	Ongoing biodiversity loss and environmental degradation, especially in food and land systems
Overlooked populations	Insufficient focus on disability inclusion and intersectional vulnerabilities
Framework adaptability	Need for balance between responsiveness to new issues and stability in monitoring

5.3 Challenges and Gaps

Any time bound policy implementation is always prone to challenges and functional gaps at different levels of its execution, the same is also viable for implementation of the sustainable development goals pertaining to agenda 2030. The challenges may be global and systematic such as slow progress and setbacks, environmental and biodiversity issues, social and economic inequalities etc. Beside this there may be data and monitoring gaps in indicator framework, data availability as well as coverage and reporting methodologies. Challenges may also evolve due to emerging and overlooked issues. Persistent issues include unequal progress, lack of financing, political instability, and data gaps. The COVID-19 pandemic further disrupted progress across nearly all goals. The chronology for the conceptualization, evolution, implementation and periodic monitoring of SDGs faces major challenges from different actors, broad challenges are as described in Table 5.

6 Toward 2030 and Beyond: Future Perspectives

The 2030 deadline is approaching and within a span of 5 years there would arise a need for future planning for new set of sustainable development goals, taking account of the attained goals and unaccomplished ones. New strategies need to be developed to accelerate SDG implementation, key strategies include:

- Strengthening multilateral cooperation and global partnerships.
- Enhancing data collection and digital infrastructure.
- Mobilizing sustainable finance and private sector investment.
- Promoting education, innovation, and youth engagement.

The post-2030 development agenda will likely build on the SDGs, incorporating lessons learned and emerging global priorities, such as digital equity and resilience to global health crises etc. The future of the SDGs will be governed by the lessons learned from the challenges met in the implementation of current SDGs. The future goals also demand for extended timelines and strengthened accountability along with better alignment with international frameworks. In context of the evolving challenges posed in implementation of the SDGs innovative concepts and more international cooperation is required to achieve the desired sustainability outcomes.

7 Conclusion

Understanding the history of SDGs, their design, implementation and ongoing challenges is essential not only for governments and academics but for all individuals, hence carefuly discussed in this chapter. SDGs are far more than a normal list of targets that protect our environment, giving all people good education, protect biodiversity. It made a deep thought into key historical milestones, who's starting held with the 1972 Stockholm Conference on the Human Environment, that first time put out environmental issues on the global map. After this the momentum continued to the 1987 Brundtland Report that introduced the powerful concept of the sustainable development, and concluding with the MDGs from 2000 to 2015, which laid very crucial groundwork by addressing poverty and basic human needs.

Beauty of the SDGs is their interconnectedness; progress in one goal usually gives advancements in others. For example, improving education (Goal 4) leads to better economic opportunities (Goal 8), which in turn contributes to reducing inequalities (Goal 10). The challenge lies in ensuring that economic progress doesn't come at the cost of environmental degradation or social inequality. It's a continuous effort, requiring cooperation across nations, sectors, and communities.

Since 2015, formal adoption of the SDGs has had a strong influence on policies worldwide. Lot of research and investigation has been done throughout the world on the different SDGs. The governments took an important step by increasingly combining them into national development plans, controlling data and evidence to inform smarter decision-making. Although, despite these positive parts, obstacles like financing that is estimated in trillions annually which hamper many of the countries' ability to implement the necessary actions (Mohanty et al., 2024). Furthermore, many nations struggle with limited capacity to gather and analyses detailed data critical for tracking progress, making it difficult to target interventions effectively.

In closing, it is very much right to quote that SDGs stand as one of the most inclusive and ambitious efforts to unite the world around a common vision for development and the sustainability. The SDGs highlight that the future is not fixed, and it depends upon our today's actions. By considering different sustainable approaches there are more chances to secure a fairer, healthier world for future generations instead of leaving them with crises caused due to the ignorance. These goals are achievable only when the meaningful action will be taken now without waiting the recommendation from others and considering these goals as own collective responsibility.

References

Ameli, M., Esfandabadi, Z. S., Sadeghi, S., Ranjbari, M., & Zanetti, M. C. (2023). COVID-19 and sustainable development goals (SDGs): Scenario analysis through fuzzy cognitive map modeling. *Gondwana Research, 114*, 138–155.

Awasthi, A., Pattnayak, K. C., & Tiwari, P. R. (2025). Application of artificial intelligence-supported process-based climate models to understand the atmosphere/weather patterns and their prediction. *Frontiers in Environmental Science, 12*, 1543951.

Awasthi, A., Sharma, A., Kaur, P., Gugamsetty, B., & Kumar, A. (2021). Statistical interpretation of environmental influencing parameters on COVID-19 during the lockdown in Delhi, India. *Environment, Development and Sustainability, 23*, 8147–8160.

Bryner, G. C. (2023). Agenda 21: Myth or reality? In *The Global Environment* (pp. 157–189). Routledge.

Cling, J. P., & Delecourt, C. (2022). Interlinkages between the sustainable development goals. *World Development Perspectives, 25*, 100398.

Curtó-Pagès, F., Ortega-Rivera, E., Castellón-Durán, M., & Jané-Llopis, E. (2021). Coming in from the cold: A longitudinal analysis of SDG reporting practices by Spanish listed companies since the approval of the 2030 agenda. *Sustainability, 13*(3), 1178.

De Jong, E., & Vijge, M. J. (2021). From Millennium to sustainable development goals: Evolving discourses and their reflection in policy coherence for development. *Earth System Governance, 7*, 100087.

Dijoo, Z. K., & Khurshid, R. (2022). Environmental degradation as a multifaceted consequence of human development. In *Environmental Biotechnology* (pp. 39–56). Apple Academic Press.

Estes, R. J. (2024). United Nations development programme. In *Encyclopedia of quality of life and well-being research* (pp. 7355–7357). Springer.

Fletcher, C., Ripple, W. J., Newsome, T., Barnard, P., Beamer, K., Behl, A., Bowen, J., et al. (2024). Earth at risk: An urgent call to end the age of destruction and forge a just and sustainable future. *PNAS Nexus, 3*(4), 106.

Golding, N., et al. (2017). Mapping under-5 and neonatal mortality in Africa, 2000–2015: A baseline analysis for the sustainable development goals. *Lancet, 390*(10108), 2171–2182.

Griggs, D., Stafford-Smith, M., Gaffney, O., et al. (2013). Sustainable development goals for people and planet. *Nature, 495*, 305–307.

Guerra, J. B. S. O. A., Hoffmann, M., Bianchet, R. T., Medeiros, P., Provin, A. P., & Iunskovski, R. (2022). Sustainable development goals and ethics: Building "the future we want." *Environment, Development and Sustainability, 32*, 1–22.

Hajian, M., & Kashani, S. J. (2021). Evolution of the concept of sustainability. From Brundtland Report to sustainable development goals. In *Sustainable resource management* (pp. 1–24). Elsevier.

Henderson, K., & Loreau, M. (2023). A model of sustainable development goals: Challenges and opportunities in promoting human well-being and environmental sustainability. *Ecological Modelling, 475*, 110164. https://doi.org/10.1016/j.ecolmodel.2022.110164

Jacob, J. (2024). A background history of the sustainable development goals. *Sustainable Development., 45*, 1154.

Kumar, R., Goel, R., Singh, T., Priyadarshi, N., & Twala, B. (2024). Incentivizing green building technology: A financial perspective on sustainable development in India. *F1000Research, 13*, 924.

Mhatre, A. S., & Shukla, P. (2024). A comprehensive review of energy harvesting technologies for sustainable electric vehicles". *Environmental Science and Pollution Research, 15*, 1–24. https://doi.org/10.1007/S11356-024-34865-8/METRICS

Mohanty, R., Samal, A., & Kumar, R. (2024). Green finance supports sustainable environment: Empirical approach towards applicability in tourism sector. *Theoretical and Practical Research in Economic Fields, 15*(1), 114–126.

Moyer, J. D., & Hedden, S. (2020). Are we on the right path to achieve the sustainable development goals? *World Development, 127*, 104749.

Pathak, V., & Deshkar, S. (2023). Transitions towards sustainable and resilient rural areas in revitalising India: A framework for localising SDGs at gram panchayat level. *Sustainability, 15*(9), 7536.

Sharma, M., Rawat, S., Kumar, D., Awasthi, A., Sarkar, A., Sidola, A., et al. (2024). The state of the Yamuna River: A detailed review of water quality assessment across the entire course in India. *Applied Water Science, 14*(8), 175.

Shukla, P., & Choudhary, S. (2022). A critical review of the impact of COVID-19 on plastic and food waste. *Natural Environmental Pollution Technology, 21*, 2283–2290.

Silva, A. L. (2021). Innovation in development cooperation: Emerging trajectories and implications for inclusive sustainable development in the 21st century. *Innovation and Development, 11*(1), 151–171.

Yang, C., & Xiu, Q. (2023). A bibliometric review of education for sustainable development, 1992–2022. *Sustainability, 15*(14), 10823.

The Urgent Need for Sustainable Development Goals in the Present Era

Subhranil Das, Rashmi Kumari, Raghwendra Kishore Singh, and Shantanu Pundhir

Abstract

The Sustainable Development Goals (SDGs) integrate different dimensions while evaluate the economic development within a nation. The objective is to examine the different patterns of variation and the relationships among various SDG indicators which offer a comprehensive interpretation across different countries. In this analysis, 40 SDG indicators have been reported by 125 countries in the 2021 Sustainable Development Report which are analyzed in order to explore these relationships. For this technique, identifying the significant connections between the indicators and the countries enable a deeper understanding of global sustainability. The simulation results indicate that countries have been in the northern regions exhibit sustainability features which would support the economic growth. However, substantial efforts are required for safeguarding the environment and encouraging the production and consumption practices. In other words, nations in southern regions where the limited purchasing power must focus on developing targeted policies and strategies which are aimed at enhancing the health and well-being of their populations that ensure the access to essential household services such as water, sanitation, electricity, and internet, and promoting social equality. By addressing these challenges, a balanced sustainable development goals has been considered as one of the important criteria.

S. Das
School of Computer Science, UPES, Dehradun, Uttarakhand, India

R. Kumari (✉) · S. Pundhir
School of Computer Science Engineering and Technology, Bennett University, Greater Noida, Uttar Pradesh, India
e-mail: Rashmi.kumari@bennett.edu.in

R. K. Singh
School of Computer Science and Artificial Intelligence, SR University, Warangal, India

T. Choudhury et al. (eds.), *Open Access as a Knowledge Platform for Implementing the United Nations Sustainable Development Goals*, World Sustainability Series,
https://doi.org/10.1007/978-3-031-99605-4_2

Keywords

Fault detection • Grey box testing • Scrutiny • Optimization

1 Introduction

The Sustainable Development Goals (SDGs) have been identified as one of the 17 global objectives which are established by the United Nations in 2015 for addressing the wide range of critical challenges facing the world which include poverty, inequality, climate change, environmental degradation, peace, and justice (Carlsen & Bruggemann, 2022; Georgeson & Maslin, 2018). With the aim of designing in finding the ultimate aim in order to achieve a better sustainable future, SDGs address global issues by considering the international action. By 2030, these goals aim to increase the overall development, environmental protection, and social equity. Goal 1 has been focused on removing poverty by giving universal access to essential resources, while Goal 2 eliminate the hunger which would help to promote sustainable agriculture for enhancing food security. Goal 3 has been targeted through the improved health outcomes by reducing child mortality and combating epidemics which ensure access to universal healthcare. Goal 4 has been emphasized on the provision of quality education and learning opportunities, whereas Goal 5 advocates for gender equality by empowering women, combating gender-based violence, and promoting equal rights and opportunities. In addition to that, Goal 6 has ensured universal access to clean water and sanitation with aim to protect the water resources from pollution, while Goal 7 has been applied as an affordable and clean energy source by promoting renewable sources and improving energy efficiency. Goal 8 has been considered as an increase in the economic growth by creating decent jobs which ensure fair wages and supporting sustainable economic innovation. In addition to that, Goal 9 promotes the development of infrastructure, sustainable industrialization, and innovation which has been aligned with the environmental responsibility. For reducing the inequalities within the countries, Goal 10 supports equal opportunities for policy support. Goal 11 supports building sustainable cities through improved urban infrastructure, pollution reduction, and affordable housing. Goal 12 is responsible for consumption and production by reducing waste which encourages recycling which promotes sustainable resource use. Here, Goal 13 addresses climate action by reducing greenhouse gas emissions and enhancing the climate impacts. Goal 14 has been focused on conserving marine resources by considering ocean pollution and protecting marine ecosystems, while Goal 15 has been emphasized on the terrestrial ecosystems which include biodiversity, forests, and other natural habitats. Goal 16 seeks to reduce violence, corruption, and human rights abuses by promoting peace, justice, and strong institutions. Goal 17 has been the main criteria for the importance of increasing the global partnerships for mobilizing resources by enhancing

Table 1 Overview of SDGs and their focus areas of 2030 agenda

Goal number	Goal title	Focus area description
1	No poverty	Eradicate poverty in all forms, with a focus on social protection and basic needs such as healthcare and education
2	Zero hunger	Achieve food security, improve nutrition
3	Good health	Ensuring healthy lives for all ages
4	Quality education	Emphasizing quality education which is learning opportunities for everyone
5	Gender equality	Achieving the equality among the genders

international cooperation. By combining all these goals, these interconnected goals have been provided with a comprehensive framework for guiding the global efforts towards building a more prosperous future. Table 1 describes some of the key goals outlined in the 2030 Agenda.

For long-term sustainability, the 17 SDGs have been subdivided into smaller and actionable fragments as one of the parts of 2030 Agenda for Sustainable Development. Progress in one area have been designed in such a way that to support and enhance advancements in others by considering the interconnected nature. The SDGs have been incorporated into both international and domestic policy frameworks to strengthen coordination, monitoring, and evaluation efforts related to sustainable development. In addition to that, they provide a global framework that fosters inclusion and promotes collaboration among nations, institutions, and stakeholders (Le Blanc, 2015). Importantly, the SDGs offer flexibility, allowing individual countries to adapt the global targets to their specific economic, cultural, and demographic circumstances by developing localized strategies within the larger 2030 Agenda (Fukuda-Parr, 2016). Each goal encompasses multiple dimensions, addressing a wide range of policy objectives that collectively shape the sustainability discourse. Monitoring efforts have revealed significant progress in many industrialized nations, largely supported by stronger economies, better infrastructure, and effective governance systems (Sachs et al., 2022). The advancements made by wealthier nations are often attributed to their higher financial capacities and their governments' enhanced ability to assess, implement, and monitor policy measures effectively. However, despite these achievements, the journey towards fulfilling the SDGs in developing countries has been impeded by systemic challenges, including limited economic resources and persistent issues such as corruption (Gupta & Vegelin, 2016).

The onset of the COVID-19 pandemic further complicated global efforts toward achieving the SDGs. As countries redirected resources and attention toward immediate public health concerns, early studies indicated that low-income regions, particularly in Africa, Asia, and Latin America, suffered disproportionately, reversing recent gains in poverty reduction and public health outcomes (Sumner et al., 2020a, 2020b). The pandemic exposed and exacerbated existing inequalities, significantly affecting access to essential services such as education, healthcare, and clean water—core areas integral to the SDG framework (Barkemeyer et al., 2020). Although preliminary assessments have been made, the full impact of the pandemic on SDG achievement is still being evaluated. According to the 2021 Sustainable Development Report, the pandemic led to a decline in the global SDG Index score, highlighting a widespread setback in progress (Sachs et al., 2022). Moreover, reporting delays have left many data indicators incomplete, suggesting that the true extent of the pandemic's effect on sustainable development may not yet be fully captured. Nevertheless, these challenges reinforce the urgent need for renewed commitment, innovative policy solutions, and strengthened international cooperation to realign global efforts with the 2030 Agenda as shown in Fig. 1.

The chapter is structured as follows: The next section outlines unprecedented global challenges such as poverty, inequality, and environmental degradation, followed by an in-depth discussion on the establishment of the United Nations 2030 Agenda and the 17 SDGs. Subsequent sections explore the interconnected nature and significance of individual SDGs, with emphasis on how they collectively address the pressing developmental needs of both developed and developing nations. A dedicated segment evaluates the impact of economic factors and corruption on the progress of specific goals in the Global South. The latter part of the paper examines strategic policy frameworks, institutional cooperation, and the role

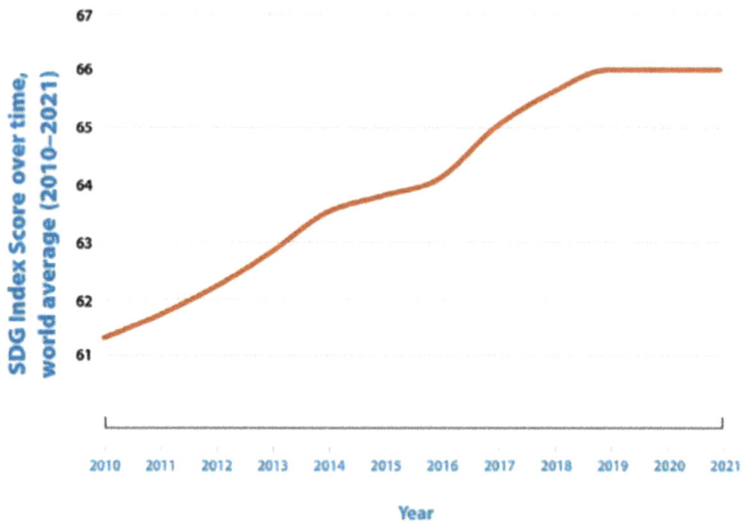

Fig. 1 SDG score over 2010–2021 years

of businesses in supporting the SDGs. Finally, the paper concludes by summarizing key insights and recommending forward-looking approaches to accelerate global sustainable development.

2 Unprecedented Global Challenges

In the evolving global landscape, unprecedented challenges are reshaping societies and economies at an accelerating pace. From the intensifying impacts of climate change to the persistent issues of poverty and inequality, these interconnected challenges demand urgent attention and coordinated global action. They not only disrupt present-day systems and livelihoods but also pose significant threats to future generations, underscoring the critical need for comprehensive and sustainable development strategies. Addressing these challenges holistically is essential to safeguard both current and future well-being, promote resilience, and ensure a more equitable and sustainable future for all.

2.1 Climate Change: Impacts and the Need for Immediate Action

Climate change stands as one of the most pressing challenges facing humanity today. Driven predominantly by human activities that have significantly increased the concentration of greenhouse gases in the atmosphere, the Earth's climate is changing at an alarming and unprecedented rate. According to the Intergovernmental Panel on Climate Change (IPCC), global temperatures have already risen by ~1.2 °C, with projections indicating the potential for even further increases if urgent action is not taken (Change, 1995). This rise in temperature is expected to result in more frequent and severe extreme weather events, including droughts, floods, hurricanes, and heatwaves, thereby intensifying environmental and socioeconomic vulnerabilities worldwide. The impacts of climate change are widespread, affecting ecosystems, economies, and communities across all regions. Polar ice caps are melting at an accelerated pace, contributing to rising sea levels that pose a significant threat to coastal cities and low-lying island nations. Agricultural productivity is being increasingly disrupted, particularly in vulnerable regions, leading to heightened risks of food insecurity and economic instability. Furthermore, climate change exacerbates existing social inequalities, disproportionately affecting marginalized and vulnerable populations. Immediate and sustained action is crucial to mitigate these adverse effects and build resilience against future climate-related risks. Strengthened international cooperation is essential for reducing greenhouse gas emissions, transitioning to renewable energy sources, and ensuring a more sustainable and equitable future for all.

2.2 Poverty and Inequality

Despite notable steps toward global economic progress, poverty and inequality continue to pose significant and persistent challenges. According to a United Nations report, over 700 million people worldwide still live in extreme poverty, surviving on <$1.90 per day (Simma et al., 2024). Inequality manifests both within and between countries, severely limiting access to basic necessities such as food, clean water, healthcare, and education. The COVID-19 pandemic has further exacerbated these disparities, pushing millions more into poverty as economic and social systems struggled to withstand the crisis.

Inequality is not merely a moral concern; it represents a major barrier to sustainable development by hindering social mobility, slowing economic growth, and perpetuating cycles of poverty across generations. Addressing these interconnected challenges requires the implementation of targeted policies that promote inclusive economic growth, reduce income disparities, and ensure equitable access to resources and opportunities for all. Improving access to quality education, comprehensive healthcare, and robust social protection systems can significantly diminish inequality, particularly among marginalized and vulnerable populations, thereby fostering a more resilient and just global society.

3 The Role of Rapid Changes and Growing Uncertainty for Sustainable Development

The accelerated pace of technological advancement, globalization, and demographic changes has introduced new challenges to achieving sustainable development (Mhatre & Shukla, 2024). While the digital revolution offers unlimited opportunities for innovation and economic growth, it also brings significant risks, including job displacement, concerns over data privacy, and the widening of social and economic inequalities. Similarly, the increasing frequency of major global crises—such as the COVID-19 pandemic, financial instability, and geopolitical conflicts—has exposed the vulnerabilities inherent in interconnected global systems, further complicating the formulation and implementation of sustainable development strategies. These developments underscore the need for adaptive, resilient policies that can harness the benefits of technological progress while mitigating associated risks. Table 2 outlines some of the major challenges currently facing the achievement of the SDGs.

In these uncertainties, the importance of sustainable development as a framework has underscored which addresses the complex challenges of the twenty-first century. By considering the long-term environmental, social, and economic goals, sustainable development goals provide a roadmap for navigating global uncertainties. This requires a shift in the development by measuring the numerous factors of economic success so that it could build sustainable societies.

Table 2 Summary of key challenges faced while developing goals in the SDG

Global issue	Key challenges	Opportunities for improvement
Climate change	Rising greenhouse gas emissions, extreme weather events	Transition to renewable energy, international cooperation
Poverty and inequality	Access to basic services, income disparity	Inclusive economic growth, targeted social policies
Education access	Inequity in educational resources, high out-of-school rates	Educational reforms, international programs like Malala Fund
Clean energy	Limited access to affordable clean energy in developing regions	Investment in renewable energy sources and energy efficiency

4 The United Nations 2030 Agenda for Sustainable Development

The United Nations' 2030 Agenda for Sustainable Development represents one of the important developments with respect to ambitious global frameworks by promoting the inclusive and sustainable growth. Adopted in 2015 by all United Nations Member States, the agenda provides the blueprint for peace and prosperity for people and the planet. There are the 17 Sustainable Development Goals (SDGs) which are central to the proposed agenda that call for that areas of critical to the humanity and the environment.

4.1 The Establishment of the 2030 Agenda

The establishment of the 2030 Agenda makes an important step while considering to the global cooperation. This establishment has been built on the success of the Millennium Development Goals (MDGs) that basically eradicates the extreme poverty, improving health, and ensuring universal education between 2000 and 2015. However, the MDGs were considered limited in scope as they primarily addressed issues in developing countries without considering the interconnectivity of global challenges such as environmental degradation and inequality.

By considering the limitations, the UN has been covered with global stakeholders, i.e., governments, civil society, and the private sector for drafting the 2030 Agenda which is designed as comprehensive and universally applicable. According to the United Nations Department of Economic and Social Affairs, this approach recognizes that poverty eradication, peace, and environmental protection are interdependent goals that must be handled together. The 2030 Agenda has been structured in such a way that the universal call to action which ensures no one is left behind.

The agenda goes beyond the economic growth by incorporating key dimensions such as environmental sustainability, social responsibilities, and institutional governance highlighting the importance of a balanced approach to development. The establishment of the agenda emphasizes on the global cooperation with commitments from both developed and developing nations for addressing local and global issues.

5 Overview of the 17 SDGs

At the 2030 Agenda, there are 17 SDGs with each represents the critical area for human and environmental development. These goals cover a wide range of issues from poverty and hunger with respect to the climate action and gender equality. The SDGs are universal and interconnected by understanding some of the important challenges such as poverty, inequality, and climate change are complex and interrelated.

The goals include No Poverty (End the poverty in all its forms), Zero Hunger (Achieve the food security and improved nutrition and promote sustainable agriculture), Good Health (Ensure healthy lives and promote well-being for all ages), Quality Education (Ensure the quality education and promote learning opportunities), Gender Equality (Achieve the equal quality of the genders), Clean Sanitation (Ensuring the sustainable management sanitation for all people),

Clean Energy (Emphases in the reasonable, dependable, and sustainable energy for all), Decent Growth (Promoting the economic growth for productive employment), Innovation Infrastructure (Constructing the best infrastructure and promoting the industrialization), Reduced Inequality (All forms of inequality is reduced across the globe), Sustainable Cities (Build smatter cities and human settlements that are safe, resilient, and sustainable cities), Responsible Consumption (Safeguarding the sustainable consumption), Climate Action (Take urgent action for challenges in the climate change and its impacts), Life Below Water (Conserve the marine resources), Life on Land (Safeguard the sustainable use of terrestrial ecosystems and decreasing the biodiversity loss), Peace Institutions (Promote strong peaceful societies and giving the right justice to the institutions), and Goal Partnerships (Improving the global partnership for sustainable development).

As noted by Sachs et al., (2020), the SDGs serve as a set of benchmarks that governments and other stakeholders can be applied to measure progress in the social, environmental, and economic challenges. Each goal has some specific targets and indicators that is used to track the progress at national, regional, and global levels. Figure 2 explains the number of publications with related to each SDG.

5.1 The Interconnected Nature of the SDGs in Ensuring Peace

One of the important features of the SDGs is their interconnected nature. The goals which can recognize the social, environmental, and economic dimensions of development that can be addressed. For example, in achieving Goal 1 (No Poverty)

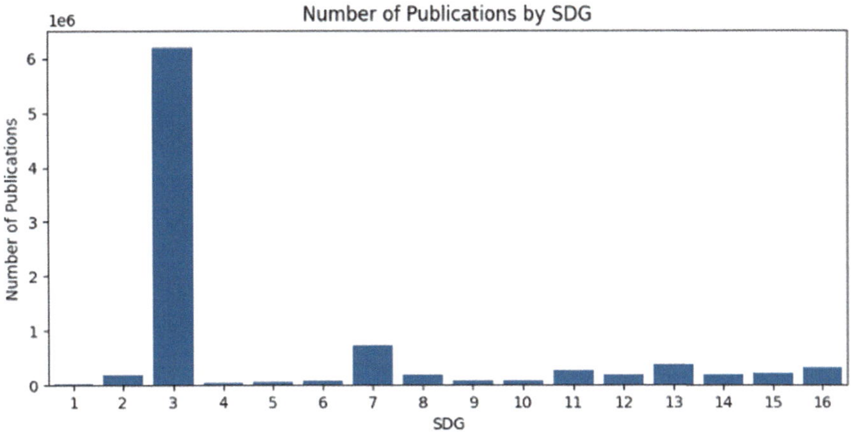

Fig. 2 Number of publications with respect to each SDG

Goal 2 (Zero Hunger), Goal 3 (Good Health and Well-being), and Goal 4 (Quality Education) has been considered for connectivity. While addressing the climate change (Goal 13), on land (Goal 15) and life below water (Goal 14) has direct connection with for protecting life.

The interconnectedness of the goals developed in the SDGs requires broad approach. According to Vladimirova and Le Blanc (2015), the combining the environmental, social, and economic targets are very critical for ensuring long-term sustainability and stability. For example, by addressing environmental degradation, efforts are established for reducing poverty and promote health that have limit access to resources.

For peace and good governance, The SDGs emphasizes the importance of Goal 16 as one of the foundational to all other goals. By not considering the institutions and peaceful societies, the progress in areas like health, education, and environmental protection is very difficult to sustain. Moreover, (Dodds et al., 2017) stated that global partnerships (Goal 17) are essential for financial, technical, and human resources which are required for achieving the SDGs. These partnerships would help in bridging gaps between the state governments that the goals are achieved through collective action. Figure 3 refers to the growth of citations per article with respect to each SDG as shown in Fig. 3.

5.2 Significance of SDG

The SDGs present a global vision for ensuring peace, prosperity, and protection of the planet for future generations. Each goal is interconnected with others which forms the comprehensive framework which covers all aspects of development from the economic growth and social responsibilities for the environmental protection. In this section, the significance of each SDG, beginning with Goal 1

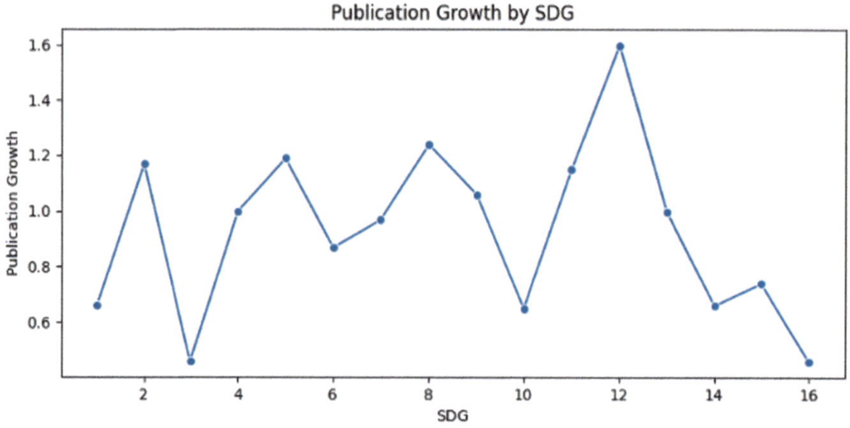

Fig. 3 Growth of publications with respect to each SDG

and then briefly summarizing the remaining goals, while highlighting how each one addresses various societal segments.

5.2.1 Goal 1: Poverty Eradication

Goal 1 focuses on ending poverty in all its forms by recognizing that poverty remains one of the greatest challenges faced by humanity. Although global poverty rates have been reduced over the past few decades, as of 2021, over 700 million people still live in extreme poverty, earning <$1.90 a day (Rocha et al., 2021). Poverty is multidimensional, involving not just income but access to basic services like health, education, and sanitation.

Achieving Goal 1 requires a multifaceted strategy that addresses both the symptoms and root causes of poverty. Strategies include expanding access to social safety nets such as conditional cash transfers, unemployment insurance, and food assistance programs, promoting inclusive economic growth and decent work, particularly in rural areas, where poverty is most concentrated.

Increasing investments in infrastructure to improve access to water, sanitation, healthcare, and education, as these are critical for poverty reduction and addressing systemic inequalities, including gender, racial, and ethnic disparities that perpetuate cycles of poverty.

The COVID-19 pandemic has exacerbated global poverty, pushing millions of people back into extreme poverty due to job losses, reduced incomes, and disruptions in access to essential services (Sumner et al., 2020a, 2020b). In response, national governments and international organizations have emphasized the importance of recovery efforts that focus on building resilience and promoting sustainable development to reduce vulnerability to future crises.

5.2.2 Goal 2–17: Brief Overview of the Remaining SDGs and Their Contributions to a Sustainable Future

Each of the remaining SDGs plays a critical role in advancing global sustainability and resilience. Below is a brief overview of these goals and their contributions to a sustainable future.

Goal 2 (Zero Hunger) aims to end hunger, achieve food security, improve nutrition, and promote sustainable agriculture. As referenced by sustainable farming practices are most essential part for feeding the growing global population by not considering climate change and environmental degradation (Boliko, 2019). Goal 3 (Good Health and Well-being) explains about healthy lives are required which is being promoted for all ages. Investment in healthcare systems especially for vulnerable populations which is one of the vital components for reducing mortality rates and combating global health crises (Bull et al., 2020). Goal 4 (Quality Education) states that the importance of education has been emphasized since it is lifelong learning opportunities for all. Access to education is the key drivers of economic growth and social mobility (Duedahl, 2016). Goal 5 (Gender Equality) elucidates promoting as one of the gender equalities. Gender equality is very much critical issue for achieving of all SDGs since education and healthcare are involved in this goal (Women, 2021). Goal 6 (Clean Water and Sanitation) elucidates the availability of water and sanitation for all is one of essential requirement in this goal. Accessing to clean water and adequate sanitation is very much essential for health and livelihoods for developing countries (Nicholl, 2019). Goal 7 (Affordable and Clean Energy) ensure access to affordable, reliable, sustainable, and modern energy is very much vital elements which are required in this goal. The transition of the clean energy transition is critical for reducing carbon emissions and mitigating climate change (Energy & Iea, 2020).Goal 8 (Decent Work and Economic Growth) explains about the sustainable economic growth, full employment, and decent work is required for promoting this goal. Environmental sustainability must be balanced with economic growth for avoiding climate change (Silva, 2022).

Goal 9 (Industry, Innovation, and Infrastructure) elucidates about the infrastructure development is essential for economic growth which has been focussed on building resilient infrastructure, particularly in rural and underserved regions (Cernas Ortiz et al., 2018). Goal 10 (Reduced Inequalities) focuses on tackling income inequality requires progressive policies, wealth redistribution, and ensuring equal opportunities for all (Clarke, 2021). Goal 11 (Sustainable Cities and Communities) seeks to make cities and human settlements inclusive, safe, resilient, and sustainable. Urbanization must be managed in a way that promotes environmental sustainability, resilience to disasters, and social inclusion (Habitat, 2019). Goal 12 (Responsible Production) encourages production patterns, particularly in high-income countries. Reducing waste and promoting circular economies are essential for achieving environmental sustainability (Macarthur & Heading, 2019). Goal 13 (Climate Action) calls for urgent action to combat climate change and its impacts. Climate change poses one of the greatest threats to sustainable development and requires global cooperation to reduce carbon emissions and adapt to its effects

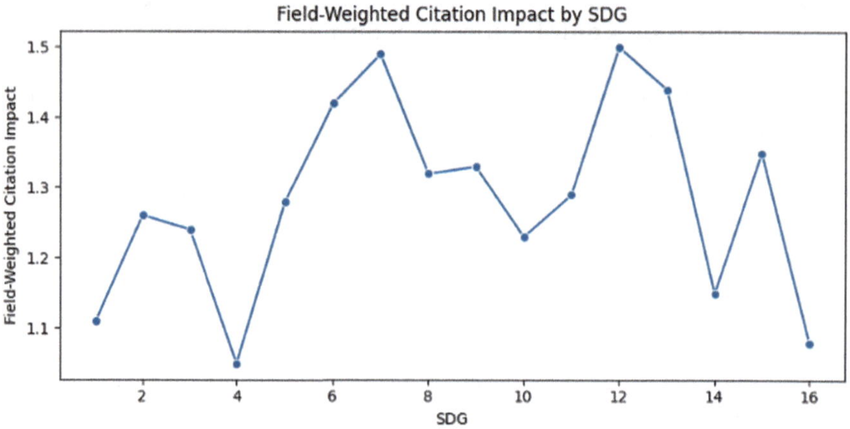

Fig. 4 Field Weighted citations impact by each of the SDG

(Masson-Delmotte et al., 2021). Goal 14 (Life Below Water) focuses on conserving and sustainably using the oceans, seas, and marine resources. Healthy marine ecosystems are crucial for global food security and biodiversity (UNDP, 2021). Goal 15 (Life on Land) aims to protect, restore, and promote the sustainable use of terrestrial ecosystems, combat desertification, halt and reverse land degradation, and halt biodiversity loss. Biodiversity is essential for ecosystem services that support agriculture, water supply, and climate regulation (Hoban et al., 2020). Goal 16 (Peace Institutions) promotes peaceful societies and giving access to the justice for all institutions. Strong institutions are critical for maintaining the rule of law in governance systems (Verger & Bonal, 2012). Goal 17 (Partnerships for the Goals) emphasizes the need for global partnerships to mobilize resources and strengthen the means of implementing the SDGs. International cooperation and financial assistance are vital for achieving the ambitious targets set out by the SDGs, particularly in developing countries (OECD, 2019). Figure 4 explains the field weighted citations impact by each of the goals in SDG.

6 Contemporary Challenges and Strategic Responses in Achieving the SDGs

6.1 Impact of the COVID-19 Pandemic on the Advancement of SDGs

The COVID-19 pandemic has significantly hindered global progress toward achieving the SDGs by pre-existing diverting resources from long-term development goals to emergency response measures. The global SDG Index score experienced a notable decline during the pandemic, particularly in health (SDG 3), education (SDG 4), poverty eradication (SDG 1), and employment (SDG

8). Disruptions in healthcare systems which lead to job losses, and increased inequality that would affect low- and middle-income countries. As emphasized in the Sustainable Development Report, the pandemic not only reversed hard-won gains in poverty reduction and healthcare delivery but also exposed weaknesses in digital infrastructure and governance, thereby delaying comprehensive SDG implementation across multiple sectors.

6.2 Policy Strategies for the Global South

Countries in the Global South must adopt tailored, equity-centered policies that address local health determinants and socioeconomic disparities. Key strategies include strengthening Primary Healthcare Systems, Community-Based Interventions, Integrated WASH Programs, Social Protection Mechanisms, and Health Education and Digital Health. Promoting awareness campaigns and investing in telemedicine to bridge accessibility gaps. These strategies align with SDG 3 (Good Health and Well-being) and SDG 6 (Clean Water and Sanitation), and can foster resilient health systems tailored to demographic and economic realities of the Global South.

6.3 SDGs as Catalysts for International Cooperation

The SDGs foster global cooperation by creating a unified agenda that transcends national boundaries. Goal 17 (Partnerships for the Goals) explicitly emphasizes the need for collaborative financial, technological, and capacity-building partnerships. Multilateral platforms such as the High-Level Political Forum (HLPF) and regional bodies (e.g., ASEAN, AU, EU) facilitate alignment of development strategies and peer-review mechanisms. Institutions such as the UNDP, World Bank, and regional development banks also provide funding, technical assistance, and policy guidance for harmonized SDG progress. Furthermore, academic, and private sector partnerships contribute to innovation diffusion and data sharing across borders, enhancing collective impact.

6.4 Global Monitoring and Assessment of SDG Progress

Progress toward the SDGs is monitored using a tiered framework of indicators established by the Inter-Agency and Expert Group on SDG Indicators (IAEG-SDGs). Each goal contains specific, measurable targets totaling 169 targets and 231 unique indicators. Global assessments are carried out via SDG Index and Dashboards Report (by SDSN), Voluntary National Reviews (VNRs), Global Sustainable Development Report (GSDR), UN DESA's SDG Indicator Database. These mechanisms enable evidence-based policymaking, regional benchmarking, and adaptive governance in SDG implementation.

6.5 Metrics for Gauging SDG Progress and Cross-Country Variations

Metrics to evaluate SDG progress are organized around quantifiable indicators such as poverty rate (% below $1.90/day)—SDG 1, Maternal and Child Mortality Rates—SDG 3, primary school enrollment rates—SDG 4, CO_2 emissions per capita—SDG 13, and proportion of population with access to electricity—SDG 7.

Variations arise due to disparities in national statistical systems, data availability, and indicator relevance. For instance, while OECD nations report real-time labor and education metrics, many low-income countries struggle with data collection and verification. Customized national indicator frameworks, like India's SDG India Index or Kenya's SDG Kenya Roadmap, are examples of context-adapted assessment tools that enhance localized monitoring and comparability.

6.6 Barriers Faced by Developed Nations in Implementing the SDGs

Despite their economic advantage, developed nations face complex challenges in fulfilling SDG obligations such as High Carbon Footprint (SDG 13), Overconsumption (SDG 12), aging populations (SDG 3 and 8), Policy Fragmentation, and Inequality and Migration Pressures (SDG 10). Unsustainable production and waste generation hinder circular economy transitions with demographic shifts stress pension systems and healthcare capacity, Socioeconomic disparities and immigration management remain contentious. These barriers require whole-of-government approaches, stronger inter-ministerial coordination, and public–private-academic partnerships to maintain SDG momentum.

7 How Each Goal Addresses Different Segments of Society

The SDGs are designed to address the needs of various segments of society, recognizing that different populations face unique challenges. For example, Goal 1 (No Poverty) targets the poorest and most vulnerable communities, aiming to ensure social protection systems for all, especially those living in extreme poverty. Goal 5 (Gender Equality) focuses on addressing systemic gender discrimination, particularly in regions where women face significant barriers to education, employment, and political participation. Goal 10 (Reduced Inequalities) seeks to address disparities not just within countries but also between countries, ensuring that marginalized groups, such as ethnic minorities and people with disabilities. Additionally, Goal 13 (Climate Action) calls attention to the disproportionate impact of climate change on vulnerable populations, particularly those living in low-lying coastal areas or regions prone to extreme weather events.

By addressing the needs of different segments of society, the SDGs promote an inclusive approach to development that leaves no one behind. As noted by Kabeer (2016), this emphasis on inclusivity is critical for ensuring that marginalized communities benefit from economic growth and social progress.

8 The Role of Economic Factors in Ensuring Environmental Sustainability

Economic factors play role in environmental sustainability. A balanced approach that fosters economic growth while preserving the environment is essential for sustainability. The integration of economic policies with environmental conservation efforts ensures that resources are used responsibly, supporting the achievement of Sustainable Development Goals (SDGs).

8.1 Impact on the Sustainable Economic Growth

Sustainable economic growth has been refereed as the economic progress which meets the needs of the present by considering the ability of future generations for meeting their own needs (Daly, 1990). Economic growth can drive environmental degradation as rapid industrialization and urbanization lead to over-exploitation of natural resources, pollution, and deforestation. However, economic growth can support environmental conservation when paired with sustainable practices. Table 3 describes about the SDG affected by the COVID-19 pandemic.

Green technologies has been one of the key drivers of sustainable economic growth. The development of renewable energy sources, energy-efficient infrastructure, and sustainable agriculture practices can lead to economic development

Table 3 SDG affected on the COVID pandemic

SDG affected	Specific impact during COVID-19	Recovery strategy suggested
SDG 1: no poverty	Increased poverty due to economic downturn	Strengthened social safety nets, economic resilience policies
SDG 3: good health	Strained healthcare systems, increased health disparities	Investment in healthcare infrastructure and universal healthcare access
SDG 4: quality education	School closures affecting millions, increase in dropout rates	Digital education platforms, support for marginalized communities
SDG 8: decent work	Job losses in vulnerable sectors	Job creation programs, support for sustainable businesses

(Kumar et al., 2024; Stern, 2006). Countries which has been invested in sustainable industries that led to growth of economic factors. For example, the European Union has consistently grown its economy while decreasing emissions due to robust environmental regulations and investments in green technologies (Petersen et al., 2014). In addition to that, sustainable economic growth includes the circular economy model, which aims to minimize waste and make the most of resources. The circular economy decouples economic growth from resource use, fostering economic development through sustainable production and consumption patterns (Korhonen et al., 2018). This approach ensures that economic activities are designed to support environmental sustainability.

9 How Economic Policies Can Support the Achievement of SDGs

Economic policies are one of the important policies that can be achieved during the goals of SDGs, by considering the environmental sustainability. Governments implement policies which can promote environmental conservation. Some of the examples are carbon pricing, renewable energy subsidy and tax incentives for companies adopting sustainable practices. For example, Carbon pricing is an effective tool for reducing carbon emissions by making it more expensive to emit greenhouse gases. By putting a price on carbon, more businesses are invested in cleaner technologies which supports both economic and environmental goals. As per (Nordhaus, 2019), indicates that carbon pricing mechanisms have led to significant reductions in carbon emissions in several developed countries.

One of the key policies is the subsidization of renewable energy. Governments provide financial support for solar, wind, and other renewable energy sources which is a gradual shift away from fossil fuels which helps in achieving SDG **7** (Affordable and Clean Energy) but also contributes to climate action under SDG 13 (Climate Action). Renewable energy stimulates economic growth by creating jobs in green industries while reducing dependence on non-renewable resources.

In addition to that, tax incentives are proposed for companies which are adopting sustainable practices that can promote environmental sustainability. For example, tax breaks has been given to those companies which can reduce waste or invest in energy-efficient technologies operating at lower costs. Also, Trade policies play a significant role in the goals in SDG. International companies produce an agreement that incorporate environmental standards into trade regulations which can ensure that global supply chains support sustainability. For example, The European Union Green Deal aims to include sustainability criteria in its trade agreements for promoting the production and consumption globally (Burki, 2021).

10 Case Studies

Several countries have demonstrated the positive impact of economic policies on environmental sustainability which are as follows in sub sections.

10.1 Costa Rica Renewable Energy

Costa Rica has been referred as a model for balancing economic growth with environmental sustainability. In this country, 99% of its electricity has been generated from renewable sources, primarily hydropower, wind, and geothermal energy. This government has invested in renewable energy infrastructure which has resulted in creating jobs in the green energy sector and reducing carbon footprint. This has been aligned with SDG 7 (Affordable and Clean Energy) and SDG 13 (Climate Action).

10.2 Sweden Carbon Tax

Sweden introduced a carbon tax in 1991 which is one of the first countries for reducing emissions. The carbon tax has been increased which resulted in increase in the businesses in cleaner technologies. As a result, Sweden has reduced its greenhouse gas emissions by nearly 27% while its economy has grown by more than 78% during the same period (Santosh, 2021). This demonstrates the optimal economic policies that can contribute to both environmental conservation and economic growth.

10.3 Germany Circular Economy Approach

Germany is one of the nations which has been implemented the circular economy principles. The country has developed a robust recycling system and promotes sustainable production practices that minimize waste. Germany's Circular Economy Act supports innovation in waste management, contributing to both SDG 12 (Responsible Consumption and Production) and SDG 9 (Industry, Innovation, and Infrastructure). Table 4 describes some of the case studies related to the SDG with its key economic policies and affected SDGs.

11 Key Focus Areas of the SDGs

The SDGs comprises of the broad range of targets which is aimed at well-being. However, several key areas of SDG has been required which require a special focus due to the impact on the planet future. These areas address issues that are fundamental to achieving sustainable development, which includes poverty eradication,

Table 4 Some of the case studies related to the key economic polices

Country	Key economic policy	Environmental impact	Related SDGs
Costa Rica	Investment in renewable energy	99% electricity from renewable sources, reduced carbon footprint	SDG 7, SDG 13
Sweden	Carbon tax	Significant greenhouse gas emission reductions	SDG 13
Germany	Circular economy approach	Reduced waste, increased recycling	SDG 12, SDG 9

climate action, education, and clean energy. Each focus area presents unique challenges and opportunities, which must be met with innovative and targeted strategies.

11.1 Poverty Eradication: Tackling the Root Causes and Providing Long-Term Solutions

Poverty eradication is one of the central issues of the SDGs and is the first goal (SDG 1). Despite the progress in reducing extreme poverty, about 9.2% of the global population still lives on <$1.90 a day, as of 2017. Poverty is one of the multifaceted issues which includes lack of access to basic services, such as education, health, and sanitation, as well as political and social exclusion.

For addressing the causes of poverty, it is essential for considering both economic growth and social policies. Various solutions are required for comprehensive approach which includes the creation of jobs, expanded the social safety nets, and improves the infrastructure in marginalized communities. For example, cash transfer programs, such as Brazil's Bolsa Família, have simulated in reducing poverty by providing financial support to low-income families for commitments related to the health and education participation.

In addition to that, poverty has been eradicated which is an intersection with other issues such as gender, race, and disability. The World Bank emphasizes that a critical long-term solution to poverty which is improving access to education, healthcare, and sustainable livelihoods.

11.2 Climate Action: Strategies for Mitigating and Adapting to Climate Change

Climate action (SDG 13) is another urgent focus area where the climate change can have potential risks to life on Earth. The increase in extreme weather events, rising sea levels, and shifting climate patterns are some of the events which can led to affect ecosystems and human societies globally. The scientific consensus which has been led by the IPCC calls for limiting global warming to 1.5 °C above pre-industrial levels for preventing catastrophic impacts (Macarthur & Heading, 2019).

By mitigating the climate change, immediate reductions in greenhouse gas emissions are required for transitioning to renewable energy sources, such as solar and wind power as well as improving energy efficiency. For example, the European Union Green Deal aims to achieve net-zero carbon emissions by 2050 which makes it a model for other regions to follow in addressing climate change. For vulnerable communities, the impacts of climate change has impacted which is equally important. Investing in climate-resilient infrastructure, increasing the disaster, and promoting sustainable agricultural practices are some of the critical measures for ensuring that societies can adapt to a changing climate. Initiatives such as the Green Climate Fund support developing countries for adapting in mitigating the climate change impacts.

11.3 Quality Education: Ensuring Inclusive and Equitable Education

Quality education (SDG 4) is one of the important SDG which is foundational for achieving all the other SDGs. Accessing to education has empowered individuals with their knowledge and skills for improving the lives and contribution to the society. However, UNESCO reports that over 258 million children, adolescents, and youth were out of school in 2018 due to the poverty, gender inequality for accessing significant barriers which remain in low-income countries (Vuong et al., 2021).

For improving education, equity has been focused on the marginalized groups, such as girls, children with disabilities, and those living in conflict zones having the access to learning opportunities. Programs like Malala Fund have supported for the education in girls which have been instrumental in raising awareness and mobilizing resources (Ikrama & Qumer, 2023).

In addition to that, education systems must be reoriented towards promoting sustainable development by incorporating climate literacy, critical thinking, and problem-solving skills where students are equipped to address global challenges. Education for Sustainable Development (ESD), endorsed by the United Nations has been aimed to achieve this by integrating the principles of sustainability into educational frameworks.

11.4 Affordable Clean Energy: Promoting Renewable Energy

Affordable and clean energy (SDG 7) is very much important for reducing poverty and climate action. With access to the sustainable energy, education, healthcare, and overall well-being is required for essential for economic development. UNICEF reported that nearly 759 million people are still lacking with electricity.

With expanded access to the affordable and clean energy, investments in renewable energy sources, such as solar, wind, and hydropower have been increased. The International Renewable Energy Agency (IRENA) has shown that renewable energy costs have been reduced significantly over the past decade which makes it more feasible for developing countries to clean energy sources. Countries like India have successfully expanded the capacity of the renewable energy such as, in solar power as part of efforts for providing the affordable clean energy to millions.

Energy efficiency is another important component of sustainable energy strategies. By improving the efficiency of energy use in buildings, industries, and transport, countries can significantly reduce their energy consumption and greenhouse gas emissions. Programs like the European Union Energy Efficiency Directive have been promoted to save the energy practices across all sectors of the economy as their main aim. In addition to that national as well as international cooperation is required for ensuring that the clean energy technologies are accessible to developing countries. The Intitative of Sustainable Energy for All (SEforALL) works to bridge this gap by mobilizing public and private financing for clean energy projects in underserved regions which ensures that no one is left behind in the energy transition.

12 Conclusion

SDGs represent a comprehensive framework which is designed to address the most challenges of our time which includes the poverty, inequality, climate change, and environmental degradation. This chapter explores the interlinked nature of the SDGs and emphasizes the urgent need for coordinated, multi-stakeholder efforts to meet these global objectives by 2030. Through the analysis of 40 key indicators from 125 countries, some developed nations exhibit robust sustainability features, significant gaps remain in the global south, particularly among countries with limited economic resources. The chapter underscores the impact of unprecedented challenge such as climate change, the COVID-19 pandemic, and geopolitical instabilities on the achievement of the SDGs, emphasizing the importance of resilience, inclusive policies, and adaptive governance. Moreover, it highlights the critical role of economic policy, green technology, and international cooperation in ensuring environmentally sustainable growth.

References

Barkemeyer, R., Faugère, C., Gergaud, O., & Preuss, L. (2020). Media attention to large-scale corporate scandals: Hype and boredom in the age of social media. *Journal of Business Research, 109*, 385–398.

Boliko, M. C. (2019). FAO and the situation of food security and nutrition in the world. *Journal of Nutritional Science and Vitaminology, 65*, S4–S8.

Bull, F. C., Al-Ansari, S. S., Biddle, S., Borodulin, K., Buman, M. P., Cardon, G., et al. (2020). World Health Organization 2020 guidelines on physical activity and sedentary behaviour. *British Journal of Sports Medicine, 54*(24), 1451–1462.

Burki, T. (2021). European commission classifies obesity as a chronic disease. *The Lancet Diabetes and Endocrinology, 9*(7), 418.

Carlsen, L., & Bruggemann, R. (2022). The 17 United Nations' sustainable development goals: A status by 2020. *International Journal of Sustainable Development and World Ecology, 29*(3), 219–229.

Cernas Ortiz, D. A., Mercado Salgado, P., & León Cázares, F. (2018). Job satisfaction and organizational commitment: Measurement equivalence test between Mexico and the United States. *Accounting and Administration, 63*(2), 145.

Change, C. (1995). Intergovernmental panel on climate change (IPCC). *CC BY, 4*, 548.

Clarke, G. (2021). The credibility of international non-governmental organizations (INGOs) and the Oxfam scandal of 2018. *Journal of Civil Society, 17*(3–4), 219–237.

Daly, H. E. (1990). Sustainable development: From concept and theory to operational principles. *Population and Development Review, 16*, 25–43.

Dodds, F., Donoghue, D., & Roesch, J. L. (2017). *Negotiating the SDGs: A transformational agenda for an insecure world.*

Duedahl, P. (Ed.). (2016). *A history of UNESCO: Global actions and impacts.* Springer.

Energy, I. W., & Iea, A. (2020). *IEA. Renewable energy market update.*

Fukuda-Parr, S. (2016). From the Millennium development goals to the sustainable development goals: Shifts in purpose, concept, and politics of global goal setting for development. *Gender and Development, 24*(1), 43–52.

Georgeson, L., & Maslin, M. (2018). Putting the United Nations sustainable development goals into practice: A review of implementation, monitoring, and finance. *Geo Geography and Environment, 5*(1), e00049.

Gupta, J., & Vegelin, C. (2016). Sustainable development goals and inclusive development. *International Environmental Agreements: Politics, Law and Economics, 16*, 433–448.

Habitat, U. N. (2019). *The strategic plan 2020–2023.*

Hoban, S., Bruford, M., Jackson, J. D. U., Lopes-Fernandes, M., Heuertz, M., Hohenlohe, P. A., et al. (2020). Genetic diversity targets and indicators in the CBD post-2020 global biodiversity framework must be improved. *Biological Conservation, 248*, 108654.

Ikrama, S., & Qumer, S. M. (2023). Malala Yousafzai: Promoting girls' education in conflict-affected countries. *The Case for Women, 36*, 1–30.

Kabeer, N. (2016). Gender equality, economic growth, and women's agency: The "endless variety" and "monotonous similarity" of patriarchal constraints. *Feminist Economics, 22*(1), 295–321.

Korhonen, J., Nuur, C., Feldmann, A., & Birkie, S. E. (2018). Circular economy as an essentially contested concept. *Journal of Cleaner Production, 175*, 544–552.

Kumar, R., Goel, R., Singh, T., Priyadarshi, N., & Twala, B. (2024). Incentivizing green building technology: A financial perspective on sustainable development in India. *F1000Research, 13*, 924.

Le Blanc, D. (2015). Towards integration at last? The sustainable development goals as a network of targets. *Sustainable Development, 23*(3), 176–187.

Macarthur, E. L., & Heading, H. E. (2019). How the circular economy tackles climate change. *Ellen MacArthur Foundation, 1*, 1–71.

Masson-Delmotte, V. P., Zhai, P., Pirani, S. L., Connors, C., Péan, S., Berger, N., et al. (2021). IPCC, 2021: Summary for policymakers. In *Climate change 2021: The physical science basis.*

Mhatre, A. S., & Shukla, P. (2024). A comprehensive review of energy harvesting technologies for sustainable electric vehicles. *Environmental Science Pollution and Research, 43*, 1–24. https://doi.org/10.1007/S11356-024-34865-8/METRICS

Nicholl, P. (2019). For every child, the right to a childhood UNICEF. *Child Care in Practice, 25*(4), 345–348.

Nordhaus, W. (2019). Climate change: The ultimate challenge for economics. *American Economic Review, 109*(6), 1991–2014.

OECD. (2019). *An OECD learning framework 2030. The future of education and labor* (pp. 23–35).
Petersen, J. E., Elbersen, B., & Fritsche, U. R. (2014). Applying resource efficiency principles to the analysis of EU-27 bioenergy options by 2020: Findings from a recent study for the European environment agency. *Biomass and Bioenergy, 65*, 170–182.
Rocha, L. A., Cárdenas, L. Q., Silva, N. G. A., & de Almeida, C. A. S. (2021). The Covid-19 pandemic and its impact on the performance of firms: An analysis based on World Bank microdata. *The Journal of Developing Areas, 55*(3), 411–433.
Sachs, J. D., Karim, S. A., Aknin, L., Allen, J., Brosbøl, K., Barron, G. C., et al. (2020). Lancet COVID-19 commission statement on the occasion of the 75th session of the UN general assembly. *Lancet, 396*(10257), 1102–1124.
Sachs, J., Kroll, C., Lafortune, G., Fuller, G., & Woelm, F. (2022). *Sustainable development report 2022.* Cambridge University Press.
Santosh, K. M. (2021). Management of environmental protection initiatives—evidence from Sweden. *GNOSI: an Interdisciplinary Journal of Human Theory and Praxis, 4*, 91–111.
Silva, V. (2022). The ILO and the future of work: The politics of global labour policy. *Global Social Policy, 22*(2), 341–358.
Simma, B., Khan, D. E., Nolte, G., & Paulus, A. (2024). *The charter of the United Nations: A commentary.* Oxford University Press.
Stern, N. (2006). *Stern review: The economics of climate change.*
Sumner, A., Hoy, C., & Ortiz-Juarez, E. (2020a). Estimates of the impact of COVID-19 on global poverty (No. 2020/43). In *WIDER Working Paper.*
Sumner, A., Ortiz-Juarez, E., & Hoy, C. (2020b). *Precarity and the pandemic.* https://www.wider.unu.edu/sites/default/files/Publications/Working-paper/PDF/wp2020-77.pdf
UNDP, S. (2021). *Creating a nature-positive future: The contribution of protected areas and other effective area-based conservation measures.*
Verger, A., & Bonal, X. (2012). "All things being equal?" Policy options, shortfalls, and absences in the world bank education strategy 2020. In *The World Bank and Education* (pp. 123–142). Brill.
Vladimirova, K., & Le Blanc, D. (2015). How well are the links between education and other sustainable development goals covered in UN flagship reports? *United Nations, Department of Economic and Social Affairs 146*, 152.
Vuong, Q. H., & Trung, T. (Eds.). (2021). *Academic contributions to the UNESCO 2019 forum on education for sustainable development and global citizenship.* MDPI.
Women, U. N. (2021). Women's representation in local government: A global analysis. *Hämtad, 27*, 2022.

Open Access and Its Role in Realising SDG 3: Pathway to Health and Well Being

Himani Upreti, Rupa Khanna Malhotra, Priti Sharma, and Tanupriya Choudhury

Abstract

United Nations has provided a global blueprint for achievement of a better and more sustainable future in the Year 2015 comprised of pressing areas of global concern categorised into seventeen Sustainable Development Goals (SDGs). Among them the 3rd SDG health and wellbeing is designed to promote healthy lives and wellbeing for everyone regardless of age. This goal embraces nine health and wellness related targets, which aims at reduction in maternity and child mortality, getting rid of infectious and non-infectious diseases, prevention of and treatment of substance abuse, reducing the deaths and injuries worldwide due to road accidents, bringing in global health coverage, provision of global access to reproductive health care services and strengthening the capacity for global health security. For achieving this ambitious goal innovative approaches are needed which can address the health challenges, reduce disparities and can foster equitable access to the healthcare information and resources globally. Open Access (OA) plays the pivotal role in realising SDG 3 by democratising knowledge, enabling the dissemination of critical health research, and easing global collaboration. This chapter explores the transformative potential of open access in advancing health and wellbeing by an in-depth analysis of the contribution of OA compared to subscription-based research in advancing SDG 3. Utilizing bibliometric data sourced from the Scopus database (2000–2024), it examines year-wise and country-wise publication trends, revealing a marked acceleration in OA publications, particularly post-2018. The chapter also examines the challenges in using open access for SDG

H. Upreti (✉) · R. K. Malhotra · Priti Sharma
Department of Commerce, Graphic Era Deemed to Be University, Dehradun, India
e-mail: himani.pant10@gmail.com

T. Choudhury
School of Computer Science, UPES, Dehradun, India

T. Choudhury et al. (eds.), *Open Access as a Knowledge Platform for Implementing the United Nations Sustainable Development Goals*, World Sustainability Series,
https://doi.org/10.1007/978-3-031-99605-4_3

3 and underscores the strategies to maximise open access impact to address the multifaceted dimensions of SDG 3.

Keywords

Open access • SDG 3 • Health • Wellbeing • Sustainable development goals

1 Introduction

Everyone has a fundamental right to good health irrespective of their ethnicity, cast, religion, political views, economic or social conditions (World Health Organisation, n.d.). Wellbeing is a multifaceted concept that encompasses the satisfaction of material, bodily, and psychological demands (Fernandez, 2020). While well-being includes more than just physical and mental health, it is obvious that physical and mental health are fundamental to the concept of well-being (SDG-Legal Guide_Chapter 3). United Nations has provided a global blueprint for achievement of a better and more sustainable future in the Year 2015 comprised of pressing areas of global concern categorised into seventeen Sustainable Development Goals. Among them the 3rd SDG health and wellbeing is designed to promote healthy lives and wellbeing for everyone regardless of age. This goal embraces nine health and wellness related targets which includes reduction in maternal and child mortality, ending infectious and transmissible diseases, safeguarding and curing substance abuse, global reduction in injuries and deaths because of accidents, ensuring universal health coverage, provision of global access to reproductive health care and strengthening the capacity for global health security (Chiu & Fong, 2023).

Achieving the targets set under SDG 3 requires prompt, affordable, equitable and quality access to healthcare services, resources and more importantly information. One of the major challenges encountered in bringing in the health equity is the existence of disparity in access to health-related information and knowledge (Fu, 2021). Informed decision making by healthcare providers, policymakers and individuals is impeded because of lack of reliable, prompt and accessible data. Researchers' ability to produce high-quality work with contemporary relevance is hampered by their limited access to the most recent and advanced research information in lower middle-income (LMIC) and upper middle-income (UMIC) nations. This obstacle needs to be removed. This disparity is filled by open access, which also promotes equity in the scientific community (www.frontiersin.org).

One of the major concerns for the global health care is the disequilibrium, termed as 10/90 disequilibrium. This disequilibrium is an unsettling fact, as barely ten percent of the 50–60 billion dollars are spent globally every single year on health-related research is focused on health issues that affect ninety percent of the global population, which requires immediate action (Chisholm, 2023).

Integrity of research is another major concern in the field of medical research and OA can have critical role in data ethics and integrity. In this regard, Pakistani clinicians defined research integrity as the ability to provide accurate and dependable data that can be adjusted to meet the health requirements of the local community and emphasised that OA can play a pivotal role in data ethics and integrity (iris.who.int).

For empowering communities, enabling evidence-based practices and for fostering innovation in the healthcare field it is imperative to address the information gap. Addressing this information gap can improve the healthcare outcomes worldwide. According to WHO open access to publicly funded research, including research data is elementary for tackling the public health challenges met worldwide in the twenty-first century (WHO, n.d.). Thus, they advocate and promote policy of Open Access as a fundamental part of their mission of improving public health and wellbeing (www.who.int).

Open Access is appearing as a transformative approach in bridging the information gap. Open Access refers to the unrestricted availability of research article online, thus easing academic and scientific findings available and accessible to all. This democratisation of information and knowledge beyond the geographical, economic or institutional barriers have a potential to drive advancement in the field of health care and research related to healthcare, particularly in underserved regions (ijmsweb.com).

The conception of Open Access (OA) has its origin in the early 2000s. The opening moves like the Budapest Open Access Initiative (2002) and then the Berlin Declaration made on Open Access to Knowledge in the Sciences and Humanities (2003) laid the framework for free and unlimited access to scholarly literature. By cutting paywalls and restrictive copyright licenses, OA appeared as a distinguished platform (Budapest Open Access Initiative, n.d.). A platform which allowed academicians, practitioners, policymakers, and the general public to access and use research products without any financial and legal barriers. In the sphere of health and medicine, OA allows for the rapid dissemination of medical research, promotes cross-border collaboration, and supports creativity in responding to health emergencies.

Regardless of these plausible benefits, substantial disparities persist in access to scientific knowledge, specifically in developing countries which face deep rooted impediments such as limited technological infrastructure, inadequate funding for research, and language barriers. This inequity hampers global progress toward achieving SDG 3. Moreover, the reliance on proprietary databases and the excessive costs associated with accessing peer-reviewed literature further limit the accessibility of vital health information.

The continuing differences in access to health-related research and information present a substantial impediment to achieving SDG 3. For enhancing the stakeholder's ability who can implement evidence-based policies, for improving healthcare practices, for effectively addressing emerging health concern it is of vital importance, that poor distribution of health research, particularly developing countries should be addressed. There is an urgent need to investigate how

Open Access might bridge this gap, encouraging fair access to knowledge and contributing to better health outcomes worldwide.

The current chapter is designed to traverse the intersection of Open Access in advancing SDG 3. Under the current chapter an attempt has been made to highlight how Open Access can function as an enabler in advancing SDG 3-health and well-being. The structure of the chapter is bifurcated into different sections, the first section is about the background and context of the Open Access and SDG 3 including the problem statement, the next section includes review of literature to provide a basis for comprehending the role of open access in achieving health and wellbeing. The next section highlights the challenges in meeting the sustainable development goal 3 and the opportunities provided by the free knowledge sharing platforms in achieving the SDG 3 additionally it also highlights that how open access can act as enabler for health and equity through various evidence across the world Utilizing bibliometric data sourced from the Scopus database (2000–2024), it examines year-wise and country-wise publication trends, revealing a marked acceleration in OA publications, particularly post-2018. Furthermore, the chapter highlights the Challenges in leveraging Open Access for SDG and suggests certain strategies which could be adopted to maximise Open Access impact.

2 Review of Literature

Jean Francois et al. (2018) The study emphasises that to address lack of adequate health related data, especially in tropical and southern countries, there is crucial need for a profound reorganisation of health monitoring and surveillance systems. The study recommends that for reorganising this monitoring system, collaboration with local and regional partners must be ensured. The study has also accentuated on adoption of One Health Eco Health approach.

Vij and Singh (2022) conducted there study on India and concluded that India has made significant progress in reaching Sustainable SDG 3. Highlighting the weaknesses of Indian Healthcare system, the study has suggested certain strategies which could enable the nation to achieve the goals stated under Health and well-being goals even earlier than the targeted dates. Rosa (2019) the study stresses that SDG 3 cannot be seen in isolation, moreover it is interconnected with other SDGs too, such as water and sanitation, nutrition, quality education which are crucial for health and wellbeing. Unrestricted access can function as a facilitator in the diffusion of knowledge in these interconnected areas and can enhance efforts to achieve Sustainable development. Latchem (2018) the study advocates that by utilising open, distance and mobile learning community health workers can be empowered, thereby improving the effectiveness of health interventions. The authors accentuates that by making wider availability of health resources, communities can be empowered to detect, prevent and treat health issues, resulting into overall wellbeing. Fu (2021) identified determinants of wellbeing, like log GDP per capita, social support, and life expectancy and concluded that these determinants significantly influence life satisfaction in all other categories, but not in lower middle-income countries. The study concluded that strategies to improve wellbeing should be tailored made and should be based on their specific income level and contextual factors. Chiu and Fong (2023) highlighting the relevance of goal 3 proposed by United Nations in the year 2015, which aims to ensure healthy lives and promotes well-being for all at all ages, the study addresses significance of various health issues Outlined by targets of SDG 3. Katsuma (2023) stressed upon the relevance of global health diplomacy a significant tool to combat transmittable diseases and for promoting global health coverage. The author also insisted that principle of "leave no one behind" must be endorsed as a part of efforts to achieve Sustainable Development Goal 3 by 2030. Weeks et al. (2023) Emphasising the significance of a holistic approach to health in the context of sustainable development, this study states that all the SDGs are interconnected, and they jointly contribute towards promotion of the health and wellbeing goal. Study also argued that SDG 3 can function as an enabler for advancing the other SDGs, suggesting that interventions in the health system can have far reaching impacts which can positively influence other sectors leading to overall societal progress.

3 Material and Methods

This chapter aims to explore how Open Access to health related research data can act as a catalyst in achieving the SDG 3 by examining the challenges that exists in meeting SDG 3, opportunities for leveraging knowledge sharing and broader role of Open Access into sustainable development strategies. By reviewing existing literature this chapter provides a comprehensive overview of how Open Access can be harnessed to support sustainable development and offers insights into the pathways for effective implementation specifically for SDG 3. Research methodology adopted is conceptual in nature which is based on secondary data and review of literature. A significant source of present study is secondary data collected from reviewing relevant documents, reports, and publications from government agencies, international regulatory bodies like World Health Organisation and United Nation Organisation. The Scopus database is used to identify the top 15 publishing countries in SDG 3 research through OA and subscription-based platforms and to illustrates the annual publication trends of Open Access (OA) and subscription-based research related to SDG 3—Health and Well-being. The data has been extracted from 2000 to 2024.

4 Results and Discussion

4.1 Understanding SDG 3

United Nations has provided a global blueprint for achievement of a better and more sustainable future in the Year 2015 comprised of pressing areas of global concern categorised into seventeen Sustainable Development Goals (Upreti &

Khanna Malhotra, 2024). Among them the 3rd SDG health and wellbeing is designed to foster health and nurture wellbeing for everyone regardless of age. This goal embraces nine health and wellness related targets, which aims at reduction in maternity and child mortality, getting rid of infectious and non-infectious diseases, prevention of and treatment of substance abuse, reducing the deaths and injuries worldwide due to road accidents, bringing in global health coverage, provision of global access to reproductive health care services and strengthening the capacity for global health security (Katsuma, 2023). These targets reflect a commitment to addressing the multifaceted nature of health and wellbeing. According to Sustainable Development Report 2024 country wise performance is being ranked for all the seventeen SDGs. Performance of every SDG is categorized under the following heads "SDG achieved", followed by "Challenges remain", followed by "Significant challenges remain" and at the end "Major challenges" remain, there are few for which information is not available thus highlighting them by stating Information unavailable. The report accentuates that majority of the nations are either ranked under the category of significant challenge remaining or major challenge remaining for SDG 3, which is a matter of global concern (dashboards.sdgindex.org).

Good health is not just an indicator of development, but it also provides foundation for progress in education, employment and equality. Although, significant advancement in the healthcare sector has been witnessed globally still millions of people lack access to needed medications, medical services and healthcare information. In 2023 World Health Organization stated that about 30% of the world population still lacks access to the basic healthcare services, which is a significant challenge need to be addressed promptly. The healthcare system around the globe is still struggling to meet the required healthcare services for their population. Many researchers and practitioners are struggling to access the groundbreaking medical studies which are hidden behind the expensive pay walls, especially in the regions with low and middle income. The fragmented knowledge across different platform is another challenge which the policy maker and practitioner has to face while undergoing the health research. Further, pandemics, chronic diseases, climate related illnesses and inflated cost of care are among few other challenges encountered by the global healthcare system. Open Access can function as a powerful solution to many of these challenges. It can bridge the gap in knowledge and can foster the collaboration by making the health research freely available to everyone. This can empower the healthcare researchers, workers and policy makers which can improve the outcomes in the healthcare sector. The Table 1 shows the challenges in meeting the sustainable development goal 3 and the opportunities provided by the free knowledge sharing platforms in achieving the SDG 3.

Table 1 highlights various Challenges in meeting SDG 3 and Opportunities for leveraging knowledge sharing. By addressing these challenges and seizing the opportunities provided by knowledge sharing, stakeholders can make significant strides towards realizing the ambitious targets of SDG 3.

Table 1 Challenges in meeting SDG 3 and opportunities for leveraging knowledge sharing

Challenges in meeting SDG 3	Opportunities for leveraging knowledge sharing
Existing inequalities in healthcare access among high- and low-income countries in terms of quality and availability of healthcare services is one of the major challenges which impedes the progress towards achieving SDG 3	Collaboration and partnership can play pivotal role in overcoming the challenges to achieve SDG 3. Global and regional collaborations can ease the exchange of best practices, innovations and resources (Weeks et al. 2023)
For addressing health challenges there is need of robust mechanism for collection and reporting of reliable data on prompt basis. Lack of accurate and prompt data which can be relied upon, acts as an obstacle in the progress towards achieving SDG 3	Promoting open access to research and data enables the dissemination of scientific advancement to wider audience, including policymakers, health care providers and general public. A study revealed that Open Access documents are 18% more likely to be cited than subscription-based articles which indicates that Open Access articles are more influential and reachable, thus can function as an enabler in conducting health care research (Piwowar et al., 2018)
For providing adequate health care services sufficient and equitable investments in healthcare infrastructure and research is required. Shortage of financial, technological and human resources particularly in rural and underserved areas is a challenge	Sharing knowledge through training, digital tools and e learning programmes can enhance the skills of healthcare workers and expand the reach of educational initiatives
	For bridging the gaps in the h-healthcare access, particularly in remote areas application of technology like tele medicine, mobile health application can be of vital relevance. Furthermore, tools like data analytics and artificial intelligence can offer new opportunities for disease surveillance, personalised treatment and operational efficiency

4.2 Open Access: An Enabler for Health Equity

Open access refers to the dissemination of knowledge without financial, technological or legal barriers. It allows free of charge access of scientific research, data and other educational material online. In context of health and scientific research OA can be crucial for fostering collaborations and addressing the global health challenges. Through open access various stakeholders including the healthcare provider, researchers, policymakers and the general public can get an equitable access to the vital information and break the obstacles that impedes the advancement of health and wellbeing.

There are various models through which open access can be implemented. Every model has its own characteristics and implications for users and contributors. Most commonly used types of open access are gold Open Access, green Open Access, hybrid Open Access, bronze open access. Under Gold open access, published articles are made available free of charges on the publisher's platform immediately upon release, the cost of publishing is borne by authors, institution or funders often through APCs article processing charges. Under green Open Access authors self-archives preprint or post print in repositories and the access is usually free in these platforms. Hybrid Open Access allows individual articles to be published open access if author pays a fee, this model combines element of both open and subscription access. Bronze OA refers to free to read articles made available by publishers, without an explicit mention to any Open Access license. This Open Access type was first suggested by Piwowar et al. (2018) (Robinson-Garcia et al. 2020).

Open access by removing paywalls can provide equitable access to scientific and medical knowledge globally. Making health related research work and relevant data related to healthcare available and accessible to all, has significant implication for health equity. Open Access ensures that healthcare professionals can have up to date excess to research and clinical guidance which can enhance their ability to evidence-based care and function as an empowering tool for then them. To design informed and effective health strategies particularly developing countries where resources are scarce, free access to health and research data can act as an enabler for the policy maker. Irrespective of institutional or financial constraints researchers worldwide can access and contribute to a shared pool of knowledge this can foster innovation and facilitate collaborative problem solving globally. Open Access democratises the information which empowers the general community and patients to engage actively in health-related decisions and campaigns. Thus, open access by eliminating the barriers to knowledge act as a catalyst for reducing the disparities on promotes informed decision making which can help in advancing the global progress towards the achievement of sustainable development goal 3 health and wellbeing.

4.3 The Role of Open Access in Advancing SDG 3: Evidence Across the World

Table 2 above highlights the comparative contribution of the top 15 publishing countries in SDG 3 research through OA and subscription-based platforms, data has been extracted from Scopus database. Under Fig. 1 the bar plot presents, the comparative contribution of the top 15 publishing countries in SDG 3 research through OA and subscription-based platforms. It reflects countries like the United Kingdom, Germany, the Netherlands, Sweden, and Spain show a higher inclination towards OA publications, while countries like India and Canada exhibit a relatively subscription-dominated pattern. The Fig. 1 highlights regional disparities and transitions towards open science practices globally.

The Fig. 2 illustrates the annual publication trends of Open Access (OA) and subscription-based research related to SDG 3—Health and Well-being, data has been extracted from Scopus database for the period 2000–2024. From 2000 to 2017, subscription-based publications dominated. However, from 2018 onwards,

Table 2 Country-wise insights (top 15 contributors to SDG 3 publications)

Rank	Country	Subscription based	Open access	Total publication
1	United States	26,406	21,459	47,865
2	United Kingdom	7,559	13,580	21,139
3	Australia	5,821	6,384	12,205
4	Canada	5,533	5,461	10,994
5	India	4,967	2,640	7,607
6	Germany	3,298	4,073	7,371
7	China	3,260	3,986	7,246
8	Italy	3,276	3,613	6,889
9	Spain	2,041	3,466	5,507
10	Netherlands	1,585	3,584	5,169
11	Undefined	3,320	1,081	4,401
12	Sweden	1,630	2,618	4,248
13	France	1,424	1,691	3,115
14	Brazil	1,109	1,944	3,053
15	Switzerland	1,053	1,922	2,975

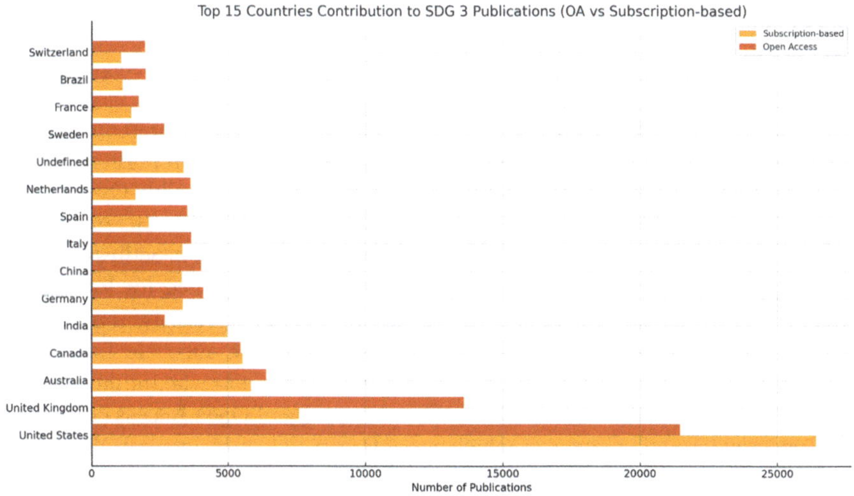

Fig. 1 Top 15 countries contribution to SDG 3 publications (OA versus subscription based)

OA publications surpassed subscription-based outputs, indicating a significant shift towards open dissemination of health-related research. The marked increase post-2018 aligns with global efforts promoting open science, particularly during the COVID-19 pandemic (Srivastava, 2021).

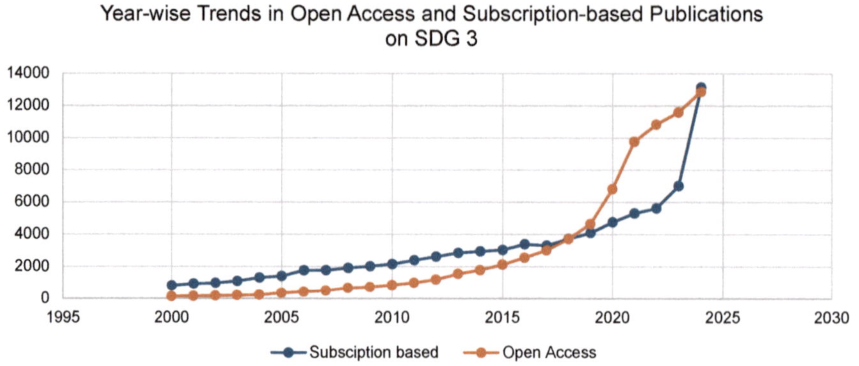

Fig. 2 Year-wise trends in open access and subscription-based publications on SDG 3 (2000–2024)

As a response to the increasing cost of scientific journal and the need for equitable access to information, the Budapest Open Access initiative was among the first to advocate for Open Access (Budapest Open Access Initiative, n.d.). The Berlin declaration gave momentum to the Open Access movement in the early 2000s (Berlin Declaration, n.d.). Efforts have been made through these initiatives to make the research freely available to everyone. In the field of healthcare, PubMed central became a game changer by offering access to millions of articles in biomedical and life sciences without any paywall (NLM Musings from the Mazzanine, n.d. https://nlmdirector.nlm.nih.gov/). The World Health Organization have also embraced Open Access to promote global collaboration, particularly with the aim to tackle the diseases which affect the poorer region disproportionately. By eliminating the pay walls and making research universally accessible Open Access play the very pivotal role in bridging the knowledge gaps. The HINARI programme common offered by World Health Organization is an exemplary in addressing these knowledge gaps, the HINARI programme is specifically tailored for low-income countries, for providing excess to thousands of scientific journals, thus benefiting health professionals an academic institution that lack adequate funding for subscription-based resources (WHO EMRO, n.d. www.emro.who.int/information-resources/hinari/). In low-income countries availability of Open Access content has bought transformative changes which provide a pathway to integrate innovative health practices hand groundbreaking research into local healthcare system. Open access has facilitated the dissemination of research on low-cost healthcare interventions like point of care diagnostics and community based maternal care models. Thus, empowering the healthcare workers empowering the healthcare workers hand professionals to implement the solution customised as per the needs of the local community. Pub Med central is another platform that offers free access to a vast repository of biomedical literature, which enables the researchers, academicians, clinicians and even students in the source limited settings to stay updated with the advancement in medicine and public health (NLM Musings from the Mazzanine, n.d.). Open access, through global collaboration has accelerated the pace of health research and innovation. Researchers across different regions regardless of the financial and the institutional constraints can get free access towards shared knowledge base. Innovative ideas of researcher can also be shared to the Open Access knowledge base. Thus, bringing in inclusivity which leads to diverse perspectives and innovative solutions to the complex health problems. Additionally, open access has facilitated the rapid dissemination of research during pandemic like COVID-19 when the whole world was facing public health crisis, when publisher temporarily made research articles freely accessible. Lot of studies has been done to understand behaviour and effect of COVID19 with respect to different parameters like metrological parameters and other variable (Awasthi et al., 2021; Mhatre & Shukla, 2024; Srivastava, 2021). Role of Open Access and its impact can be recognised from the outcomes of COVID-19 pandemic (Fraser et al., 2021). The freely accessible databases and preprint repositories such as BIOS RX IV and Med RX IV facilitated rapid sharing of research on the SARS covert to virus which led to breakthrough in vaccine development, making out public health strategies, and bringing in treatment protocol (Fraser et al., 2021).

While crafting effective and informed health related policies Open access can act as facilitator for the policymakers. Free of charge access to the health-related research and data enables evidence-based decision making. This evidence-based support becomes crucial for addressing health challenges persisting developing countries. A study published in The Lancet Planetary Health reported that air pollution in India led to ~1.67 million deaths in 2019, accounting for 17.8% of all deaths in the country. The economic losses associated with these health impacts were estimated at $36.8 billion, representing 1.36% of India's GDP (Boston College, n.d.). Similarly, a study published in Pub Med highlighted that the deaths attributable to ambient particulate matter pollution has increased by 115.23% between 1990 and 2019. A review in frontiers in sustainable cities identified the fifth leading cause of death in India is outdoor air pollution (Brown et al., 2022). Critical data like this, enables the policy makers to understand the severity of air pollution and its health impact in India. This understanding helps in gauging the severity of the health risk associated with the polluted air and thus proves to be essential in formulating and implementing the policies aimed at mitigating the health risks associated with increased air pollution (Gautam et al., 2024; Roy et al., 2024). Benefits of really available resources are not limited to policy makers and researcher but, they also act as an empowering tool for individuals and communities. Open Access to community health guidelines and patient education enables individuals to make informed decisions about their health and wellbeing. Free access to information helps individuals to comprehend their medical condition, treatment options, unpreventable care practices in a better way. For fostering healthier behaviour, reducing diseases related stigma and strengthening grassroot advocacy for better health services, awareness is must platforms like medicine and health information for all acts as a tool to enhance health literacy. Hence it can be culminated that free access to health-related knowledge can act as a cornerstone in bridging the knowledge gaps, driving innovations in forming policies and enhancing health literacy, making it an indispensable tool for advancing SDG 3.

4.4 Challenges in Leveraging Open Access for SDG

As discussed above, free access to health-related knowledge has significant potential in bridging the knowledge gaps and fostering innovation but still there are several challenges which impede the widespread adoption and effectiveness of Open Access to health knowledge. The funding models, copyright issues, high article processing charges, resistance from traditional publishing systems are certain such obstacles which are required to be addressed to realising Open Access transformative potential for health equity and wellbeing.

Shifting towards Open Access publishing often requires robust financial support, but sustainable funding models are not available for the same. Majority of the Open Access journals depend on Article Processing Charges (APC) which allows the leaders to gain access to the content for free but shifts the financial burden how the authors. This creates financial challenge for authors, specifically belonging from underfunded institution or low-income regions. Similarly, governments and institution from low resources may also like the funds to support Open Access programs thus making the adoption more difficult.

Another significant barrier to open Access implementation is the existing complexities of copyright laws. Majority off researchers are unaware of their rights about the sharing of their work or choosing appropriate creative common licences. Father publishing agreements often restrict author from depositing their work in the Open Access repositories, this also acts as a hindrance in getting access to the critical health research related data.

High APCs charged by many of the leading Open Access journal, creates inequalities in publishing opportunities, specifically for the researchers from low- and middle-income countries. This leads to under presentation of local health challenges and Indigenous knowledge system existing developing countries. This imbalance perpetuates a one size fits all approach to health solutions, undermining efforts to address region specific health disparities.

Open Access model is viewed as a threat by many traditional publishers, thus resulting into resistance from them. The traditional publishing system is deeply entrenched in profit driven models which rely upon subscription fees. Resistance from such stakeholders slows down the broader adoption of Open Access And creates a fragmented publication landscape.

Due to the perceived prestige and influence in academic evaluation system, many researchers and institutions prioritise publishing in high impact and subscription-based journals. This perception slows down the adoption of Open Access and limits the availability of groundbreaking health research to broader audience.

Strategies to Maximise Open Access Impact

These challenges are needed to be addressed, potential solutions may include the development of innovative funding models to be created by governments, philanthropic organization and international coalitions, who can invest in Open Access infrastructure and can subsidise the APCs for the researchers from underfunded organizations. Inclusive Open Access platform can be promoted which can give fair access and representation of diverse research topics. Governments and funding bodies may mandate Open Access publication for public funded research projects. Policies related to counteract copyright reception can also be formulated by the governments. Further, academic institutions can value research on the basis of broader dissemination of knowledge through Open Access platforms.

To harness the full potential of Open Access in advancing SDG 3, it is essential to adopt targeted strategies that address existing challenges while maximising its impact. Robust policies and sustainable funding mechanism are critical for fostering a thriving open access ecosystem. Thus, promoting such policies and funding mechanism which can support Open Access publishing is the need of the hour. Governments and funding agencies can frame out such policies which require researchers to publish publicly funded studies in open access journals or repositories. The National Institute of Health (US) and Plan S in Europe prove that how mandates can drive open access adoption globally.

Further, implementation of fee waivers or discounts by the journals can ensure inclusivity for authors coming from the low on middle income countries and for those institutions who lack the resources. Collaboration across the sectors can also accelerate the adoption and effectiveness of Open Access. Public private partnership can help in creating affordable and sustainable Open Access infrastructure. HINARI Access to research programme demonstrate the partnership between The WHO and technology companies, which reflects the impact of cross sector collaboration. Formation of global consortia by universities and research institutions can help in negotiating with the publishers, sharing of open access infrastructure and can also advocate for the fair publishing practices. For addressing the local needs effectively global knowledge hubs can be formed. These global knowledge hubs can also help in prioritising multilingual content and region-specific health research which can address the local needs effectively.

For meeting out the needs of underserved and marginalised population inclusivity in Open Access resources has to be ensured so that it can help everyone. Thus, for insuring inclusivity in Open Access resources it is essential to publish health materials in diverse languages and contexts. Culturally relevant content and translations can make open excels more meaningful for non-English speaking researchers, cool healthcare workers and general community.

Further it is very much essential that Open Access initiative must be created as per the tailored needs of marginalised group and local community. Technology accessibility is another aspect which has to be taken care off. Many low-income regions lack stable Internet connectivity, which hinders the access to Open Access content. Downloadable repositories, USB based Open Access libraries or mobile apps should be designed for such areas which lacks this stable Internet connectivity (Strydom et al., 2022).

Multifaceted approach is needed for maximising the impact of Open Access in transforming the health equity and achieving the ambitious targets of SDG 3.

5 Conclusion

Through the above discussion it can be culminated that Open Access has the potential to accelerate the pace of health research and innovation and can function as an enabler in advancing SDG 3. At the heart of Open Access is the principle of equity, this helps in prioritising fair knowledge sharing facilitating health and well-being for all without any disparities. Democratisation of health research through Open Access accelerates adoption of evidence based practices, enhances innovation and empowers stakeholders across all sectors to address pressing health challenges. Open Access ensures that knowledge is not a privilege but a shared resource accessible to all. Open Access can play a pivotal role in achieving health equity, where policymakers of low- and middle-income nations can frame out their decisions based on cutting edge data, and where every individual can have access to the knowledge needed to lead the healthier and better life. Open Access breaks the paywalls and goes ahead with a vision to provide the information to every individual regardless of their socio-economic status and geographical boundaries.

This vision underscores the transformative power of open access in shaping a more inclusive, healthier and sustainable world.

However, to harness the full potential of Open Access in advancing SDG 3, it is essential to adopt targeted strategies that address existing challenges while maximising its impact. Robust policies and sustainable funding mechanism are critical for fostering a thriving open access ecosystem. Thus, promoting such policies and funding mechanism which can support Open Access publishing, is the need of the hour.

References

Awasthi, A., Sharma, A., Kaur, P., Gugamsetty, B., & Kumar, A. (2021). Statistical interpretation of environmental influencing parameters on COVID-19 during the lockdown in Delhi, India. *Environment, Development and Sustainability, 23*, 8147–8160.

Berlin Declaration. (n.d.). Retrieved January 12, 2025, from Max Planck Society. https://openaccess.mpg.de/Berlin-Declaration

Boston College. (n.d.). *Air pollution in India.* https://www.bc.edu/bc-web/bcnews/nation-world-society/international/air-pollution-ininda.html

Brown, P. E., Izawa, Y., Balakrishnan, K., Fu, S. H., Chakma, J., Menon, G., & Jha, P. (2022). Mortality associated with ambient PM2.5 exposure in India: Results from the million-death study. *Environmental Health Perspectives, 130*(9), 538.

Budapest Open Access Initiative. (n.d.). *Read the declaration.* https://www.budapestopenaccessinitiative.org/read/

Chisholm, A. (2023). The role of open access publishing in addressing the 10/90 disequilibrium. In *Open pharma: Innovations in medical publishing*. Retrieved January 12, 2025, from https://www.openpharma.blog/blog/accessibility/open-access/the-role-of-open-access-publishing-in-addressing-the-10-90-disequilibrium/

Chiu, W.-K., & Fong, B. Y. F. (2023). Sustainable development goal 3 in healthcare. In *Environmental, social, and governance and sustainable development in healthcare* (pp. 33–45). Springer.

Fernandez, R. M. (2020). SDG 3 good health and well-being: Integration and connection with other SDGs. In: Leal Filho, W., Wall, T., Azul, A. M., Brandli, L., Özuyar, P. G. (Eds.), *Good health and well-being. Encyclopedia of the UN sustainable development goals*. Springer. https://doi.org/10.1007/978-3-319-95681-7_64

Fraser, N., Brierley, L., Dey, G., Polka, J. K., Pálfy, M., Nanni, F., & Coates, J. A. (2021). The evolving role of preprints in the dissemination of COVID-19 research and their impact on the science communication landscape. *PLoS Biology, 19*(4), e3000959. https://doi.org/10.1371/journal.pbio.3000959

Fu, G. (2021). Toward achieving sustainable development goal 3: Determinants, innovations, and reactions from 110 countries with different income levels. *Sustainable Development, 29*(4), 607–623. https://doi.org/10.1002/sd.2162

Gautam, S., Gautam, A.S., Awasthi, A. N. R. (2024). The nature of air pollution. In: *Sustainable air. Springer briefs in geography* (pp. 29–37). Springer. https://doi.org/10.1007/978-3-031-77057-9_5

Jean-François, G., Gerardo, S., Séraphin, K.-C., Nkoko, B., & Jean-Paul, M. (2018). Sustainable development goal 3: Health and well-being and the need for more integrative thinking. *Veterinaria Mexico, 5*(2), 443.

Katsuma, Y. (2023). Global health diplomacy to combat communicable diseases and to promote universal health coverage in achieving sustainable development goal 3. In *Sustainable development goals series* (pp. 17–35). Springer.

Latchem, C. (2018). Health care, childcare, safe water, sanitation, and hygiene. In *Open and distance non-formal education in developing countries* (pp. 121–130). Springer Singapore.

Mhatre, A. S., & Shukla, P. (2024). A comprehensive review of energy harvesting technologies for sustainable electric vehicles. *Environmental Science Pollution Research, 43*, 1–24. https://doi.org/10.1007/S11356-024-34865-8/METRICS

NLM Musings from the Mezzanine. (n.d.). Retrieved January 12, 2025, from National Library of Medicine. https://nlmdirector.nlm.nih.gov/

Piwowar, H., Priem, J., Larivière, V., Alperin, J. P., Matthias, L., Norlander, B., & Haustein, S. (2018). The state of OA: A large-scale analysis of the prevalence and impact of Open Access articles. *PeerJ, 6*, e4375. https://doi.org/10.7717/peerj.4375

Robinson-Garcia, N., Costas, R., & van Leeuwen, T. N. (2020). Open Access uptake by universities worldwide. *PeerJ, 8*, e9410. https://doi.org/10.7717/peerj.9410

Rosa, M. (2019). SDG 3: Good health and well-being—integration and connection with other SDGs. In *Encyclopaedia of sustainable development* (pp. 629–636). Springer.

Roy, A., Mandal, M., Das, S., Kumar, M., Popek, R., Awasthi, A., et al. (2024). Non-exhaust particulate pollution in Asian countries: A comprehensive review of sources, composition, and health effects. *Environmental Engineering Research, 29*(3), 598.

Srivastava, A. (2021). COVID-19 and air pollution and meteorology-an intricate relationship: A review. *Chemosphere, 263*, 128297.

Strydom, A., Mellet, J., Van Rensburg, J., Viljoen, I., Athanasiadis, A., & Pepper, M. S. (2022). Open access and its potential impact on public health: A South African perspective. *Frontiers in Research Metrics and Analytics, 7*, 975109. https://doi.org/10.3389/frma.2022.975109

Upreti, H., & Khanna-Malhotra, R. (2024). Bridging the urban-rural education gap in India through corporate social responsibility (CSR) initiatives: A conceptual study with special reference to sustainable development goal 4 (Quality Education). *E3S Web of Conferences, 556*, 01032. https://doi.org/10.1051/e3sconf/202455601032

Vij, D., & Singh, H. (2022). Sustainable development goal 3 strategies to take a leap. In *Research anthology on measuring and achieving sustainable development goals* (pp. 335–352). IGI Global.

Weeks, W. B., Weinstein, J. N., & Lavista, J. M. (2023). All sustainable development goals support good health and well-being. *International Journal of Public Health, 68*, 1606901. https://doi.org/10.3389/ijph.2023.1606901

WHO EMRO. (n.d.). Information resources for HINARI. https://www.emro.who.int/information-resources/hinari/

WHO. (n.d.). *Open access policies*. https://www.who.int/about/policies/publishing/open-access

World Health Organization. (n.d.). Constitution of the World Health Organization. http://www.who.int/governance/eb/who_constitution_en.pdf

Future Trends and Opportunities in Robotics and Automation in Different Sectors Under Sustainable Development Goals

Shristi Garg, Satyam Prakash Srivastava, Rupa Khanna Malhotra, Pooja Kanojia, and Tanupriya Choudhary

Abstract

In recent years, there has been extensive adoption of robotics and artificial intelligence in multiple sectors, wherein one of the sectors, the hotel industry, has brought a paradigm shift in switching from a manual interface to a technology-driven interface. The study has been conducted to showcase how businesses function by integrating technology-driven systems. Further, the extent of robotics adoption depends upon organizational preferences, maintenance, competitive factors, etc. This study not only comprehends how the different sectors are embracing the adoption of robotics in their service wing but also examines the consequences on employees' job prospects, retention, and customer satisfaction. With rapid advancement, revolutionary changes seem to be happening in most sectors. The automation market has consistently grown in recent years despite a downturn in the global economy. The coordination of AI and machine learning with Robotics can leverage many promises to adapt to a changing environment, being more decisive in situations like medical healthcare, automatic vehicles, and agriculture. The study reflects the benefits, limiting factors, and future possibilities of its adoption in different sectors.

Keywords

Automation • Challenges • Opportunities • Robotics • Sectors • Trends

S. Garg (✉) · S. P. Srivastava · R. K. Malhotra · P. Kanojia
Graphic Era Deemed to be University, Dehradun, Uttarakhand, India
e-mail: shristigarg800@gmail.com

T. Choudhary
School of Computer Science, UPES, Dehradun, Uttarakhand, India

T. Choudhury et al. (eds.), *Open Access as a Knowledge Platform for Implementing the United Nations Sustainable Development Goals*, World Sustainability Series,
https://doi.org/10.1007/978-3-031-99605-4_4

1 Introduction

The pace of advancements in the last many years has changed people's mindsets and forced them to bring changes to their work culture environment by incorporating revolutionary changes across industries through adopting robotics and automation. This chapter explores the integration of robotics, automation, and challenges has been done in different sectors, and the prospects for its sustainability have been made. The entire robotics and automation market has shown significant growth with the most promising future trends. The research forecasting states that globally, there can be an upsurge of $ 156 billion by 2025, shimmering a compound annual growth rate of 15% from 2019 to 2025. As per the statement by the International Federation of Robotics (IFR), an upsurge of 13% was found in the sales of robots globally in 2020. The industrial robotics market prediction seems to be staying positive, with an annual growth rate of 3.6% from 2024 to 2028. America, as such, has reported sluggish sales conditions in the automotive sector. The overall market growth represents a staggering trajectory with global production figures.

Figure 1 explains market prediction for industrial robots in terms of revenue. The chart shows the trend from 2019 to 2028, wherein revenue attainment was attained at the highest level of 34.9%. It further slashed to 14.9% in 2022 and reached the lowest level of 2.6% in 2023. The revenue perspective seems to stay stable between 2024 and 2028, ranging from 3.4 to 4.5% or up to 3.5%. According to a London report and global market research of Interact Analysis done on 5th June 2024, around 500,000 units of industrial robots have been shipped globally. The average cost of industrial robots has declined in the last year. Mexico has much dependence on the automotive sector, representing its strong contribution to industrial robot sales figures. The American market accounted for 18% of the global industrial robot sales. America in the post-COVID-19 period, both in the automotive and non-automobile industries, wherein producers were trying to improve their manufacturing procedures and mitigate the overall cost by adopting such technologies.

The quick progression of technology, in particular in the areas of robotics and artificial intelligence, is producing sweeping shifts that have never been seen before across all sectors of the economy. These developments are extremely significant. Working together, industrial and service robots are revolutionizing operations across various industries, from healthcare to manufacturing. This revolution is driving a revolution in operations. In the hospitality and travel industry, where robots enhance the quality of the customer experience and increase the efficiency of operations, this wave of innovation is also redefining the landscape of the industry. Within the hospitality sector, robots are utilized for a wide range of tasks, including interacting with guests, cleaning, and delivering items, among other things. In addition to addressing changing problems, such as those brought about by the COVID-19 pandemic, their deployment is done to improve both the efficiency of the services offered and the quality of those services. However, as technological improvements continue, questions are being asked about individuals' role in

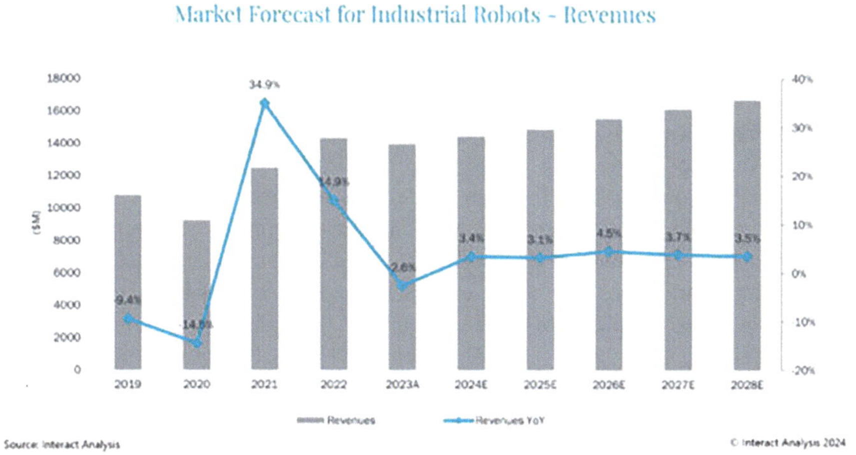

Fig. 1 Market prediction for industrial robots in terms of revenue. *Source* Statista, www.statista.com, 2023

this world that is becoming increasingly automated. When it comes to the optimization of processes and the provision of personalized services, Industry 5.0 emphasizes the collaboration between humans and robots (Brougham & Haar, 2017; Brown et al., 2005). Even though robots are particularly adept at completing repetitive tasks, humans bring to the table the ability to think critically and creatively to deliver personalized services. Even though robotics and artificial intelligence have the potential to be advantageous, their acceptance and incorporation are dependent on several circumstances. These considerations include the compatibility of the technology, the organisation's readiness, and the stakeholders' viewpoints. This paper analyses the behaviour of hotel managers in adopting robotic technology. Throughout the aim of the above inquiry, the Technology-Organization-Environment (TOE) paradigm will act as a framework. This research investigates technical, organizational, and external variables that impact administrators' decisions regarding whether or not to adopt robotics. Furthermore, the research considered the moderating impact of creativity on such aspirations. In addition to the academic knowledge about the hospitality sector, this research also contributes to the practical application of that knowledge. Additionally, it accomplishes this by addressing individuals' issues regarding the research. Furthermore, it provides a structure to evaluate an organization's availability and the methods for making decisions associated with the use of robotics. In addition, it provides knowledge about the factors that, given the administration's perspective, determine whether or not robots are adopted. The all-embracing goal of this study is to improve the competitiveness and financial viability of hospitality organizations, in addition to assisting in generating decisions based on accurate information regarding those organizations. This study will be carried out.

To draw attention to the extent of their effect. We examine the studies on accepting robots and the repercussions of this adoption in Sect. 2. A conversation of real-world applications, adoption models, and theoretical frameworks is obtainable in Sect. 3. The fourth section deliberates on the advantages, disadvantages, and applications exclusive to the sector. In Part 5, we will examine the relationship between robots and the Sustainable Development Goals (SDGs) made by the United Nations. Finally, the chapter delivers an overview of the most critical aspects, probable ramifications, and optional areas for further research.

A researcher from the Centre for Pew Research shows that artificial intelligence is changing workplaces and communities throughout the globe. The study surveyed 20 national people from different countries, including the Americas, Europe, Russia, Canada, and Asia–Pacific. Between 2019 and 2020, researchers surveyed people about their thoughts on the societal effects of artificial intelligence; 53% had a good impression, while 33% had an unfavourable one. It is worth mentioning that most Asians have positive views on AI, which highlights how important it is becoming in our daily lives. On top of that, opinions vary on whether human workers will soon be replaced by robots, yet nearly half of those surveyed saw automation of work as an improvement.

Multiple approaches to improve worker performance are indications that organizations have adopted information and communication technology to increase worker productivity and effectiveness (Jonker-Hoffren, 2020; Ruiz, 2020). Technostress, described by clinical psychologist Brod (1984), is the stress people experience when they cannot manage their health using information and communication technologies. This is only one of many present-day issues brought about by the widespread adoption of ICTs. According to Rabenu et al. (2017), the pressure that employees experience at work has an effect on their health as well as their standard of life in general. Although businesses are aware of the risks to workers' well-being, they should pay greater consideration to the risks to workers' mental welfare and implement treatments to improve their emotional well-being.

There has been a lack of in-depth research on how artificial intelligence affects worker productivity, even though it is becoming more popular in many industries, including the hospitality industry. Moore and Benbasat (2000) and Tarafdar et al. (2007) are just two examples of the positive and negative ways that adopting artificial intelligence might affect HR. One of the worries is the potential for technological stress between workers. Nevertheless, there is a clear lack of knowledge concerning the complicated impacts of AI on worker productivity, which calls for more investigation into the pros and cons of AI, including the risks associated with technological anxiety against the background of an environment that is getting more and more progressively mechanized.

1.1 Strategic Adoption and Integration of Robots and Automation in Hotels

As computing power, networking, and other related technologies grow, many robotic and AI-powered machines are finding their way into every aspect of human life. A robot is a self-contained device that may be programmed to execute a complex sequence of tasks. replace people and carry out the same tasks that people do. Several sectors are making use of robots for industry, which have been surrounding for more than 30 years (Belk, 2017; Bae & Chung, 2019; Maurer et al., 2016), which includes production, shipping, farming, medical care, transportation, and De Koning and Rodenburg (2004).

Additionally, interactive robots have entered the academic sector (de Kervenoael et al., 2020) and the elderly care sector (Čaić et al., 2018). The tourist and hospitality sectors are not a unique case (Cha, 2020). According to Collins (2020), numerous organizations in the travel and hospitality sector have already begun using robotic applications in different settings. Several changes are coming to the hotel and tourist sectors due to artificial intelligence's growing capacity to provide robots with the capacity to make decisions (Buhalis, 2020; Buhalis et al., 2019). An example is the initial robotics hotel, the Henn-na in Japan, which offers entirely robotic client service (Buhalis & Leung, 2018). A recent report from the accounting firm Price Water estimates that by 2030, the global economy will have been affected by AI to $15.7 trillion. Of this total, $6.6 trillion will be attributable to improvements in efficiency, while $9.1 trillion will be linked to effects on consumer spending (PwC2017). Artificial intelligence acquisitions were twenty-six times bigger in 2017 than in 2015, based on an estimate by the Economist (2018). Recent developments in artificial intelligence have prompted and facilitated this slew of mergers (Awasthi et al., 2024; Furman & Seamans, 2019). Organizations worldwide are pouring more and more money into technological innovation (IT) because it helps businesses make more money, which keeps the economy going. Widespread adoption of the innovation and additional investments are necessary to sustain and increase this impact, but this can only happen if the factors that influence the acceptance of technology are properly identified and accounted for (Oliveira & Martins, 2011).

Academics continue to be captivated by the deployment and uptake of automation by both appropriate industries and consumers (Makridakis, 2017). However, the opinions of operations and those making decisions have been the subject of very few investigations. Technological acceptance models (TAMs) (Davis, 1989), theory of reasoned action (TRA) (Ajzen & Fishbein, 1980), and innovation of diffusion theory (IDT) (Rogers, 2003) are examples of conventional frameworks that focus on how consumers use technological advances. However, these theories are largely technocentric (Awa et al, 2017). Originally proposed by Tornatzky and Fleischer (1990), the technology-organization-environment (TOE) framework identifies three elements at the organizational level that impact the acceptance and execution of innovations in technology. Therefore, this study set out to examine how hotel managers interact with and use robots and technological advances in robotics. Drawing on the Theory of Constraints (TOC) framework, The present research aimed to comprehend the influence of the following variables on hotel management's decisions to employ robots: technological advancements (i.e., dependability, complexity, advantageous position, and anticipated cost), organizational (i.e., backing from executive levels, organizational preparedness), and environmentally friendly (i.e., affordable stress).

Based on data acquired from hotel management throughout 11 nations, the research reveals extremely important revelations—a massive effort on an unequalled scale. That is why this investigation is crucial as a preliminary step towards delivering consequences for politicians and technological managers. Furthermore, this research adds substantial conceptual weight to the ongoing research into using robots in rooms.

In addition, the study's real-world implications include a structure that hotel management and others making decisions may use to assess their business's procedures and practices about robotic acceptance, allowing them to make educated choices. Finally, the study's seven-dimensional structure for analyzing organizational robotic adoption of technology offers beneficial information for technology providers in the hotel industry.

The entire chapter pertains to covering future trends and opportunities in robotics and automation across different sectors within the ambit of the Sustainable Development Goals. The purpose is to showcase its role in different sectors. It also helps how robotics and automation are bridging the gap of service lending companies. It also explains how work efficiency can be ensured by engaging robotic services in different arenas. Moreover, it is also mitigating the cost factor for most of the industries that are only dependent upon a manual interface.

2 Literature Review

There are a lot of obstacles that come along with the introduction and implementation of machinery, and those who oversee enterprises in the hospitality and tourism sector must thoroughly consider both the advantages and disadvantages of utilizing robotics in contrast with human employees navigating these difficulties. These elements have already been extensively examined in earlier research (Dogan & Vatan, 2019; Kuo et al., 2017; Lee et al., 2021), which may be accessed here. These papers examine these topics in great detail. A few of the many benefits that customer service robotics give include the capacity to operate 24 h a day, complete tasks in a precise and timely manner, broaden the scope of their employment via the implementation of appropriate updates, and deliver excellent customer service. Dogan and Vatan (2019) state that they are dependable with occupations that are repetitious, harmful, etc., causing employees to get weary, walk off occur, etc., and become unwell. Additionally, they were loyal and unsafe.

According to Dogan and Vatan (2019) and Kazandzhieva and Filipova (2019), certain limits should be taken into consideration. These limitations include the deficiency of emotional intelligence, personal touch, and creativity and the inability to effectively manage circumstances that are not structured. Even with these challenges, advancements in artificial intelligence and robotics can minimize certain downsides and improve others, which could result in increased application in the business world. Ivanov and Webster (2019) state that professions that include human emotions or physical presence are more difficult to automate than jobs that involve supplying information, cleaning, housekeeping, or handling luggage. This is because human elements are more difficult to replicate.

According to Osawa et al. (2017) and Ivanov and Webster (2019), introducing service robots into the workforce has beneficial and negative repercussions for employee jobs. These consequences can be categorized as either positive or negative. According to Ivanov et al. (2017), robots can increase operational capacity, simplify operations planning, and need mobile-friendly infrastructure, which results in increased expenditures. Several factors influence decisions regarding operations, including the logistics of maintenance and repairs and the possibility of vendor lock-in scenarios (Farrell & Klemperer, 2007). Ivanov and Webster (2019) point out that leasing models, such as robot-as-a-service, can address these concerns by moving the responsibility for maintenance to the provider. According to Choi et al., and and's (2020) research, robots have the potential to significantly improve marketing campaigns by creating one-of-a-kind experiences for customers. This can boost the possibility of guests returning and disseminating positive evaluations online.

According to Yadegaridehkordi et al. (2018), several non-robotic factors influence consumers' choices regarding hotels. Three of these factors include location, service value, and price. When deciding whether to invest in robots, to be successful from an economic perspective, they must be able to increase their productivity and earn more money. To justify spending, it is necessary to carry out extensive cost-effectiveness assessments considering all elements. In 2019, Ivanov and Webster Robotics performed repetitive tasks, which could lead to a decline in the costs associated with operations. The method for arriving at decisions throughout the COVID-19 pandemic is impacted by different factors, including the capacity to adapt to viral events, protection, and privacy. This assertion was demonstrated by Seyitoglu and Ivanov in the year 2020, in addition to Zeng et al. in the same year. When determining whether or not the implementation of service robots is achievable, consider several essential factors, such as personal and organizational features. When it comes to making judgments on creativity, the organizational atmosphere is influenced by factors like the organization's scope, the structure of leadership, and the immediately available finances. Sexuality, age, and years of employment are examples of personal variables that impact this environment (Venkatesh et al., 2003; Tornatzky & Fleischer, 1990). One additional example of the interaction between the two types of variables is offered by the simple fact that decisions about implementing technological improvements are impacted by outside influences such as competition applied by opponents (Baker, 2012). This is an additional demonstration of the relationship between the two types of variables. Several elements must be carefully considered when discussing the effective incorporation of service robotics into the hospitality sector. Technology abilities, operating transportation, advertising strategies, monetary consequences, and organizational interactions are some aspects that fall under this category. Regarding the hotel industry, the incorporation of robotics has to be the company's objective and be mindful of the constantly shifting circumstances in the marketplace to make the most of the opportunities that robots provide. As a result of robotics' multiple positive effects regarding the accuracy of operations and the overall satisfaction guests have, such is the situation that comes about.

This has been said by Brown et al. (2005) that the creation of UGC on websites like YouTube has evolved into a strong instrument that can be used for advertising purposes and for acquiring knowledge regarding the perspectives of buyers.

YouTube, one of the most widespread platforms for sharing videos, makes it simple to publish videos linked to tourism and encourages user interaction through comments and likes (Khan, 2017). Users communicate their views and feelings on YouTube, affecting how others perceive them (Waters & Jones, 2011). Businesses utilize YouTube to promote their products and services. Reino and Hay (2016) state that the interactive aspects of YouTube contribute to the construction of consumers' views of items or locations and the disclosure of their preferences. When researching the presentation of service robots in hotel operations, it is vital to have a solid understanding of customers' feelings. Research indicates that individuals may have a sense of unease when they are in the presence of robots exhibiting human-like features due to the "uncanny valley" result (Mori, 1970). Strait et al. (2017) state that this phenomenon occurs when a robot's characteristics are excessively similar to those of actual people without being realistic. According to Nowak and Biocca (2003), individuals have greater hopes for humanoid robots, and if those hopes are not realized, it might cause a serious problem. As a result, individuals have greater demands. They have set higher standards for themselves. On the additional hand, Tung and Au (2018) state that the collaboration of extraordinary experiences with robotics may have a favorable effect on individuals' perceptions of an item or service. This is because robots can generate. To acquire a more comprehensive understanding of how consumers engage with service robotics, examining remarks made on the web is imperative.

According to Walther et al. (2010), remarks can be utilized for collecting data regarding customers' thoughts, emotions, and viewpoints that, consequently, can affect how other individuals view the advances in technology. According to Lee et al. (2008), these claims provide organizations with information regarding their consumers that they would not normally divulge, impacting their consumers' choice to make an order. The evaluation of the public's impression via the feedback feature of YouTube is a crucial step in developing advertising tactics and improving the quality of service. In the same spirit, the technological affordance hypothesis and the unexplained valley concept are two theories that aim to explain how guests will respond to the service provided by robotics in the hospitality business. Both of these ideas are related to the concept of affordance. In the words of Norman (1988), the innovation opportunities theory examines how humans interact and become acquainted with technological advancements. On the other hand, the strange valley concept describes how robotics with human-like features might cause uncomfortable. The two approaches investigate how individuals engage with advancing technical developments. By utilizing these structures, academics investigate service robotics in hospitality contexts and how customers perceive human traits. Robots in the hospitality industry offer staff members a wide range of benefits, which may be broken down into several categories. In the case of McLeay et al.'s research from 2020, these opportunities are categorized based on the functions they can carry out and the human qualities they contain. For example, robotics that offers hotel service can assist with workouts, and robotics that offers personal assistance may offer mental support. Each type of robotics has a specific set of benefits. Workers have a choice for tangible opportunities with automated machinery, as stated by Huang and Rust (2018), due to their belief in the effectiveness of these robots compared to different kinds of robotics whenever performing jobs.

This preference influences their purpose to use robots in service delivery. Additionally, the relative advantage that service robots provide, which is mediated by trust, is a significant element in adopting these robots. According to Davis (1986), workers are more willing to put their faith in robots when they demonstrate a substantial improvement over the methods that are now in place. According to Mooradian et al. (2006), when employees trust one another, it results in more positive attitudes and behaviors in the workplace, impacting their inclination to adopt robotic technologies. This is because increased trust leads to more positive attitudes and behaviors. Regarding the hotel business, using user-generated content on websites such as YouTube can provide us with a wealth of information regarding how people feel about service robots. By utilizing theoretical frameworks and acquiring a more in-depth understanding of the affordances provided by robots, businesses can improve client service and speed up the adoption of robotic technology (Fig. 2).

2.1 Role of Artificial Intelligence Robots in the Hotel Sector

Intelligent machines and robotics are interchangeable terms. Robotic devices, both human and non-humanoid, that mimic behavior from humans and have the potential to improve and expand the abilities of humans demonstrate this trait. Effectiveness in business operations (Russell & Norvig, 2016). Even while artificial intelligence has been around since the 1950s, it has only been in the last few years that it has started to make a splash in academia. Finance, medical transport, employment services, travel, and the hospitality sector are just a few examples of the countless service industries that can greatly benefit from the widespread use of AI. (for example, see Buhalis and Leung (2018), Kim (2011), and Yu and Schwartz (2006)). In their 2009 operationalization of AI, Russell Stuart and Norvig took a multi-pronged approach, focusing on reasoning, human-like behaviour, and

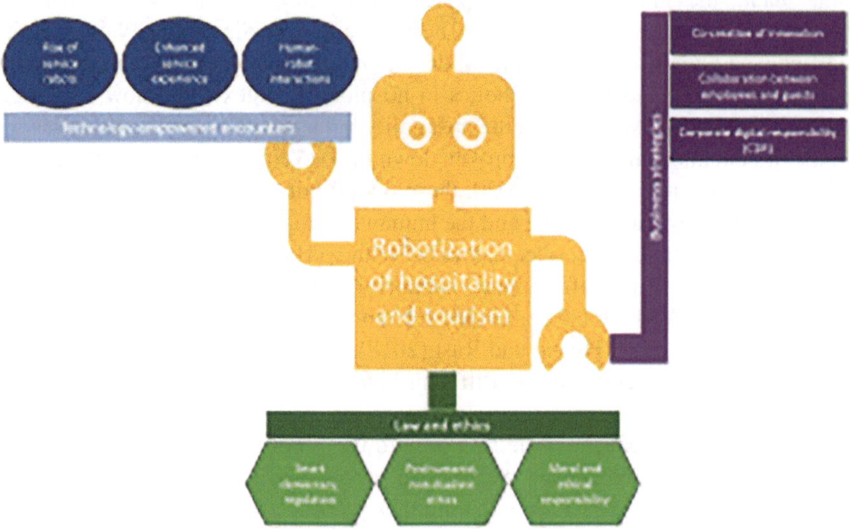

Fig. 2 Robotics in hospitality and tourism (https://www.mdpi.com/2673-5768/2/1/3)

human-like thought. According to Rijsdijk et al. (2007), six characteristics of artificial intelligence have been proposed: independence, responsiveness, cooperation, character, and the capacity to learn. The evolution and use of artificial intelligence also allow for its categorization. According to Kaplan and Haenlein (2019), there are three primary kinds of artificial intelligence: specific artificial intelligence, wide artificial intelligence, and extraordinary intelligence. The original AI, artificially restricted intelligence, was designed to carry out very particular activities.

While AI has undeniable benefits for businesses in terms of effectiveness, there is growing worry that technology may supplant humans in the workforce (Larivière et al., 2017). The "Momentum Machines Plan" has supposedly brought fast-food industry employment down a notch, as stated by Robinson (2017). Artificial intelligence is predicted to replace around a quarter of tasks in the hospitality industry (Chui et al., 2016). In 2016, the OECD found that nine per cent of occupations in twenty-one nations were amenable to automation. In addition to the 5% reduction in employment predicted by artificial intelligence in a 2017 McKinsey analysis (Manyika et al., 2017), a study out of Oxford University has projected that by 2033, 47% of occupations could be replaced by machines (Ramaswamy, 2017). Study Online reports that 72% of the US population is concerned that AI will eventually displace humans in the workforce (Smith & Anderson, 2017).

Nonetheless, research suggests that AI is limited to performing menial tasks at a low level (Morikawa, 2017; Smith & Anderson, 2017). The overall AI characteristics mirror how workers perceive the overall level of AI. Forsgren et al. (2016) and Wixom and Todd (2005) are only a few examples of studies that found that these data and quality of system factors can improve workers' perspectives towards technologies and their performance on duties. Artificial intelligence can greatly assist staff by translating client inquiries into human-friendly language, exploring company information systems, and creating responses (Kirkpatrick, 2017). In the travel business, AI may 'additionally offer data on price modifications and timetable issues. Analyzing artificial intelligence may enhance duties, as Serbanescu and Necsulescu (2013) show. Artificial intelligence (AI) has generated disagreement among scholars and practitioners alike. Adapting laws and regulations, ensuring the security of automobiles, and replacing hundreds of thousands of jobs are only a few of the new problems that artificial intelligence (AI) could bring about, according to Calo (2015).

There are a lot of potential dangers associated with using artificial intelligence to build HR services, such as the potential for employees to be too complicated, overlooked, or even taken over by robots (Reilly, 2018). Additionally, Nilson (2006) asserts that artificial intelligence has the potential to assist organizations in enhancing their efficiency and effectiveness. In 2019, 40% of modernization efforts will employ artificial intelligence services, and by 2021, 75% of commercial apps will do the same (Crews, 2019). The standard definition of excellent service is the gap amid the actual and expected levels of service provided by an organization. Parasuraman et al. (1988) proposed a theoretical description of client service quality as an overarching evaluation or boldness related to the service's overall distinction or dominance. The Evaluated Performance model (Caro, 2007) is based on Teas (1993) and illustrates the discrepancy between the quantity of an optimum attribute and the number viewed as effective. In 2007, Chaudhary and Prakash revisited the topic of customer service metrics.

2.2 Organizational Adaptation to New Technologies

The implementation of technological advances in work environments may result in significant improvements. Corporate efficiency. On the other hand, workers will experience the implications of this advancement in the form of changes to employment patterns and, unfortunately, the loss of some jobs (Noël, 2017). Consequently, it is logical to suppose that certain staff members feel anxious about transformation and resist it (Schneider et al. 2018). The degree to which employees can adjust to and think about the complex technological advances and the severe nature of the transition itself may also play a role in determining whether the shift is successful (L. Aiman-Smith and S. G. Green). Technological advances in work environments are becoming increasingly prevalent, which is a further problem that has the possibility of the well-being of employees (Brinzer & Banerjee, 2018). Technological stress has repercussions for both people and businesses. Technostress can also negatively affect wellness and satisfaction with work and commitment (Atanasoff et al. 2017).

Additionally, implementing fresh innovations within a company may be met with a serious challenge in organizational shifts among employees. There is a possibility that organizations will face challenging shortages of labour, in addition to the increasing responsibilities for hiring, educating, and retaining employees

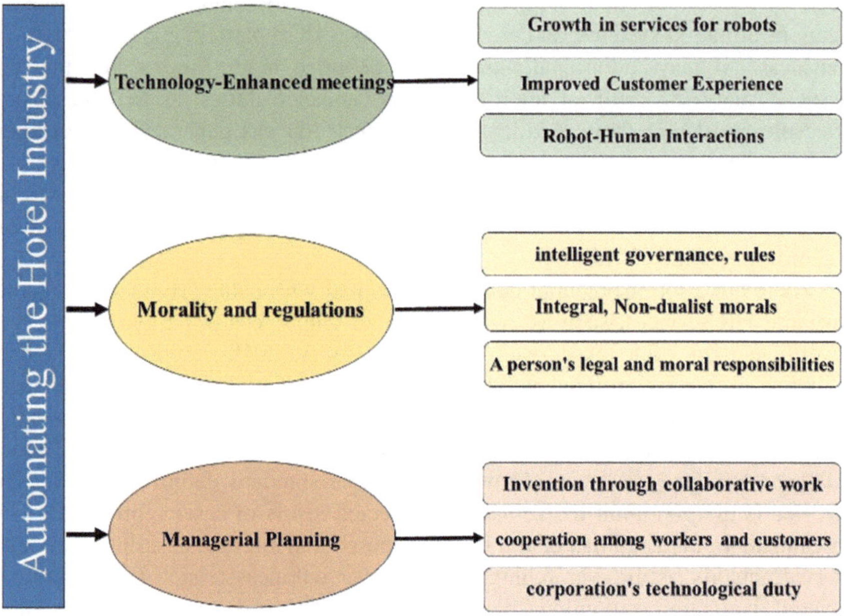

Fig. 3 Automation in the hotel industry

(Sivathanu & Pillai, 2018). Furthermore, there is an extensive variety of personal, social, sex, and age disparities in the abilities that managers prioritize in applicants (Kerpen et al., 2016). These variances can be found in employment opportunities. It is possible that older workers have different demands on education and training, for example, or that they may be reluctant to use advances in technology (Khan et al., 2016). Given the background information, multiple avenues of investigation were covered.

Figure 3 explains that Technology-enhanced meetings, Morality and Regulations, and Managerial Planning are the three primary domains that make up the model depicted in the image, which aims to automate the hotel sector. The following are some of the aims and factors to be considered for each segment.

2.3 Meetings Enhanced by Technology

1. Robotic Service Expansion: Highlights the growth of robotic service expansion in hotel operations.
2. Enhanced Customer Satisfaction: Centers on utilizing technology to boost visitor happiness. The integration and interaction of robots with human staff/customers is addressed in Robot-Human Interactions.

2.4 The Law and Morality

1. Smart Government, Regulations: Supports establishing regulatory frameworks and smart governance systems.
2. Integrated, Non-Dualist Morals: Advocates for a comprehensive view of ethics and the promotion of integrated moral principles.
3. The Moral and Legal Obligations of an Individual: Stresses the significance of being aware of and following one's ethical and legal responsibilities.

2.5 Managerial Planning

1. Invention Through Collaborative Work: Promotes creativity by bringing people together in a team setting.
2. Worker-Customer Cooperation: Highlights the value of staff-customer cooperation in running a business efficiently.
3. The Technological Responsibility of Corporations stresses that businesses must adopt and oversee new technologies.

This model provides a thorough framework for directing the hotel industry's automation process, with an equal emphasis on technical, ethical, and managerial considerations.

3 Research Questions

i. What technical aspects impact hotel managers' decisions to use robots?
ii. To what extent do hotel managers' intentions to implement robotic technology depend on organizational elements?
iii. What external elements impact hotel managers' decisions to implement robotic advances in technology?
iv. Is there a curbing effect of creativity on the links between ecological, administrative, technical, and managerial aspects and the intention of managers in hotels to implement robotic technology?

3.1 Research Objectives

i. To study the role of robotics and automation in different sectors.
ii. To study factors that influence the adoption of robotics in the different sectors.
iii. To study how robots and automation influence management effectiveness, customer satisfaction, staff positions, and adaptation.
iv. To study the challenges involved in robotics and automation in various segments.

3.2 Dimensions of Robotics

i. Technological Reliability: One of the most important elements impacting acceptance was the predictability and uniformity of robotic technologies' functioning. Hoteliers must have a strong technology foundation, as they are using Artificial Intelligence (AI)-driven robots for client satisfaction and productivity improvements, among other things.
ii. Organizational preparation refers to how open hotel executives and staff are to embracing new technologies. For new technology to be smoothly integrated into current activities, there must be successful backing from upper management and organizational flexibility.
iii. The Competitive Environment: To stay ahead of the opposition, hotels implement robotics to improve efficiency, enhance client happiness, and reduce costs. Being an early adopter of robots, the Japanese Henn-na Hotel has proven that they can increase client happiness.
iv. Limiting the Effects of Innovation: The concept of innovation has a crucial regulatory function in determining how effectively hotels are creating and using robotics. Hotels can set themselves apart from the competition and satisfy the changing demands of their customers by being innovative with the gadgets they use.

Figure 4 describes the conceptual framework for adopting robotics in hotels. It proposes that several factors impact a hotel's willingness to usage artificial intelligence (AI) and robotic technology (Results), as explained below. The difficulty, cost, precision, and advantages in the competition are some of the technical aspects that impact how appealing and feasible certain inventions seem. Organizational factors, such as approval from management and readiness, dictate the level of support and the ability to effectively execute these technological advances. Hotels may feel pressured by outside sources, such as competition, to use artificial intelligence and robotics to maintain their competitive advantage. The constraint of original thought and imagination increases the likelihood that hotels will successfully incorporate AI and robotics into their everyday operations. The model aims to offer a systematic way to comprehend and forecast the use of artificial intelligence and robots in hospitality environments by including these aspects in the Theory of Constraints paradigm. To effectively execute aims, it is crucial to connect technological resources with organizational readiness and external forces. The hotel's potential for creativity acts as a moderator in this process.

3.3 The Automated Waiting Staff at Eateries

2015 marked the year in which the Henn Na Hotel in Nagasaki, Japan, made history as the first global hotel to function largely by robotics (Alexis, 2017). Since the robotics at the hotel is responsible for seventy per cent of the duties, just 7 individuals are required to administer the whole facility. The technological entities

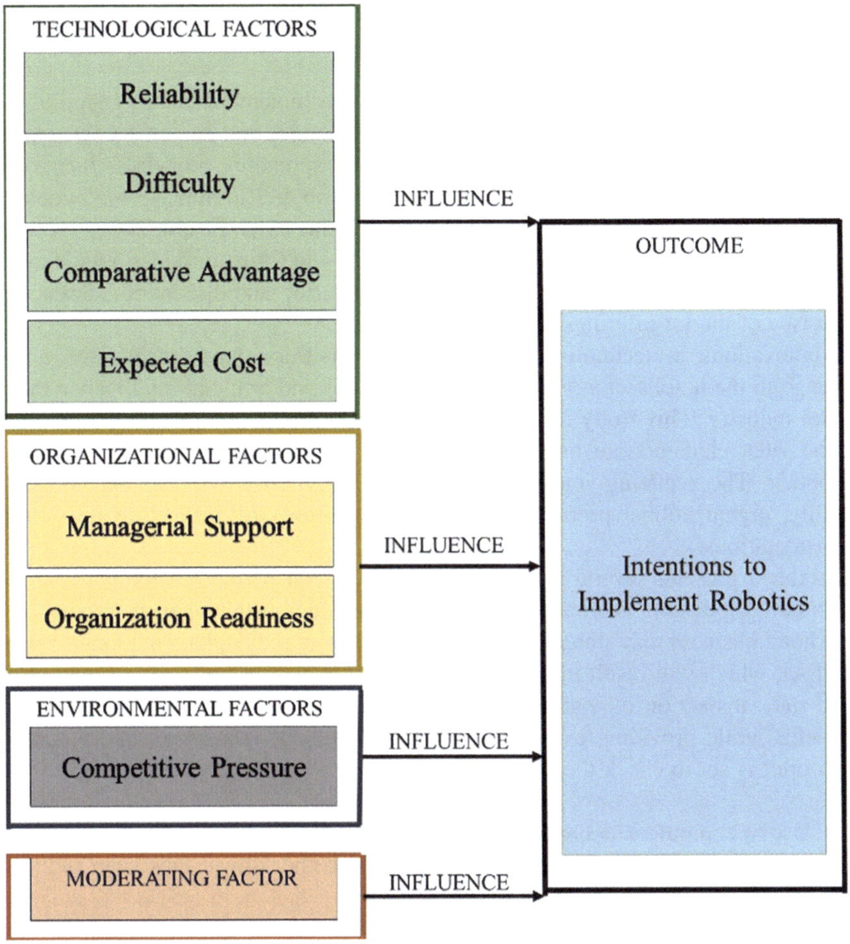

Fig. 4 Conceptual framework for robotics adoption in hotels

in question could assist clients with their luggage and guide them through every service offered by the staff at the front desk. Robotics was also responsible for completing tasks such as vacuuming surfaces and preparing drinks. Stated that the Hen Na Hotel swiftly ceased fifty per cent of its robotic services shortly after opening (Bhimasta & Kuo, 2019). This was done due to the hotel's lackluster track record of success. The "Connie Robot" at the Hilton Hotel is more fortunate than other robots. Connie, a concierge who was developed by IBM, assists guests in identifying products within the hotel premises in addition to suggesting restaurants and activities in the surrounding area (Trejos, 2016). In the wake of the pandemic, smart robots, capable of thinking and judging to a certain extent, are discovering new applications in the hospitality industry and other areas. It is possible for service robotics equipped with AI to provide clients with safe assistance regarding queries, cleaning, food pickup, and entertaining when they need assistance.

During the COVID-19 pandemic, dining establishments have also begun utilizing robotics to transport meals from storage to vehicles and customers (Demaitre, 2020). This is done to reduce the likelihood of spreading the virus. Increasing numbers of individuals are placing their trust in robotics to manage food cooking due to worries regarding the security of the pandemic (Meisenzahl, 2020). According to Bhardwaj (2018), Alibaba's Hema restaurant in China provides a wide range of services, including robotic fried rice manufacturing and distribution. These are just two of the lot offerings.

Innovations in technology and intense competition have accelerated a sea change in the hotel sector, with the rise of robotics and artificial intelligence in the hotel industry. This study uses the Theory of Constraints (TOC) model to determine what elements are most important to hotel managers when implementing robotics. The reducing impact of originality on innovation, technical dependability, organizational preparedness, and competition are all part of this set of considerations.

Table 1 presents several applications for robots and Artificial intelligence in the hospitality industry. The above table of contents demonstrates, on its whole, how the hotel business may stand to gain a great deal from the implementation of robots and AI, which will result in an increase in efficiency, a reduction in expenditures, and the satisfaction of visitors. These applications illustrate these technologies' benefits while providing examples of how they might be implemented when the autopilot is set to check the following areas.

(a) Use of computers and robots: To automatically assign guests the check-in time. One of the benefits is that the procedure for checking in has been reduced,

Table 1 Real-world applications of AI and robots in hotels

Application	Example	Benefits
Automated check-in	Use of kiosks and robots for guest check-in	Reduces wait times, enhances guest experience
Concierge services	AI-driven robots provide information and recommendations	Personalized service, improved guest satisfaction
Room service	Robots delivering food and amenities to guest rooms	Increased efficiency, reduced labor costs
Housekeeping	Robots cleaning rooms and public areas	Consistent cleaning quality frees up staff for other tasks
Maintenance and security	Robots monitor hotel facilities and perform maintenance	Enhanced safety, proactive issue resolution

which has resulted in shortened wait times and significantly increased the overall experience for guests.

(b) High-end services suppliers: Robotics equipped with AI and provide both advice and feedback. Benefits include: Since robotics can respond to inquiries while making precise recommendations, they can provide consumers with personalized attention, which raises the customers' satisfaction with their overall experience.

(c) Easy services for rooms provided by hotels: Robots that supply room service and refreshments to guests. It is possible to increase profitability while reducing the amount of money spent on labour by automating the process of delivering things to hotel rooms.

(d) Maintenance and washing: Robots can vacuum areas such as rooms and public areas. Benefits include ensuring that the highest levels of hygiene are maintained while allowing personnel to focus on other tasks and increasing production.

(e) Maintenance and Security: A good example of this would be robots that monitor hotels and perform routine maintenance on them. Advantages include: It always keeps a watch on problems and enables you to rectify them before they get more severe, which enhances protection.

3.4 Applications in Real Life

This report offers significant knowledge to help managers of hotels, lawmakers, and IT companies through the maze of robot and artificial intelligence acceptance. The results can be used by hotel managers to create guidelines for evaluating and deploying robotic devices, ensuring they align with the business's goals and plans to compete. By utilizing these findings, governments can create beneficial rules that help hotels and restaurants responsibly integrate artificial intelligence and robotics, while striking a balance between revolution and safety issues. The hospitality sector has unique requirements, and companies specializing in robotics and artificial intelligence can improve their products to match those demands.

4 Applications of Robotics and Automation

There could be several tasks that can be triggered by involving robotics, but the most common are discussed below:

4.1 Safety and Security

This is one of the most relevant applications of robots that every firm entity would look forward to having in their system process or procedures. To get the enriched experience of safety and security against all external threats and protect human life from vulnerable conditions. One of the companies in the United States, Knightscope, has provided cutting-edge security by helping security guards effectively. These robots are mapped to guard all areas of concern related to robbery, violence, and burglaries, furnish real-time intelligence as and when required, and take stringent actions to impediment misleading or fraudulent practices.

4.2 Healthcare

Robotics has assisted the healthcare sector tremendously in achieving a better future tomorrow. It helps doctors and experts develop patients' theories and more effectively diagnose conditions. One of the robots, da Vinci, can handle complex surgeries without hassle and sort out issues related to a heart ailment, body parts or sensitive areas. Devices like exoskeletons furnish additional support for patients who are under rehabilitation and suffering from spinal injuries, heart strokes, and kidney, and liver issues. In the medical sector, robots have transformed the pattern of performing surgeries, streamlining the supply chain and sparing more time caring for patients for better treatment.

Intel provides the leverage of technology that assists surgical equipment, developing robots for automated treatments. In the 1980s, surgical help was obtained through robotic arm technologies. They are not limited to such treatments but are also used in clinical trials to increase the care and concern for the patient's healthcare. Hospitals consistently deploy them for better access and to minimize exposure to pathogens during uncertainties like COVID-19.

Over the years, Artificial intelligence-driven vision and analytics have enriched the capabilities of medical robots, extending their support in handling other medical healthcare facilities. The robots are being involved in laboratories to overcome the issues of being cumbersome and handling repetitive tasks. Eventually, robots can keep the room neat and look after all maintenance required during the patient's room stay. It can also look after all the requisites prevailing in the infectious room. AI-enabled robotic software minimizes the time spent identifying and matching medical pills based on the diagnoses given by experts and doctors.

Figure 5 elaborates on robotic process automation in the healthcare sector from 2023 to 2028. The data reflect the progression over the span of years. It was $1.71 billion in 2023. Further, it increased to $1.97 billion in 2024 and is predictable to show a trajectory CAGR of 15.1% over the years. In the year 2028, it is anticipated to attain $ 3.46 billion (Table 2).

4.3 Agriculture

This is one of the sectors on which human life depends. However, it is dependent upon the weather conditions. Ideally, the farm produce is cultivated only when the farmer has some indicators of good monsoon conditions. Incorporating the role of

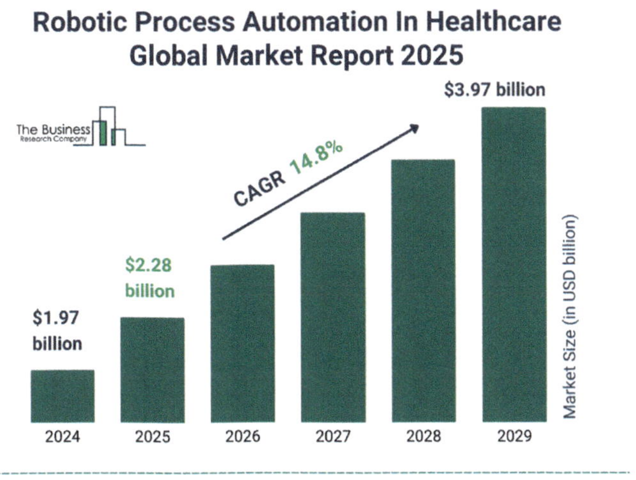

Fig. 5 Robotic process automation in healthcare—Global Market Growth (2023–2028). *Source* Adapted from The Business Research Company (2024)

Table 2 Report attribute and respective details

Report attribute	Details
Revenue forecast In 2033	\$3.46 billion
Market size value in 2024	\$1.97 billion
Growth rate	CAGR of 15.1% from 2024 to 2033
Actual estimates/historical data	2018–2023
Base year for estimation	2023
Market representation	Revenue in USD Billion and CAGR from 2024 to 2033
Forecast period	2024–2028–2033
Segments covered	(1) By solution: Software, Services, Implementation, Support and Maintenance, Training and Consulting (2) By Operations- Rule based operation, Knowledge based operation (3) By application: Claim Management, Clinical Documentation, Billing and compliance management, Appointment Scheduling, Workflow Management
Regional scope	Asia–Pacific, Western Europe, Eastern Europe, North America; South America, Middle East, Africa
Country scope	Australia, Brazil, China, France, Germany, UK, USA, Canada, Italy, Spain, India, Indonesia etc.

Source https://www.thebusinessresearchcompany.com/

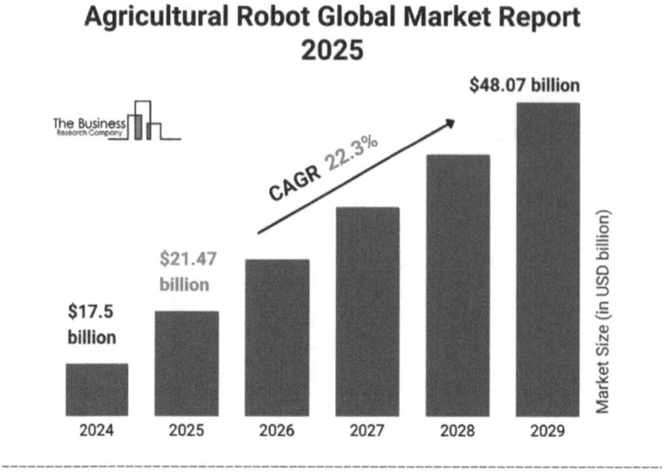

Fig. 6 Robotic process automation in agriculture—Global Market Growth (2023–2028). *Source* Adapted from the Business Research Company (2024)

robotics in this field, there seems to be a lot of scope for farmers to accomplish their functions properly.

In Fig. 6, the robotics and automation market was $ 6.71 billion in 2023. It further scaled up to $ 8.37 billion in 2024 and is anticipated to rise with a CAGR of 23.5% between 2025 and 2027 and could be touching the height of $ 19.45 billion in 2028. The developments in this sector of robotics show the following areas of concern.

4.3.1 Growth Trajectory

There seems to be an exponential growth trajectory in the next many years. The chances are bright that it will increase from $ 6.81 million in 2023 to $ 8.47 billion in 2024 at a CAGR of 24.7%. The important growth can be credited to the low cost of labour, incentives and subsidies by government entities, and the shortage of labour. It is anticipated that the agriculture market will grow $ 19.20 billion in 2028 at a CAGR of 24%. The development in agriculture is the result of robots involved in weeding operations, robotic soil management, harvesting through robots, and automated tractors. These robots are embedded with inbuilt cameras that hover all around the cropping area and function to capture all issues which crops need to be catered from any issues which may be pertinent and likely a matter of concern for farmers to work out to keep them healthy and alive. Robots also help to check the quality of soil and advise adding nutrients as and when required.

4.3.2 Upsurge in Agriculture Automation

Automation involves procedures with minimal manual intervention, using automated instructions for executing the given task and for better farm management and carrying out routine nature of tasks which are more laborious and fatiguing.

This is how productivity is enhanced in terms of a variety of crops multiple times in a year. According to Formstack, a United States software company, in June 2023 states that around 75% of organisations have automated their workforce, automated data and reporting are used by 59% for better planning and simultaneously complying with regulations is used by 35% of firms. This is the reason for the consistent growth observed in the agriculture robotics market.

4.3.3 Global Prediction for Robots

Several companies operating in the market are consistently engaging their minds in escalating research and development activities and increasing their sustainability through strong positions. In March 2021, an autonomous harvesting machine, Carry, was introduced by Future Acres, a US-based company, to transit hand-picked crops involved in delicate picking crops on farming land. For example, grape farming needs to be picked this way and overcome the issues of moving crops around the field. It is accomplished of moving crops up to 500 lbs in different weather conditions and enhances efficiency by 32% in farming land area.

4.3.4 Autonomous Monitoring Crops

Companies working in this sector are developing new technologies to lend dependable services to consumers. Solix is one of the autonomous crop-nursing robot that checks plants individually and maintains their productivity. It is embedded with an inbuilt camera and paired with artificial intelligence technology to check and detect unhealthy plant conditions and provide support for their better health.

4.3.5 Smart Agro Robot Market Merger and Acquisitions

AGCO Company, a US-based manufacturer, acquired Farm Robotics and Automation to enhance its capacity by extending innovations in farm areas and assisting its customers to attain great profits besides improving the welfare of live stocks. S.L Faromatics is a Spanish company working in Clever Robotics for animal welfare.

4.3.6 Policy Developments for Agro Robots Market

The largest area known for agriculture robots was North America in 2023. Europe is envisaged to be the fastest country known for the global agriculture robots marketplace. It covers areas like Eastern Europe, North America, South America, Western Europe, the Middle East, and Africa. Such robots market comprises organizations that earn revenue by furnishing satellite images, GPS agriculture, software used for land cultivation, and data sets in online mode. It also includes the sale of different types of drones like fixed-wing drones, hybrid drones, and rotary-blade drones (Table 3).

4.4 Exploration of Space

Astronauts have to face many challenges while exploring space. It is difficult to roam all around Mars and collect soil samples from the surface or repair spaceships

Table 3 Market prediction for agriculture robots

Report attribute	Details
Revenue forecast in 2033	$19.45 billion
Market size value in 2024	$8.37 billion
Base year for estimation	2023
Growth rate	CAGR of 23.5% from 2024 to 2033
Forecast period	2024–2028–2033
Actual estimates/historical data	2018–2023
Market representation	Revenue in USD billion and CAGR from 2024 to 2033
Segments covered	(1) By offering: hardware, software, services
	(2) By type: milking robots, UAVs or drones, automated harvesting systems, driverless tractors, other types
	(3) By farm produce: fruits and vegetables, field crops, dairy and livestock, other farm produces
	(4) By farming environment: indoor, outdoor
	(5) By application: harvest management, field farming, dairy and livestock management, soil and irrigation management, other applications
Regional scope	Asia–Pacific; Western Europe; Eastern Europe; North America; South America; Middle East; Africa

Source https://www.thebusinessresearchcompany.com/

in space. In such cases, robots can easily be positioned to handle such concerns as this will save human life from any risky conditions. NASA engages robots in areas of concern where humans cannot be involved. For example, the Mars rover is a robot capable of collecting pictures that support life on Mars and revert to scientists for further investigation and study. The global market size of space robotics was valued at USD 4.50 billion in 2022, and it is predictable to grow CAGE by 8.9% from 2023 to 2030. Efficient repair and maintenance of geo satellites has increased the demand for space robots. Effectiveness in terms of cost, enriched productivity, and conditioning to work in severe environments are sufficient factors that scale up the demand for space robots. For the implementation of space programs, orbit manufacturing, and assembly in the Space station, the initiative taken by NASA needs robots to meet such requirements. NASA has come across various kinds of robotics that would set the roadmap for 2035. European Space Agency have also set out plans for running projects that are related to space robotics. Countries like Russia, China, India, and Japan have already been notified about space robotic technologies. Artificial intelligence and Deep Learning are involved in robotics and automation. Several entities are designing robots that provide AI robots with more exploration advantages of mobility and capturing data that could be relevant to decision-makers. These robots can work in space with minimum human interface and can handle situations that could be complex, involving a lot of time. In June 2023, the Japanese Aerospace Exploration Agency developed Int-Ball, which is used to shoot videos and pictures in the shortest time. It moves automatically to the expected location and captures data in zero time. An increase in satellite launches raises the demand for orbit services, thereby upscaling the services of robots involved in space. They are also used to remove scrap happening in the Earth's surrounding space environment. Robotic manipulators are used to de-orbit redundant and operational satellites. The demand for such space robotics is anticipated to increase in the coming time. A single robot cannot be sufficient to collect and capture every corner, so using multiple robots would require better exploration and accomplishing planetary missions. Such multiple robots' engagement will drive the entire market very fast. With great dynamism in advancements, the space expedition has become complex and to get the condition of atmosphere, soil, temperature, and surface layer, exploration should not be done at the cost of human life. This is the biggest reason robots come into the picture.

4.5 Deepwater Exploration

Robots can now explore deep into the ocean and collect information beyond human beings. There is a lot of pressure in the ocean, which is difficult for humans to manoeuvre. Submarines can also go to a certain extent. In situations where deepwater exploration is required, robots can easily approach and collect data about unique undersea creatures in the phase of extinction.

4.6 Music and Entertainment

Robotics applications are used in the entertainment industry. They can be used in place of actors and actresses in serials and movies to make a special impact on the background. They are involved in all kinds of boring, repetitive tasks and are not favourable for human beings. They can easily do stunts and actions that are threatening to human life. Disney World is engaging robots to showcase more flavour of creativity and immense magical experiences to customers who prefer to stay connected with them.

4.7 Production Process

In the production process, most aspects are repetitive and redundant for the labour to perform the same task repeatedly. Packing, assembling, or repairing the material through welding is a common function of any process and involves multiple areas. The robots can be mapped to perform monotonous tasks and kill humans' time. This can ease the production manager's ability to pitch in efforts in those areas, which can lead to productive hours and fruitful outcomes.

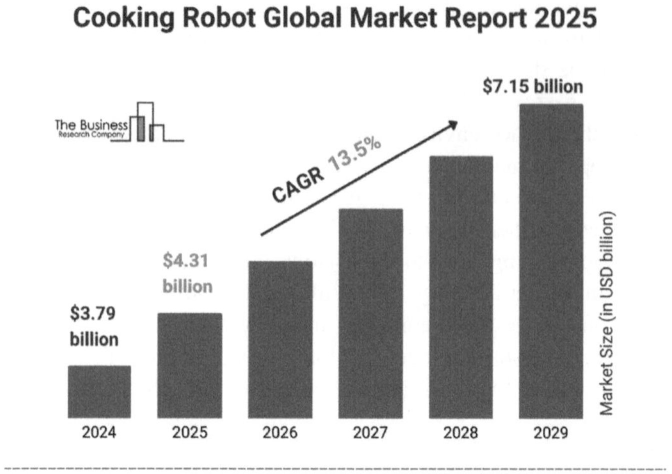

Fig. 7 Robotics adoption forecast across industries. *Source* Adapted from Statista (2024)

4.8 Preparation of Food

Some robots can cook food without any hassle. Robotic kitchens can be created to cook meals by using different recipes. Moley Robotics is one of the companies that has developed a robotic kitchen that can cook like a master chef.

Figure 7 shows a bar chart of cooking robots in the global market in 2024. The status of cooking robots in 2023 was $3.34 billion, and the data seems to be progressive from 2024 to 2028. In 2024, it further scaled up to $3.79 billion with a CAGR of 13.6% and is anticipated to be $6.31 billion.

4.9 In Military

In the military, the applications of robotics and automation to maintain discipline and take stringent measures to control crime and offence are being taken into consideration to a large extent. Use of Military robotics like MAARS (Modular Advanced Armed Robotic System), which releases tear gas in case of exigencies and is even capable of launching grenades in situations which can be threatening for them. DOGO is also a spying camera capable of shooting the enemy from a 9-mm pistol.

4.10 Customer Support

Few robots have been designed that resemble human beings. They are designed to lend customer service just to create more visibility in the service area. One of the robots designed in Singapore recognises visitors from the last visit by shaking hands, making eye contact, and having a consistent conversation, keeping in view the last visit. Junko Chihira, designed in Japan, works at a tourist information centre in Aqua City Odaiba.

In 2020, the automobile sector embraced robot absorption in its industry and showed impressive growth of 1150 robots per 10,000 employees. Moreover, the healthcare sector has again shown significant growth in caring for patients and engaging in surgery, diagnostics, and rehabilitation. The logistics and storage houses have shown exponential growth in demand for robotics.

4.11 Advantages of Robotics and Automation

Concierge, room service, cleaning, and greeting robots are examples of robotic services at hotels. These services are intended to improve visitor experiences, expedite hotel processes, and lighten hotel employees' workloads. The deployment of robotic services significantly impacts employee impressions. Some workers worry about their employment prospects and the possible loss of human contact in the service process. In contrast, others see these services as increasing efficiency and reducing repetitive chores. Robotic services can impact employee happiness by lowering manual labour and freeing up staff members to concentrate on more interesting and challenging duties. The degree of management assistance, the employees' capacity to adjust to new technologies, and the incorporation of these amenities into regular business processes could all influence employee satisfaction. Training and development programs are essential to the widespread use of robotic services so that employees can collaborate effectively with robots. This integration can create a more fulfilling and productive work environment when executed with practical training, support systems, and clear communication.

4.12 Limitations of Robotics and Automation

Although there are many benefits to implementing robotics and automation, there are also some obstacles that businesses must overcome. The huge outlay of capital needed at the start is a significant obstacle. Aside from the initial investment in hardware, additional costs are associated with training, development, and integrating robotic systems into preexisting processes when putting them into action. Businesses, especially those in the SME sector, may find these expenses especially onerous, which may cause a delay in the ROI. And a staff well-versed in cutting-edge innovation is essential for robot integration to succeed. The effective operation, maintenance, and updating of robotic systems requires employee training. The continuous requirement for training necessitates substantial organizational resources and dedication. It could be difficult for businesses to reap the advantages of automation if their employees aren't adequately trained. It is not always easy to incorporate new robotic systems into preexisting infrastructure. Thorough preparation and the creation of strong procedures to handle possible problems are necessary to guarantee compatibility between new technology and legacy systems. Organizations need to be ready to troubleshoot and address integration issues to keep operations running smoothly to keep operations running smoothly. To sum up, organizations need to think about and solve the financial, technological, and human resource problems involved in adopting robotics and automation to be more efficient and competitive.

5 Robotics and Sustainable Development Goals (SDGs)

Robotics and automation technologies are extremely significant in accomplishing several Sustainable Development Goals (SDGs) recognized by the United Nations. These technologies not only advance efficiency and advance but also contribute to safer, more sustainable, and more inclusive growth across all industries.

5.1 Robotics Supporting Sustainable Development Goals (SDGs)

Robotics contribute to understanding numerous Sustainable Development Goals (SDGs) by enhancing productivity, security, and innovation. For instance, SDG 8 (Decent Work and Economic Growth) is bolstered by the automation of hazardous and repetitive tasks, which decreases the risk of injury and improves employment efficiency. SDG 9 (Industry, Innovation and Infrastructure) is eased by robotic progressions in smart factories, while SDG 11 (Sustainable Cities and Communities) is developed through intelligent trash systems, traffic management, and smart infrastructure. Robots also contribute to SDG 13 (Climate Action) by dipping carbon emissions using energy-efficient operations (UN DESA, 2021; Ivanov & Webster, 2019).

5.1.1 Importance of Eco-Friendly Growth in Determining Robotic Answers

Sustainable development is essential when designing, organizing, and assessing the long-term effects of robotic solutions in various industries. More and more, the design and application of robotic systems are being shaped by the fundamental concepts of sustainability, which include preserving the environment, being socially accountable, and being economically viable. When designing robotic systems, developers consider how well they work, how little influence they will have on the environment and how much value they may deliver in the long run (Dwivedy & Mittal, 2010).

The cornerstones of environmentally sustainable design are using recyclable or recyclable materials and energy-efficient components like low-power sensors and actuators. With their linked design, robotic systems are becoming more popular because they can be easily mended, upgraded, and reused, which decreases electronic waste and increases product lifecycles. Lowering the frequency of replacements and guaranteeing healthier utilisation of resources contributes to Sustainable Development Goal 12 (Responsible Consumption and Production) (Bocken et al., 2016).

Robotics development aligns with SDG 9 (Industry, Innovation and Infrastructure) when green manufacturing technologies, such as additive industrial that uses recycled materials or production driven by renewable energy, are combined. According to Bechar and Vigneault (2016), these methods help businesses become cleaner while plummeting emissions.

Designing robots to supplement human work rather than supplant it is one approach to social sustainability; this will ensure that automation improves working conditions and fosters inclusivity. Goal 3 (Good Health and Well-being) and Goal 10 (Reduced Inequalities) can be progressive with the use of assistive robots in the healthcare and hospitality industries, which can advantage older people and persons with disabilities (Ivanov et al., 2019).

To sum up, sustainable development promotes eco-conscious, resilient, comprehensive, and functional robotic innovation. By coordinating their efforts this way, robotics can meaningfully impact the circular cheap and the planet's long-term viability.

5.2 Robotics for Sustainable Agriculture and Food Security

In agriculture, robotics progressed SDG 2 (Zero Hunger) and SDG 12 (Responsible Consumption and Production) by easing precision farming, minimising chemical application, and enhancing resource efficiency. Technologies such as drone seeders, autonomous harvesters, and AI-driven crop screens reduce waste, address labour lacks, and improve yield while safeguarding ecosystems (Bechar & Vigneault, 2016; Pedersen et al., 2006).

5.3 Reducing Waste and Improving Efficiency in Industries

Using proactive upkeep, lean robotics, and intelligent scheduling, robots reduce working costs and increase equipment lifespan, contributing to Sustainable Development Goals 7 (Affordable and Clean Energy) and 13 (Climate Action). Precision and quality pledges operated by robotic systems help minimise waste from defective products (IFR, 2020; Bogue, 2018). Robotics in manufacturing improve resource use and reduce material waste.

5.4 Enhancing Workplace Safety and Well-Being

Collaborative robots improve workplace well-being, reduce ergonomic strain, and contribute to the establishment of safer and more comprehensive workspaces

(International Labour Organisation, 2020; Toyota Annual Report, 2019). Robotics enhance work-related safety by replacing human services in high-risk industries such as mineral extraction, structure, and chemical handling, aligning with Sustainable Development Goal 3 (Good Health and Well-being) and Sustainable Development Goal 8 (SDG 8).

5.5 Ensuring Environmental Sustainability in Robotics

While robotics provide efficiency, their environmental influence throughout the life cycle must be addressed. This encompasses energy use, material procurement, and left-over disposal. Implementing eco-friendly robotics design, repurposing components, and adhering to rules such as RoHS and WEEE mitigate detrimental belongings. Instruments such as Life Cycle Assessment (LCA) are essential for evaluating environmental properties throughout all phases (Dwivedy & Mittal, 2010).

5.6 Promoting Sustainability in Hospitality and Retail Supply Chains

Robotics helps recover supply chain sustainability in the hospitality and retail industries by automated inventory control, order fulfilment, and waste reduction. (Ivanov et al., 2020) Skills such as automated guided vehicles (AGVs), robotic cooking staff members, and intelligent inventory bots donate to Sustainable Development Goal 12 by minimising excessive industrial deterioration and energy consumption, which is especially significant during peak demand.

5.7 Addressing Ecological Consequences of Robotics in Urban Settings

Even though robotics advances urban functionality, they can also pose ecological risks if not adequately controlled. These risks involve increased waste from electronics, interference from electromagnetic waves, and higher energy stresses. Noise pollution and the improper disposal of robotic parts damage biodiversity and urban ecosystems. Therefore, urban planning policies must incorporate Environmental Impact Assessments (EIAs) and reuse and recycling systems to ensure they align with the Sustainable Development Goals (UNEP, 2022).

Table 4 GRIA—Green robotics impact assessment framework

Dimension	Criteria	Evaluation Metrics
Environmental	Energy consumption, material use	LCA score, carbon footprint reduction
Social	Workforce enhancement	Number of jobs supported, inclusivity rate
Economic	Operational and cost-efficiency	ROI, total cost of ownership (TCO)
Ethical/CSR	Compliance, social responsibility	ISO 26000 adherence, CSR alignment

Source Author's own formulation based on LCA, ESG, SDG, and ISO 26000 frameworks (adapted from Bocken et al., 2016)

5.8 Aligning Hospitality Robotics with Corporate Social Responsibility (CSR)

Hospitality companies can use robotics to endorse CSR goals and SDGs 10 (Reduced Inequalities) and 12. For example, Marriott International uses robotics to track vigour use and make services more accessible for disabled visitors. Robots also cut down on water and power use in cleaning. Aligning technology with CSR promotes operational competence and ethical accountability (Marriott International, 2021).

5.9 Frameworks to Evaluate Robotics' Impact on Sustainability

Companies can use Life Cycle Assessment (LCA), ESG metrics, and SDG indicators to measure the sustainability effect of robotics. The deployment of robotics is aligned with global sustainability prospects thanks to standards like ISO 14001 (environmental management) and ISO 26000 (social responsibility), which provide openness, comparison, and leadership for long-term positive impact (Bocken et al., 2016).

Although gadgets such as Life Cycle Assessment (LCA), Environmental, Social, and Governance (ESG) metrics, and Sustainable Development Goal (SDG) indicators are presently employed, a tailored methodology can provide a more sector-specific analysis of robotics. The optional Green Robotics Impact Assessment (GRIA) paradigm offers a thorough method to evaluating sustainability performance in organizing robotics across various industries.

Table 4 presents the Green Robotics Impact Assessment (GRIA) framework for evaluating the sustainability impact of robotics. This approach allows administrations to assess robotics on economic performance and comprehensive sustainability and social equality results. Organisations can include GRIA in present reporting and certification frameworks to monitor sustainability progressions.

6 Conclusion

Robotics and automation are progressively reshaping numerous sectors by growing efficiency in operations, productivity, and service value. Their cutting-edge products and services help the hotel sector simplify operations and provide superior service to customers. Glitches with emotional labour, human connection in service distribution, and job security are major issues brought up by their integration. This chapter stresses the significance of a balanced strategy combining technological growth with human-centred design to promote diversity and trust. By harmonizing with the SDGs, robotics can automatically contribute to ecologically friendly, ecologically conscious, and economically sound development. Sustainable robotic incorporation across industries requires organizations to invest in upskilling, adaptable policies, and influence evaluation frameworks.

References

Ajzen, I., & Fishbein, M. (1980). *Understanding attitudes and predicting social behavior*. Prentice-Hall.

Awasthi, A., Pattnayak, K.C., Dhiman, G., & Tiwari, P.R. (Eds.). (2024). *Artificial intelligence for air quality monitoring and prediction*. CRC Press.

Awa, H. O., Ojiabo, O. U., & Orokor, L. E. (2017). Integrated technology-organization-environment (TOE) taxonomies for technology adoption. *Journal of Enterprise Information Management*.

Bae, J., & Chung, W. (2019). Efficient path planning for multiple transportation robots under various loading conditions. *International Journal of Advanced Robotic Systems, 16*(2), 1729881419835110. https://doi.org/10.1177/1729881419835110

Baker, J. (2012). The technology–organization–environment framework. In *Information systems theory* (pp. 231–245). Springer.

Belk, R. (2017). Consumers in an age of autonomous and semi-autonomous machines. In *Contemporary consumer culture theory* (pp. 5–32). Routledge.

Bechar, A., & Vigneault, C. (2016). Agricultural robots for field operations: Concepts and components. *Biosystems Engineering, 149*, 94–111. https://doi.org/10.1016/j.biosystemseng.2016.06.014

Bocken, N. M. P., Short, S. W., Rana, P., & Evans, S. (2016). A literature and practice review to develop sustainable business model archetypes. *Journal of Cleaner Production, 65*, 42–56. https://doi.org/10.1016/j.jclepro.2013.11.039

Brod, C. (1984). *Technostress: The human cost of the computer revolution*. Addison Wesley.

Brougham, D., & Haar, J. (2017). Smart technology, artificial intelligence, robotics, and algorithms (STARA): Employees' perceptions of our future workplace. *Journal of Management & Organization, 24*(2), 239–257. https://doi.org/10.1017/jmo.2016.55

Brown, J. S., Denning, S., Prusak, L., & Groh, K. (2005). *Storytelling in organizations: Why storytelling is transforming 21st-century organizations and management*. Routledge.

Buhalis, D. (2020). Technology in tourism—from ICTs to smart tourism. *Tourism Review, 75*(1), 267–272. https://doi.org/10.1108/TR-06-2019-0258

Buhalis, D., & Leung, R. (2018). Smart hospitality—Interconnectivity and interoperability towards an ecosystem. *International Journal of Hospitality Management, 71*, 41–50. https://doi.org/10.1016/j.ijhm.2017.11.011

Buhalis, D., Harwood, T., Bogicevic, V., Viglia, G., Beldona, S., & Hofacker, C. (2019). Technological disruptions in services: Lessons from tourism and hospitality. *Journal of Service Management, 30*(4), 484–506. https://doi.org/10.1108/JOSM-12-2018-0398

Čaić, M., Odekerken-Schröder, G., & Mahr, D. (2018). Service robots: Value co-creation and co-destruction in elderly care networks. *Journal of Service Management, 29*(2), 178–205. https://doi.org/10.1108/JOSM-07-2017-0179

Cha, S. S. (2020). Customer's intention to use robot-serviced restaurants in Korea. *International Journal of Contemporary Hospitality Management, 32*(9), 2947–2968. https://doi.org/10.1108/IJCHM-01-2020-0046

Choi, Y., Oh, M., Choi, M., & Kim, S. (2020). Exploring the influence of culture on tourist experiences with robots. *Current Issues in Tourism.* https://doi.org/10.1080/13683500.2020.1735318

Chui, M., Manyika, J., & Miremadi, M. (2016). *Where machines could replace humans—and where they can't (yet).* McKinsey Quarterly.

Collins. (2020). Improving human-robot interactions in hospitality. *International Hospitality Review, 34*(1), 61–79. https://doi.org/10.1108/IHR-09-2019-0019

Davis, F.D. (1986). Technology acceptance model (Doctoral dissertation, MIT).

Davis, F. D. (1989). Perceived usefulness, perceived ease of use, and user acceptance of information technology. *MIS Quarterly, 13*(3), 319–340.

Dwivedy, M., & Mittal, R. K. (2010). Challenges faced by e-waste management initiatives in India: an overview. *Waste Management & Research, 28*(11), 1100–1107. https://doi.org/10.1177/0734242X10382978

De Kervenoael, R., Hasan, R., Schwob, A., & Goh, E. (2020). Leveraging HRI in hospitality: Perceived value and empathy. *Tourism Management, 78*, Article 104042. https://doi.org/10.1016/j.tourman.2019.104042

De Koning, K., & Rodenburg, J. (2004). Automatic milking in Europe and North America. In *Automatic milking: A better understanding* (pp. 27–37).

Farrell, J., & Klemperer, P. (2007). Coordination and lock-in. In *Handbook of industrial organization* (Vol. 3, pp. 1967–2072). Elsevier.

Furman, J., & Seamans, R. (2019). AI and the economy. *Innovation Policy and the Economy, 19*(1), 161–191. https://doi.org/10.1086/699936

Huang, M.-H., & Rust, R. T. (2018). Artificial intelligence in service. *Journal of Service Research, 21*(2), 155–172. https://doi.org/10.1177/1094670517752459

International Federation of Robotics. (2020). Executive summary world robotics 2020. https://ifr.org

International Labour Organisation. (2020). *World employment and social outlook: Trends 2020.* ILO Publications.

Ivanov, S., Gretzel, U., Berezina, K., Sigala, M., & Webster, C. (2019). Progress on robotics in hospitality and tourism. *Journal of Hospitality and Tourism Technology, 10*(4), 489–521. https://doi.org/10.1108/JHTT-08-2018-0087

Ivanov, S., Webster, C., & Berezina, K. (2020). Adoption of robots and service automation in tourism. *International Journal of Contemporary Hospitality Management, 32*(6), 1861–1879. https://doi.org/10.1108/IJCHM-04-2020-0330

Ivanov, S., & Webster, C. (2019). Conceptual framework of the use of robots, artificial intelligence and service automation in travel, tourism and hospitality companies. In S. Ivanov & C. Webster (Eds.), *Robots, artificial intelligence and service automation in travel, tourism and hospitality* (pp. 7–37). Emerald Publishing Limited.

Ivanov, S., & Webster, C. (2019). Robots in tourism: A research agenda. *Annals of Tourism Research, 76*, 105–109. https://doi.org/10.1016/j.annals.2019.03.002

Ivanov, S., Webster, C., & Berezina, K. (2017). Adoption of robots by tourism companies. *Revista Turismo & Desenvolvimento, 27*(28), 1501–1517.

Kaplan, A., & Haenlein, M. (2019). Siri, Siri, in my hand: Who's the fairest in the land? *Business Horizons, 62*(1), 15–25. https://doi.org/10.1016/j.bushor.2018.08.004

Kazandzhieva, V., & Filipova, H. (2019). Customer attitudes toward robots. In S. Ivanov & C. Webster (Eds.), *Robots, AI, and service automation* (pp. 79–92). Emerald Publishing.
Khan, M. L. (2017). Social media engagement on YouTube. *Computers in Human Behavior, 66*, 236–247. https://doi.org/10.1016/j.chb.2016.09.024
Kim, W. G. (2011). Effects of hotel information quality on electronic loyalty: Application of TAM. *International Journal of Hospitality Management, 30*(3), 528–536. https://doi.org/10.1016/j.ijhm.2010.08.001
Kuo, C.-M., Chen, L.-C., & Tseng, C.-Y. (2017). Innovative service with hospitality robots. *International Journal of Contemporary Hospitality Management, 29*(5), 1305–1321.
Lee, J., Jin, M. H., & Ryu, G. (2021). Public service motivation and knowledge sharing. *Sustainability, 13*(11), 6286.
Lee, J., Park, D. H., & Han, I. (2008). Effect of negative reviews on attitude. *Electronic Commerce Research and Applications, 7*(3), 341–352. https://doi.org/10.1016/j.elerap.2007.05.004
Makridakis, S. (2017). The forthcoming AI revolution. *Futures, 90*, 46–60.
Manyika, J., Chui, M., Miremadi, M., Bughin, J., George, K., Willmott, P., & Dewhurst, M. (2017). *A future that works: Automation, employment, and productivity*. McKinsey Global Institute.
Marriott International. (2021). Marriott deploys robots to improve guest services. Press Release. https://news.marriott.com
Maurer, M., Gerdes, J. C., Lenz, B., & Winner, H. (2016). *Autonomous driving: Technical, legal and social aspects*. Springer.
McLeay, F., Osburg, V. S., Yoganathan, V., & Patterson, A. (2020). Replaced by a robot. *Journal of Service Research, 24*(1), 104–121.
Moore, G. C., & Benbasat, I. (2000). Adoption of information technology: A framework for research. *ACM SIGMIS Database, 26*(2–3), 60–85. https://doi.org/10.1145/382948.382953
Mori, M. (1970). The uncanny valley. *Energy, 4*(7), 33–35.
Norman, D. A. (1988). *The psychology of everyday things*. Basic Books.
Nowak, K. L., & Biocca, F. (2003). Presence in virtual environments. *Presence, 12*(5), 481–494. https://doi.org/10.1162/105474603322761289
Oliveira, T., & Martins, M. F. (2011). IT adoption models at firm level. *Electronic Journal of Information Systems Evaluation, 14*(1), 110.
Osawa, H., Ema, A., Hattori, H., Akiya, N., Kanzaki, N., Kubo, A., Koyama, T., & Ichise, R. (2017). Robot hotel analysis. In *IEEE RO-MAN 2017*.
Parasuraman, A., Zeithaml, V. A., & Berry, L. L. (1988). SERVQUAL: A multi-item scale for measuring consumer perceptions of service quality. *Journal of Retailing, 64*(1), 12–40.
Pedersen, S. M., Fountas, S., Blackmore, B. S., Gylling, M., & Pedersen, J. L. (2006). Adoption and perspectives of precision farming in Denmark. *Acta Agriculturae Scandinavica, Section B – Soil & Plant Science, 56*(2), 73–82. https://doi.org/10.1080/09064710510008961
Rabenu, E., Tziner, A., & Sharoni, G. (2017). Work-family conflict and stress. *International Journal of Manpower, 38*(8), 1143–1156. https://doi.org/10.1108/IJM-01-2014-0014
Rijsdijk, S. A., Hultink, E. J., & Diamantopoulos, A. (2007). Product intelligence: Its conceptualization, measurement and impact on consumer satisfaction. *Journal of the Academy of Marketing Science, 35*(3), 340–356. https://doi.org/10.1007/s11747-007-0031-8
Rogers, E. M. (2003). *Diffusion of innovations* (5th ed.). Free Press.
Ruiz, A. C. (2020). ICT usage and skills matching. *International Journal of Manpower*. https://doi.org/10.1108/IJM-03-2020-0103
Russell, S., & Norvig, P. (2016). *Artificial Intelligence: A Modern Approach* (3rd Ed.). Pearson Education.
Smith, A., & Anderson, J. (2017). AI, robotics, and the future of jobs. Pew Research Center. https://www.pewresearch.org
Strait, M., Canning, C., & Scheutz, M. (2017). Robot communication strategies. *HRI Conference*, 409–417. https://doi.org/10.1145/2909824.3020258
Tornatzky, L. G., & Fleischer, M. (1990). *The processes of technological innovation*. Lexington Books.

Tarafdar, M., Tu, Q., Ragu-Nathan, T. S., & Ragu-Nathan, B. S. (2007). Technostress and productivity. *Journal of Management Information Systems, 24*(1), 301–328. https://doi.org/10.2753/MIS0742-1222240109

Tung, V. W. S., & Au, N. (2018). Customer experiences with robotics. *International Journal of Contemporary Hospitality Management, 30*(7), 2680–2697. https://doi.org/10.1108/IJCHM-06-2017-0322

United Nations Department of Economic and Social Affairs. (2021). World social report 2021: Reconsidering rural development.

United Nations Environment Programme. (2022). Emissions gap report 2022.

Venkatesh, V., Morris, M. G., Davis, G. B., & Davis, F. D. (2003). User acceptance of information technology: Toward a unified view. *MIS Quarterly, 27*(3), 425–478. https://doi.org/10.2307/30036540

Waters, R. D., & Jones, P. M. (2011). Branding via nonprofit YouTube videos. *Journal of Nonprofit & Public Sector Marketing, 23*(3), 248–268. https://doi.org/10.1080/10495142.2011.594774

Yadegaridehkordi, E., Noor, N. M. M., Ayub, M. N., & Affal, H. (2018). A conceptual model for the successful implementation of customer relationship management in banking sector. *Journal of Relationship Marketing, 17*(1), 45–65. https://doi.org/10.1080/15332667.2018.1440130

Yu, G., & Schwartz, Z. (2006). Forecasting tourism demand with AI. *Journal of Travel Research, 45*(2), 194–203. https://doi.org/10.1177/0047287506291594887

Advancing Global Education Through Open Educational Resources for SDG 4

Geshwaree Huzooree and Mohit Yadav

Abstract

Open Educational Resources (OER) have the potential to significantly contribute to achieving Sustainable Development Goal 4 (SDG 4) by expanding access to education and promoting global educational equity. This chapter, which draws on a review of primary and secondary literature, examines how OER initiatives support marginalized communities and regions with limited infrastructure by providing open access to high-quality educational materials. These initiatives foster inclusive, learner-centered education and facilitate lifelong learning, making education more adaptable and accessible across diverse contexts. The chapter also explores the sustainability of OER, addressing key factors such as user-created content, quality assurance processes, cost-effectiveness, and the role of open licensing in maintaining the relevance and accessibility of resources. Given the challenges surrounding the widespread adoption of OER, this chapter examines key issues such as disparities in digital access, concerns about resource quality, and the need for effective instructional approaches. It also proposes strategies for addressing learners' individual needs and highlights the critical importance of designing rigorous assessments to accurately measure the impact of OER.

Keywords

Open educational resources • Sustainable development goal 4 • Global education • Educational equity • Lifelong learning

G. Huzooree (✉)
The Business School, Edinburgh Napier University, Edinburgh, UK
e-mail: a.huzooree@napier.ac.uk

M. Yadav
Human Resource Department, O.P. Jindal Global University, Sonipat, India
e-mail: mohitaug@gmail.com

T. Choudhury et al. (eds.), *Open Access as a Knowledge Platform for Implementing the United Nations Sustainable Development Goals*, World Sustainability Series,
https://doi.org/10.1007/978-3-031-99605-4_5

1 Introduction

Education is a foundational element of sustainable development, enabling individuals to overcome social and economic barriers, foster personal growth, and contribute meaningfully to their communities. The United Nations established Sustainable Development Goal 4 (SDG 4) to "ensure inclusive and equitable quality education and promote lifelong learning opportunities for all," acknowledging the important role of education in fostering sustainable development (UNESCO, 2019). SDG 4 emphasizes education not merely as a right but as an essential factor for addressing global challenges such as poverty, gender inequality, and social exclusion. To achieve these goals, countries are urged to provide quality education that adapts to evolving societal learning needs and leverages the expanding capabilities of digital technologies, particularly information technology and social software. Despite global progress in expanding access to education, significant disparities persist, particularly in low-income and marginalized communities. Millions of children, youth, and adults continue to face barriers that prevent them from accessing quality education, including inadequate infrastructure, insufficient learning materials, and a lack of trained educators (Huang et al., 2020). The digital divide further intensifies these inequalities, with limited access to technology and internet connectivity disproportionately affecting rural and underserved populations (Coleman, 2021). The COVID-19 pandemic has intensified these challenges, disrupting traditional education systems and highlighting the need for innovative, scalable solutions.

The rise of Open Educational Resources (OER) over the past decade has been seen as a significant step toward addressing disparities such as inequalities and socio-economic barriers (Mishra, 2017). Accessible, high-quality educational content is now available through a variety of open initiatives, including open textbooks, digital content repositories, and platforms, benefiting a global community of educators, students, and lifelong learners (Menzli et al., 2022; Wiley et al., 2014). Education has evolved to embrace cost-effective learning and knowledge-sharing models, with academic institutions globally adopting open access, distribution, and collaboration. This shift has led to the growing integration of openness, collaboration, and volunteerism within educational environments (Butcher & Moore, 2015). The importance of OER has been recognized globally through initiatives like UNESCO's OER Recommendation, which calls for member states to adopt supportive policies, build capacity, and promote the use of open resources (UNESCO, 2019). These efforts point out the potential of OER to bridge gaps in education, enhance teaching practices, and contribute to the achievement of SDG 4.

Given the global recognition and growing adoption of OER, this chapter explores their potential to advance SDG 4 by expanding equitable access to quality education worldwide. It examines how OER initiatives can address educational disparities, particularly in regions with limited infrastructure and resources, while promoting learner-centered and inclusive pedagogies.

This chapter is organized to provide a thorough examination of OER. It begins with an introduction that highlights the importance of OER in advancing educational equity worldwide. Following this, the chapter explored the key characteristics of OER, including their definition and different types. It then discusses the role of OER in promoting educational equity, highlighting how they facilitate access to education, support inclusive and learner-centered practices, and encourage lifelong learning. The chapter also addresses the sustainability of OER, addressing critical factors such as user-created content, quality assurance, and open licensing. The chapter also tackles the challenges associated with the adoption and implementation of OER, including issues like digital access gaps, concerns over resource quality, and the need for effective instructional strategies, offering practical solutions to mitigate these challenges. Finally, the chapter concludes with key insights and practical recommendations for educators, institutions, and policymakers to maximize the potential of OER in creating more equitable educational systems.

2 Understanding Open Educational Resources (OER)

OER have emerged as a transformative concept in global education, fostering accessibility and innovation. The term was first officially recognized during UNESCO's Forum on the Impact of Open Courseware for Higher Education in Developing Countries (UNESCO, 2002). At this forum, OER was initially defined as "open provision of educational resources, enabled by information and communication technologies, for consultation, use and adaptation by a community of users for non-commercial purposes" (UNESCO, 2019). This early understanding positioned OER as a tool for ensuring equitable access to quality higher education, particularly in developing countries where resource constraints limit participation in rapidly evolving global educational systems. At the time, OER was primarily associated with freely available university-level course materials.

Over the years, the scope and definition of OER have broadened significantly, reflecting the dynamic nature of educational practices and technological advancements. The William and Flora Hewlett Foundation, a pioneer in supporting the OER movement, now defines OER as encompassing a diverse array of materials, including "courses, textbooks, streaming videos, exams, software, and any other materials or techniques supporting learning tools and content" (COL, 2017; Hewlett Foundation, 2013). This expanded definition underscores that OER are not only content repositories but also include platforms and tools that enable learners to engage in deeper learning, emphasizing skills such as critical thinking, effective communication, and problem-solving.

One of the defining characteristics of OER is their adaptability. Unlike traditional educational resources, OER can be edited, modified, and remixed to suit specific cultural, institutional, or pedagogical needs. This flexibility has been described by the OECD as accumulated digital assets that can be adjusted (OECD,

2007). Such adaptability is essential in ensuring that educational materials are relevant and effective across diverse learning contexts. Moreover, the emphasis on openness and sharing fosters a culture of collaboration and innovation, empowering educators and learners alike to contribute to and benefit from collective knowledge.

The significance of OER goes beyond their definition; it lies in their potential to address systemic educational challenges. As UNESCO (2005) highlighted during its Second Global Forum on International Quality Assurance, Accreditation, and the Recognition of Qualifications in Higher Education, OER have evolved from being static resources to becoming tools for supporting learner communities. This shift reflects a growing recognition of the need for education to be inclusive, interactive, and socially connected.

OER play a crucial role in promoting equity and lifelong learning, aligning with the broader objectives of global educational initiatives such as the Sustainable Development Goals (SDGs) (Swain & Pathak, 2024). Tonegawa (2022) further reinforced this perspective that SDG 4 seeks to ensure inclusive and equitable quality education for all and promote lifelong learning opportunities. OER align with this vision by offering free, high-quality resources that can be accessed anytime, anywhere, enabling learners to overcome barriers related to geography, socioeconomic status, or institutional limitations. This is particularly significant in an era marked by workforce mobility and the increasing need for flexible learning pathways. OER enable learners to continue their education without interruption, even when traditional face-to-face learning environments are not feasible.

Moroever, OER contribute significantly to the development of key competences essential for success in a globalized and knowledge-driven world. Scholars emphasize the importance of lifelong learning and the creation of a "knowledge triangle" that links education, research, and innovation (Marczewska et al., 2024). Through opportunities for collaborative and interdisciplinary learning, OER support this framework by facilitating the acquisition of skills crucial for personal and professional development, such as adaptability, teamwork, and digital literacy.

However, the global OER movement is not without its challenges. One key issue is the dominance of initiatives from English-speaking, developed countries, which has raised concerns about inequities in the production and consumption of resources. The global OER ecosystem should not function as a one-way street where developed countries produce resources and developing countries remain passive consumers (King et al., 2018). Instead, there is a need to empower institutions and communities in less-developed regions to contribute to OER creation, ensuring that resources reflect diverse perspectives and contexts.

The active promotion of free and open sharing remains central to the OER movement. This ethos is grounded in the belief that education should be a collaborative endeavor, where resources and practices are shared to facilitate change and innovation (OECD, 2007). By fostering a culture of openness, OER have the potential to reshape not only educational practices but also broader societal structures, promoting equity, inclusion, and sustainability.

3 The Role of OER in Promoting Educational Equity

OER present a transformative solution to the educational challenges faced across both developing and developed regions by offering an innovative way to deliver high-quality education with fewer resources (Hewlett Foundation, 2013). The ongoing resource constraints in higher education systems, combined with a growing preference for technology-driven solutions, have created fertile ground for the widespread adoption of OER as a means of enhancing educational outcomes.

3.1 Democratizing Access to Education

A fundamental benefit of OER is its ability to democratize access to education by breaking down traditional barriers. Unlike proprietary educational resources that require significant financial investment, OER is free and openly accessible to anyone with an internet connection. This accessibility is a critical step toward bridging the equity gap in education, enabling more students to engage with educational content regardless of their geographic location, socioeconomic status, or institutional affiliations (Costa et al., 2022). For example, in remote or underserved areas where physical educational infrastructure may be limited or nonexistent, OER such as MOOCs provides an alternative pathway for learners to access high-quality materials (Barger, 2020). This ensures that education is no longer confined to traditional settings but extends to marginalized communities that have historically been excluded from mainstream educational opportunities.

The success of OER in these settings, however, is contingent upon addressing persistent digital access inequalities. Many learners in low-resource contexts lack stable internet connectivity or appropriate devices, making offline or low-bandwidth versions of OER critical for equitable access. Localized, culturally relevant content also plays a pivotal role in ensuring that learners can meaningfully engage with materials. OER initiatives that integrate local languages, contexts, and educational norms not only enhance relevance but also foster a sense of ownership and community engagement, key factors in long-term adoption and impact.

Moreover, the global reach of OER aligns with the principles of equitable education, as it allows learners in low-income countries to access the same materials used in well-resourced institutions (Mncube et al., 2021). Initiatives such as the Massachusetts Institute of Technology's OpenCourseWare (OCW) have demonstrated how openly sharing high-quality course content can help close the educational divide and inspire innovation worldwide (Abelson, 2008).

3.2 Fostering Inclusive and Learner-Centered Education

OER's adaptability makes it a powerful tool for fostering inclusive education. Unlike traditional resources, OER is not static but can be customized to meet the specific needs of diverse learners and contexts. The ability to modify and translate content into multiple languages supports inclusivity by making educational materials accessible to non-native speakers and culturally diverse populations (Fatayer, 2024).

Inclusive education also benefits from OER's potential to support learners with disabilities. By providing resources in multiple formats, such as text, audio, and video, OER ensures that learners with varying needs can engage with content in ways that suit them best. For instance, visually impaired learners can use screen readers to access textual content, while auditory learners can benefit from audio recordings (Zaid et al., 2024).

The adaptability of OER also fosters learner-centered pedagogies, enabling educators to tailor materials to their students' interests, prior knowledge, and learning preferences (Zhou et al., 2019). This personalization empowers students to take charge of their learning journey, fostering deeper engagement and a sense of ownership. Research has shown that when learners engage with materials that reflect their experiences and aspirations, they are more likely to remain motivated and achieve positive outcomes (Huzooree & Doargajudhur, 2023).

3.3 Supporting Lifelong Learning

Another critical dimension of OER's role in promoting educational equity is its capacity to support lifelong learning (Onopriienko et al., 2023). As economies evolve and the demand for new skills emerges, access to high-quality, flexible learning resources becomes indispensable. OER facilitates continuous education by providing opportunities for learners of all ages to acquire new knowledge and skills at their own pace (Nykyporets et al., 2023).

In a rapidly changing world, where upskilling and reskilling are essential for staying competitive, OER offers an inclusive and affordable solution. Adults who may have missed out on formal education or are seeking to switch careers can access OER to bridge skill gaps or explore new fields of interest. Additionally, many OER platforms, such as Khan Academy and Coursera, are designed with self-directed learning in mind, further enhancing their utility for lifelong learners (Arnavut et al., 2019).

For professional development, OER enables educators and practitioners to stay current with the latest advancements in their fields. Open resources like research articles, case studies, and interactive tools provide cost-effective ways to ensure continuous growth without the need for expensive conferences or subscription-based journals.

4 The Sustainability of OERs

As technology continues to evolve, the demand for accessible and quality educational resources grows worldwide, especially in the context of expanding and diverse educational landscapes. Resource-producing organizations that contribute to OER are increasingly challenged to keep pace with these rapid developments while ensuring sustainability and maintaining high-quality content (Baker & Dutton, 2017). OER initiatives face the dual challenge of expanding access to education while addressing issues of financial stability, content quality, and user engagement in an environment that is constantly shifting (Farrell et al., 2022). While OER platforms promise to democratize education by providing freely available resources, they also require long-term planning to ensure their continued viability, particularly as funding sources become more uncertain (Kitchin et al., 2015; Ndebele et al., 2023). As the use of OER grows, a key concern for organizations and stakeholders is how to maintain the quality and relevance of user-generated content, which often raises questions regarding accuracy, reliability, and educational effectiveness (Naab & Sehl, 2017; Nilashi et al., 2022). Additionally, there is the ongoing challenge of scaling these initiatives to meet the needs of diverse learner populations, including those in under-resourced areas or in regions with limited access to digital infrastructure (Kukulska-Hulme et al., 2024; Zhang, 2023). This necessitates an approach that balances accessibility with sustainability, ensuring that resources are not only freely available but also adaptable to local contexts and needs (Curtis, 2021). Moreover, as OER initiatives continue to expand globally, the need for ongoing advocacy around the importance of open licensing and copyright frameworks becomes more pressing, as these legal considerations play a critical role in protecting both creators and users while supporting the free exchange of educational content.

4.1 User Created Content and OERs

The rise of OER has transformed the traditional landscape of educational content creation, offering new opportunities for diverse individuals and communities to participate in the development and dissemination of learning materials. Unlike conventional educational resources, which are predominantly produced by experts or institutions, OERs allow users from various backgrounds to create, modify, and share content freely. This collaborative approach aligns with the core principles of openness, accessibility, and inclusion that define the OER movement. However, the involvement of non-expert contributors introduces both significant opportunities and challenges, particularly in terms of ensuring the quality, reliability, and long-term sustainability of OER initiatives.

One of the primary advantages of user-generated content in OER is the potential to increase access to educational resources across a wide range of contexts. As Perryman and Coughlan (2014) emphasize, social curation, where users actively curate and share content can significantly enhance the visibility and discoverability of educational materials. This model allows content to be more widely accessible, particularly in regions where educational resources may be limited or costly. The openness of OER platforms enables anyone with internet access to contribute, making educational content available to a broader audience, regardless of geographic location or economic status.

However, while user-generated content can help broaden access to educational materials, it also raises concerns about the quality and reliability of these resources. McShane (2017) points out that there is often a perception that OER created by volunteers or non-experts may not meet the same standards as resources developed by recognized scholars or academic institutions. This can undermine trust in the content, especially in disciplines where accuracy and credibility are paramount, such as science, medicine, or law. Despite these challenges, research in other fields, such as open-source software development and the gaming industry, shows that user-generated content can be both innovative and high-quality when appropriately managed (Munir et al., 2018). These industries provide valuable insights into how user contributions can enhance the overall value of a product or resource, even when created outside traditional professional structures.

To address concerns about the quality of user-generated content in OER, several strategies have been proposed (Ganapathi, 2019). Peer review, for example, is a widely recognized method for ensuring that materials meet academic standards. By engaging subject matter experts to review and validate content before publication, OER platforms can help maintain the accuracy, reliability, and educational value of resources. Additionally, the collaborative nature of many OER platforms allows for the continuous refinement of content. Users can adapt, update, and improve existing materials, ensuring that they remain current and relevant to changing educational needs. This iterative process of collaboration and feedback is one of the key strengths of OER, enabling resources to evolve and improve over time.

In addition to peer review and ongoing collaboration, the success of user-created content in OER depends on fostering active and engaged communities of contributors. Research from the National University of Distance Education (UNED) in Spain illustrates the potential of user-generated content in higher education (Rodrigo & García-Serrano, 2015). In this model, students were trained to create and share educational materials, which significantly increased both the quantity and quality of resources available. The involvement of users in content creation not only enhances the value of the materials but also fosters a sense of ownership and engagement, which is crucial for the long-term sustainability of OER projects.

However, there are ethical and practical challenges associated with user-created content, particularly regarding the issue of unpaid labor. While volunteer contributions are common in OER initiatives, concerns about exploitation can arise if contributors are not adequately supported or compensated for their work. As Terras (2015) notes, unpaid labor in the digital sphere can lead to ethical issues, especially if volunteers are not formally recognized or rewarded for their contributions. To mitigate these challenges, some OER organizations, such as Siyavula, have implemented formal agreements with volunteers, establishing clear expectations and providing support for contributors (Lambert, 2019). These agreements help ensure that volunteers are treated fairly and that their work is valued, while also maintaining high standards for the quality of the content produced.

4.2 Quality Assurance and OERs

Quality assurance (QA) plays a pivotal role in maintaining the effectiveness and long-term success of OERs. As these resources are designed and shared in an open, collaborative environment, ensuring their quality is vital not only for their credibility but also for their educational impact. With the rapid growth of OERs, where digital content is continually created and shared globally, QA has become an increasingly important factor in determining the utility and sustainability of these resources (Connell & Connell, 2020). Effective QA systems not only measure the quality of resources but also ensure their continuous improvement over time.

The process of ensuring quality in OERs begins at the creation stage. Just as traditional educational resources undergo rigorous development processes, OERs require a structured approach to ensure they meet high educational standards. This includes adhering to pedagogical guidelines, ensuring content accuracy, and aligning resources with the needs of learners and educators (Meng, 2023). Research indicates that the success of OER platforms depends heavily on the quality of the materials they host, and platforms that fail to maintain high standards often struggle to attract and retain users (Perifanou & Economides, 2023). The creation of OERs should thus be guided by clear criteria that help contributors focus on educational objectives, technical quality, and relevance. This is especially critical in regions with limited resources, where poorly curated materials can undermine learner confidence and educational outcomes.

In many OER platforms, expert review processes are key to ensuring that resources meet the required standards before being published. Platforms like Siyavula (Lambert, 2019) and Pratham Books (Ganapathi, 2019), for example, rely on expert committees to assess the content. These committees are responsible for evaluating the materials for pedagogical quality, accuracy, and appropriateness for the intended audience. However, expert review systems are not without challenges. They can be resource-intensive, particularly for non-profit platforms that may not have the budget to maintain a large team of experts. In these cases, alternative approaches to QA, such as crowd-sourced or collaborative review systems, offer viable solutions. These more participatory models not only reduce costs but also enable locally relevant insights to shape content quality.

The cultural relevance of OER content is another critical dimension of quality. Materials that align with local languages, values, and social contexts tend to see higher rates of acceptance and usage. Ignoring cultural nuances can lead to disengagement or even rejection of otherwise high-quality materials. Collaborative transparency has become a significant method for managing quality in OERs (Ramirez-Montoya, 2020). Many platforms now involve users in the process of quality review, allowing educators, students, and other stakeholders to provide feedback, suggest improvements, and report errors. This approach helps maintain the quality of the resources while also fostering a sense of ownership and engagement within the OER community. It also allows for the continuous updating of resources, ensuring that they remain current and relevant without requiring extensive resources from the hosting platform (Schnackenberg, 2019). By encouraging user participation, OER platforms can benefit from diverse perspectives and collective expertise, which enhances the reliability and educational value of the resources.

Transparency in the QA process also plays a crucial role in building trust (Luo et al., 2020). When users and contributors are aware of the standards and review processes behind the resources, they are more likely to trust the quality of the content. For example, platforms that attribute content to specific authors and contributors increase the credibility of their resources, as attribution helps ensure accountability and recognition. This trust-building mechanism is essential for sustaining an active and committed user base, as evidenced by Pratham Books, which has seen an increase in volunteer contributions by acknowledging the efforts of its community (Ganapathi, 2019). Such transparent practices help reinforce the value of the OER platform and motivate further contributions.

While expert committees are often a gold standard in QA, they are not always feasible for OER platforms operating on tight budgets. Collaborative models, where users contribute to the review and improvement of resources, offer a practical solution. These models engage the broader educational community, fostering a sense of shared responsibility for maintaining the quality of the resources. Moreover, the reputation system where contributors are acknowledged, and their work attributed also serves to motivate ongoing participation and commitment to high-quality content creation. Community engagement in these models is particularly valuable in underprivileged regions, where local educators and learners can shape resource quality based on lived experience and context.

4.3 Cost, Sustainability and OERs

The financial aspects of OERs are critical to their long-term viability, especially for organizations operating within the constraints of social business models (Darwish, 2019; Muthanga et al., 2023). While OERs are designed to be freely accessible, their creation, distribution, and maintenance incur substantial costs that need to be addressed for these resources to remain sustainable over time. A key challenge in this regard is securing consistent and diverse sources of funding to cover these costs. As discussed in the literature, many OER initiatives rely on a combination of government funding, private sector support, and philanthropic donations to maintain their operations (Hoque & Abdullah, 2021; Rohayati et al., 2016). However, this reliance on external funding poses both opportunities and challenges that significantly impact the financial sustainability of OERs.

Creating and distributing high-quality OERs requires significant investment in both human and technological resources. Content creation demands the expertise of subject matter experts, instructional designers, editors, and graphic designers, all of whom require compensation for their time and skills. For organizations like Khan Academy and OpenStax, the cost of employing skilled professionals can be prohibitive, especially given their non-profit status. To alleviate these financial constraints, both organizations supplement their operations with volunteer labor (Holl et al. 2024). Volunteers play a critical role in producing and curating educational content, which helps reduce costs. However, the reliance on volunteers also introduces challenges related to the quality of the materials produced. Volunteers may not always have the same level of expertise or commitment as paid professionals, and this can affect the QA processes that are vital for maintaining high standards in OER production.

Philanthropic fundraising is a common strategy used by OER providers to support the creation of materials and the infrastructure needed to disseminate them. As Rohayati et al. (2016) note, higher education institutions and organizations operating within the OER space often face increasing financial pressure, with government funding becoming less predictable. For these organizations, securing donations and grants from foundations, individual donors, and corporate sponsors is essential for their survival. However, while such fundraising efforts are crucial, they can also limit the autonomy of OER providers. Many funding organizations have specific expectations about how their money should be spent, and they often direct funding towards projects that align with their own priorities. This dynamic can constrain the flexibility of OER providers, forcing them to prioritize projects that may not be aligned with their core mission or goals.

Moreover, relying heavily on external funding sources can lead to financial instability. There is no guarantee that funding will continue indefinitely, and organizations must constantly seek new sources of income to stay afloat. Funding organizations may choose to support only "rewarding" projects, which can limit the ability of OER providers like Khan Academy to manage the direction of their initiatives (Lambert, 2019). This dependence on external stakeholders for financial support can create tensions between maintaining an open, flexible business model and meeting the expectations of funders.

The relationship between funding and content quality is another critical aspect of OER sustainability. For organizations like Khan Academy and OpenStax, employing professional illustrators, designers, and other experts is essential to producing high-quality educational materials (Holl et al. 2024; Luo et al., 2020). However, without sufficient funding, these organizations must rely on volunteers or less experienced contributors, which can compromise the overall quality of the resources. This reliance on volunteer labor can also create a two-tiered system within OER production, where some materials are produced by highly skilled professionals funded through external sources, while others are created by volunteers with limited experience or resources. This disparity in content quality can raise concerns about the overall effectiveness and credibility of OERs.

As Ahlf et al. (2024) argue, financial sustainability is a crucial but often neglected issue in the OER literature. Many OER initiatives focus on making educational materials available at no cost, but fail to consider the financial models needed to ensure the ongoing production and improvement of these resources. The case study organizations, however, have adopted a more proactive approach to financial sustainability, seeking multiple funding streams and business models to support their operations. For instance, OpenStax, backed by the philanthropic foundation of Rice University, uses a combination of grants, corporate sponsorships, and partnerships to fund their work (Stafford & Flatley, 2018). This diversified approach to funding allows them to maintain their open access model while still covering the costs of content creation, curation, and distribution.

However, there is a downside to this approach: the dependence on external funding can influence the direction of content creation. As scholars highlight, corporate sponsors often seek to align the educational materials with their branding or business interests, which can undermine the open and non-commercial nature of OERs (Holl et al. 2024; Lambert, 2019). In some cases, this can lead to the development of a two-tiered system in which well-funded materials are of high quality, while less-funded resources produced by volunteers may be of lower quality. This creates a potential imbalance in the resources available to learners, as well as concerns about the integrity and impartiality of the content.

To mitigate some of the financial and quality-related challenges, many OER providers, such as Khan Academy and OpenStax, utilize open licensing systems, such as Creative Commons, to manage the distribution and use of their resources. Open licenses allow these organizations to share their materials freely while still retaining some control over how they are used and adapted. This approach also facilitates collaboration with other educational institutions, organizations, and individuals, enabling them to contribute to the ongoing development and improvement of the resources.

Moreover, open licensing can help organizations overcome the challenges of copyright, especially when using public domain materials or content whose copyright has expired (Otero et al., 2019; Norris et al., 2023). By leveraging open licensing systems and blockchain technology, OER providers can expand the range of materials available to users and encourage other organizations to adopt and distribute their resources (Jones, 2018). This collaborative model of content creation and sharing also helps to build a community of educators, students, and content creators who can collectively support the ongoing development of high-quality, open resources.

However, while open licensing provides an effective way to share content, it also introduces the challenge of ensuring that all contributors adhere to quality standards. To address this, OER providers often rely on community-driven approaches or OER-enabled pedagogy project to content creation, which involve educators and users in the process of reviewing, revising, and improving resources (Williams & Werth, 2021). This collaborative approach helps ensure that OERs remain relevant and up-to-date while also fostering a sense of ownership and engagement among users.

The challenge of maintaining a sustainable business model for OERs is particularly acute in developing countries, where access to technology and infrastructure can be limited. For organizations like Khan Academy and OpenStax, OERs provide a way to expand access to quality educational materials, but sustaining this model requires ongoing investment in content creation, technology, and community engagement. The reliance on volunteers, while cost-effective, raises questions about the consistency and quality of the materials produced. Similarly, external funding sources, while necessary, can limit the autonomy of OER providers and create pressures to align content with the interests of funders.

4.4 Open Licensing and OERs

Open licensing plays an essential role in shaping the OER landscape by providing a legal framework that enables the free sharing, adaptation, and distribution of educational materials. As Butcher and Moore (2015) emphasizes, open licenses are the critical tools that define OERs, allowing them to be openly accessed and reused in ways that traditional copyright models do not permit. The flexibility of open licenses, especially those offered by Creative Commons (CC), has been pivotal in enabling educational institutions, organizations, and individuals to participate in the OER movement by offering resources that are both cost-effective and scalable. The main objective of OERs is to make educational content accessible to a wider audience, and open licensing is a key mechanism that facilitates this goal.

The permissiveness of open licenses is central to their success in the OER movement (Imberman & Fiddler, 2019). One of the most significant advantages of CC licenses is that they allow users to modify and create derivative works from the original content. This ability to adapt materials to specific needs such as translating them into different languages or tailoring them to suit local curricula enhances the inclusivity of educational resources. For instance, OER providers can upload textbooks or instructional materials that others can then alter, remix, or expand upon, making them more relevant to different cultural and educational contexts. This feature is especially important in countries with diverse linguistic and cultural landscapes, where one-size-fits-all resources may not be effective in reaching all learners. By making materials adaptable, open licensing helps overcome language barriers and cultural differences, allowing educators to customize resources for their specific students.

Moreover, open licensing can breathe new life into older, outdated educational materials (Herrera-Cubides et al., 2022; Schuwer & Baas, 2023). Traditional textbooks and other educational resources often become obsolete due to curriculum changes or advances in the field. However, through the use of open licenses, these materials can be updated, re-purposed, or republished, ensuring their continued relevance and usefulness. For example, institutions and educators can republish older content in new formats such as digital texts, audio, or video, which makes it accessible to a broader audience. This process of reviving and updating resources helps ensure that valuable educational content is not lost and that learners continue to benefit from it. Libraries, museums, and archives have increasingly adopted CC licenses to make public domain materials freely available online, demonstrating how open licensing can help preserve and distribute educational content that might otherwise remain inaccessible.

The affordability of OERs is another key benefit of open licensing (Mullens & Hoffman, 2023). Unlike traditional educational materials, which are often costly due to publishing fees, royalties, and licensing charges, OERs can be distributed without the financial burdens associated with traditional copyright protection. This cost-effective nature of open licensing allows educational resources to be made available to a broader audience, particularly in resource-constrained environments where purchasing expensive textbooks or subscribing to educational databases is not feasible. As Bammel (2014) notes, the creation of derivative works is often more accessible and cost-efficient than developing entirely new content. This makes OERs a sustainable option for educators, especially in low-resource settings, where budget limitations can severely restrict access to high-quality materials.

The flexibility of CC licenses also enables OER providers to maintain control over how their content is used and distributed. OER providers can choose from various CC license options to specify the level of freedom they wish to grant users. For instance, a CC BY (Attribution) license allows users to modify, adapt, and redistribute content, as long as they provide proper attribution to the original author. On the other hand, a more restrictive CC BY-ND (No Derivatives) license prevents users from creating derivative works, ensuring that the original content is preserved in its entirety. The ability to select the most appropriate license for specific contexts allows OER providers to balance openness and control, ensuring that content is used ethically while still encouraging innovation and collaboration.

While the advantages of open licensing are clear, there are also challenges associated with its adoption. One significant barrier is the lack of awareness and understanding of CC licenses, particularly among educators. Many educators and institutions are still unfamiliar with the concept of open licensing and may be hesitant to release their content under open terms due to concerns about losing control or being taken advantage of. As Veletsianos (2020) observes, despite the widespread availability of CC materials, only a small percentage of educators actively create and publish OERs under open licenses. This lack of engagement with open licensing can limit the growth and impact of the OER movement, as it restricts the availability of freely accessible resources.

Moreover, concerns about copyright infringement and the potential misuse of open content persist. Although CC licenses provide clear guidelines about how content can be used, there is still confusion among some users regarding the permissions and restrictions associated with different licenses. For example, the CC BY-ND license, which restricts the creation of derivative works, may be misunderstood by users who are unaware that they cannot modify or remix the content. Therefore, it is essential for OER providers to educate users about the specific terms of the licenses they use and ensure that the licensing conditions are clearly communicated to avoid confusion and legal issues.

Another challenge is the reluctance of some governments and organizations to embrace open licensing fully (Khayyat & Bannister, 2015). In some cases, governments may prefer more restrictive licensing models, such as CC BY-ND, because they provide greater control over the materials and limit the potential for misuse or unauthorized modifications. This hesitation to adopt open licenses may stem from a lack of understanding of the potential benefits of OERs or a desire to maintain control over educational content. However, such restrictive approaches can hinder the broader goals of the OER movement, which aims to create a more open and accessible educational ecosystem.

Despite these challenges, the continued adoption of open licensing is crucial for the growth and sustainability of the OER movement. As Ajiga et al (2024) and Soncin et al. (2022) argue, CC licenses offer a flexible, cost-effective, and scalable approach to sharing educational resources. By allowing content to be freely used, adapted, and distributed, open licensing enables OERs to reach a wider audience, cater to diverse educational needs, and remain relevant in a constantly evolving educational landscape. For OER initiatives to thrive, it is essential for educators, institutions, and governments to increase their awareness of the benefits of open licensing and to support the widespread adoption of CC licenses as a standard for educational materials.

4.5 Copyright and OERs

The relationship between copyright law and OERs is critical in ensuring that educational materials are accessible while respecting the rights of creators (Norris et al., 2023; Otero et al., 2019). Copyright law serves as a legal mechanism to protect intellectual property, ensuring that creators have control over the use, reproduction, and distribution of their work. Traditionally, this has meant that educational content is often subject to restrictive usage rights, limiting access and redistribution. However, OERs aim to challenge these barriers by promoting the free use, modification, and distribution of educational materials. This paradigm shift requires a delicate balance between the traditional protection of creators' rights and the mission to make knowledge freely available.

For many OER providers, copyright restrictions can be seen as an obstacle to achieving their goal of universal access to educational resources (Luo et al., 2020). Copyrighted works often come with conditions that limit how and where they can be used, creating unnecessary hurdles for educators, students, and institutions. Moreover, educational materials that are locked behind paywalls or proprietary licenses are inaccessible to many who would benefit from them, particularly in underserved communities or regions with limited access to resources (Maristela, 2024). OERs, on the other hand, aim to dismantle these barriers by offering educational content that is freely accessible and reusable. CC licenses have emerged as the key tool enabling this transformation. By providing a clear and flexible licensing system, CC licenses allow creators to share their work openly while still maintaining control over certain aspects, such as attribution and modifications.

CC licenses are an attractive alternative to traditional copyright because they provide creators with the ability to choose how their work is used without the need for complex legal processes. These licenses come in various forms, ranging from the most permissive, such as the CC BY (Attribution) license, which allows anyone to use, adapt, and share the work as long as they credit the original creator, to more restrictive options like the CC BY-ND (No Derivatives) license, which prohibits any changes to the work. For OER providers, this flexibility is invaluable.

By using CC licenses, creators can make their educational resources available to a wide audience, while also ensuring that they receive proper recognition and that their intellectual property is protected.

Despite the advantages of CC licenses, challenges arise in how these licenses are understood and applied, especially among those who are not familiar with copyright law. Copyright can be a complex and confusing concept, particularly for individuals who are not trained in intellectual property law. This lack of understanding can lead to misuse or misattribution of works, which may inadvertently violate copyright law or undermine the creator's rights. Moreover, in the case of OERs, many creators, particularly those who are new to the concept, may not be aware of the different types of CC licenses and the specific permissions they grant. This can create confusion about what is allowed under each license and can lead to unintentional infringement.

For example, many educators and content creators may not fully understand the implications of releasing their work under a CC license (Fleming et al., 2021). While the intention is often to make resources more widely available, the fear of losing control over how the content is used can deter creators from adopting open licenses. The emotional attachment that creators have to their work may make them hesitant to release it under a license that allows others to remix or adapt it. This attachment is often tied to a sense of ownership and authorship, which is a central concept in copyright law. When creators release their work under an open license, they are giving up some degree of control, which can be a significant hurdle.

Additionally, the lack of education and awareness around copyright and licensing among creators can hinder the growth of OER initiatives (Oriakhogba & Ndlovu, 2024). Many creators may not be aware of the potential benefits of open licensing, such as increased visibility, collaboration, and the opportunity to have their work adapted and reused in diverse contexts. Without a clear understanding of how open licensing works and the benefits it can provide, creators may be reluctant to participate in the OER movement. This highlights the need for OER platforms and institutions to invest in educating creators about the value of open licenses and how they can use these licenses to share their work freely while still protecting their rights.

One of the key challenges for OER providers is ensuring that the creators of educational resources understand the terms of Creative Commons licenses and how to choose the appropriate license for their work (Imberman & Fiddler, 2019). Different types of licenses allow for different levels of use and modification, and it is essential that creators understand the implications of each choice. In some cases, platforms may need to offer guidance or resources to help creators make informed decisions. This could involve providing clear explanations of the various CC license options, offering training on copyright and open licensing, and even working directly with creators to ensure that they understand how to apply the licenses correctly.

For OERs to be successful, it is crucial that both creators and users understand how to navigate the complexities of copyright law and open licensing (Costa et al., 2022; Luo et al., 2020; Mncube et al., 2021). This requires a concerted effort by OER platforms, educational institutions, and policymakers to raise awareness about the importance of copyright and the value of open access. Educating creators, users, and institutions about the potential of open licensing and OERs can help to dispel fears and misconceptions about copyright infringement and encourage broader participation in the OER movement (Swain & Pathak, 2024). As more creators and users understand the benefits of open licensing, the adoption of OERs is likely to increase, further promoting access to educational materials and fostering collaboration across educational communities.

5 Challenges and Strategies in OER Adoption and Implementation

5.1 Access to OER Does Not Guarantee Learning Success

While OER and Open Educational Practices (OEP) are frequently seen as key solutions for promoting equitable access to education, the mere availability of OER does not guarantee an improvement in learning outcomes. Research has shown that transitioning from proprietary to open educational content does not inherently lead to enhanced student achievement (Rodríguez-Abitia et al., 2020). Simply making resources freely available, without addressing the broader context in which they are used, may not yield significant educational benefits. This is especially true in marginalized communities where digital divides persist, learners may lack access to stable internet, digital devices, or the necessary digital literacy to navigate open content effectively.

A more holistic approach is necessary for realizing the full potential of OER. This includes not only changing the licensing model but also revising instructional methods, adapting the design of the materials, and considering the needs and preferences of learners. Studies have highlighted that factors such as the quality of the OER, the effectiveness of the teaching strategies employed, and individual learner characteristics play a crucial role in determining the impact on learning outcomes. Addressing digital access inequalities through community-based internet initiatives, mobile-compatible content, and low-bandwidth versions of OER can improve inclusivity and support learning in resource-constrained settings.

For instance, OERs must be pedagogically sound, engaging, and aligned with curriculum goals (Zhou et al., 2019). Without these elements, the mere shift to open access may not result in more effective learning. Additionally, learners' ability to engage with and benefit from OER depends on factors like digital literacy, access to technology, and personal motivation. Efforts to provide digital skills training and localized support networks can help bridge these gaps and create more enabling environments for learners. In short, the potential of OER to enhance learning requires a comprehensive and thoughtful implementation strategy that goes beyond just making content freely available.

5.2 Quality of OER and Its Impact on Learning Outcomes

The quality of OER plays a critical role in determining their effectiveness as tools for enhancing student learning. High-quality OER, which are accurate, up-to-date, and designed with clear learning objectives, have been shown to positively impact student outcomes. Resources that are well-aligned with the curriculum and designed to engage learners tend to support deeper understanding and more effective skill acquisition. When students have access to well-structured OER that provide meaningful learning experiences, their engagement and achievement levels tend to improve.

However, it is essential to recognize that the quality of OER can vary significantly. Not all open educational materials meet the necessary standards of clarity, coherence, and pedagogical soundness. Some OER may lack essential elements such as appropriate instructional design or may not align well with the learning goals they aim to support. This variability can lead to inconsistent learning experiences and, in some cases, diminish the potential impact of OER on student achievement. To address this issue, it is important to establish robust QA processes, such as peer review and ongoing evaluation, to ensure that resources meet high standards of educational quality (Soncin et al., 2022; Tonegawa, 2022).

Furthermore, the concern that OER are inherently of lower quality due to their open nature is often unfounded. Research has shown that when OER are created by skilled educators, instructional designers, or experienced publishers, there is no significant difference in quality between open resources and traditional proprietary content (Luo et al., 2020). This study supports the view that, under similar production conditions, OER can be just as effective as closed resources in promoting learning outcomes.

5.3 Instructional Approach and Its Role in OER Effectiveness

The instructional approach used alongside OER is a key factor influencing their impact on learning outcomes. While OER offer broad access to educational content, their effectiveness in improving student learning depends heavily on how they are integrated into teaching practices. Simply providing open access to resources does not guarantee that students will benefit from them. Effective implementation requires thoughtful instructional strategies that engage students and encourage active learning.

Research has shown that educators who adapt their teaching methods to incorporate OER in meaningful ways, such as through collaborative learning, problem-based learning, or project-based assignments, tend to see better outcomes in terms of student engagement and achievement (Zhang, 2023). In contrast, if OER are used in a passive, one-size-fits-all approach, the learning experience may fall short of its potential. This highlights the importance of moving beyond merely adopting OER to also focusing on how these resources are used to support active learning and deeper understanding.

However, the widespread adoption of Open Educational Practices (OEP), which include these innovative instructional approaches, remains limited (Tlili et al., 2021). Several barriers, such as lack of training, institutional support, and resistance to change, prevent educators from fully embracing OEP. Consequently, without a shift in instructional practices, the full potential of OER to enhance learning outcomes may not be realized. Therefore, the effective use of OER requires not just access to quality resources, but also the application of pedagogical strategies that foster interaction, critical thinking, and collaboration among learners.

5.4 Learners' Individual Factors

The effectiveness of OERin enhancing learning outcomes is not solely dependent on the quality of the resources themselves but also on the individual characteristics of learners. Key factors such as prior knowledge, motivation, and learning preferences play a significant role in determining how effectively students engage with and benefit from OER. As highlighted by Tlili and Burgos (2022), learners come from diverse backgrounds with varying levels of competency and readiness for open education. This diversity means that a one-size-fits-all approach to OER may not be effective in meeting the needs of all students. Demographic variables such as age, gender, cultural background, and socioeconomic status can further influence learners' engagement with and perceptions of OER. For instance, adult learners may prioritize flexibility and practical application, while younger students might benefit from more structured or gamified content.

Motivation is another critical factor that influences the use and impact of OER. Students who are intrinsically motivated to learn and who perceive OER as relevant and valuable to their educational goals are more likely to engage deeply with the materials. However, students who are less motivated or who do not see the immediate value of OER may not benefit as fully from these resources. Additionally, prior knowledge plays an essential role in how learners approach new content. Those with a solid foundation in the subject matter may find it easier to navigate and make use of OER, while students with limited prior knowledge might require additional scaffolding and support.

Importantly, students with disabilities or special learning needs must be explicitly considered in OER design and implementation. Ensuring that OER are accessible, through features like screen reader compatibility, closed captioning, and adaptable formats, is essential to creating inclusive learning environments. Without such measures, these learners may be inadvertently excluded from fully participating in open education. Learners from disadvantaged backgrounds, particularly those with limited access to technology or essential digital skills, may struggle to make the most of OER, further widening educational gaps (Lambert, 2019; Ndebele et al., 2023; Oriakhogba & Ndlovu, 2024). Therefore, addressing these individual factors and offering personalized support is crucial to maximizing the potential of OER for all learners. Tailoring content delivery, providing mentorship, and fostering learner autonomy are some ways to enhance the effectiveness of OER for diverse populations.

5.5 Adequate Experimental Design Is Crucial for Accurately Measuring Learning Achievement

A critical issue in assessing the impact of OERon learning achievement lies in the experimental design employed in the studies. The majority of studies on OER adoption rely on quasi-experimental designs, as random assignments are often not feasible in open education contexts (Griggs & Jackson, 2017; Gurung, 2017). The lack of randomization can create difficulties in accurately measuring the specific effects of OER on learning outcomes, potentially leading to biased or incomplete conclusions.

Moreover, separating the effects of OER from other interventions that may accompany their use is another significant challenge. For example, faculty members may receive support from instructional designers or adopt new teaching methods, such as flipped classrooms, simultaneously with the introduction of OER. In such cases, it becomes difficult to isolate whether improvements in student learning outcomes are due to the OER themselves or the other pedagogical changes (Tlili & Burgos, 2022). This overlap of variables makes it harder to determine the direct impact of OER on achievement.

Furthermore, the use of final exam scores or grade point averages as the primary measures of learning outcomes raises concerns about the reliability and validity of these assessments. The design of exams may vary based on the subject matter and course objectives, leading to inconsistencies in how learning achievement is measured. As such, employing standardized instruments to assess learning outcomes across different studies would provide more robust and comparable results. This could enhance the credibility of findings and support the validation of alternative credentials in open education (Griffiths et al., 2022).

6 Conclusion and Implications

This chapter examined the transformative potential of OERin advancing Sustainable Development Goal 4: Quality Education. OER has emerged as a pivotal tool for democratizing learning by providing open access to high-quality educational materials, reducing barriers to education, and fostering lifelong learning opportunities. By making education more accessible and equitable, especially in underserved regions and marginalized communities, OER initiatives address critical global challenges, such as disparities in educational infrastructure and access to resources. However, this chapter underscores that realizing the full potential of OER requires a comprehensive approach that extends beyond resource availability.

Key findings from this analysis reveal that the adoption of OER, while promising, does not inherently guarantee improved learning outcomes. The impact of OER on educational achievement is mediated by various factors, including the quality of resources, the instructional approaches used, and the diverse needs of learners. High-quality OER must meet rigorous standards of accuracy, relevance, and cultural adaptability to ensure effectiveness. Without addressing these aspects, the adoption of OER risks perpetuating inequities rather than resolving them.

Moreover, this chapter identifies significant challenges in OER implementation, such as digital access disparities, varying levels of technological infrastructure, and the need for capacity building among educators. These challenges highlight the importance of systemic interventions, including policy support to fund and promote OER initiatives, targeted investments in digital infrastructure, and professional development programs for educators. Equipping teachers with the skills to effectively integrate OER into learner-centered pedagogies is essential for leveraging its benefits fully. Collaborative efforts among governments, educational institutions, and non-profit organizations can further enhance the scalability and impact of OER initiatives.

In addition to these challenges, the diversity of learners' backgrounds and educational contexts must be prioritized in OER design and implementation. Effective OER strategies recognize that learners come from varied cultural, linguistic, and socio-economic settings. Culturally relevant and contextually adaptable resources can significantly improve engagement and learning outcomes, ensuring that OER contributes to inclusivity and equity.

The implications of this study extend beyond the adoption of OER itself. They emphasize the need to rethink educational practices to align with principles of openness and adaptability. OER provides a unique opportunity to foster innovative pedagogies that place learners at the center of the educational process. When integrated with strategies such as flipped classrooms, blended learning, or personalized instruction, OER can transform the learning experience and bridge gaps in traditional educational models.

Ultimately, the findings presented in this chapter underscore the importance of viewing OER as part of a larger educational ecosystem rather than an isolated solution. By addressing barriers to adoption and implementation through targeted policies, partnerships, and investments, stakeholders can harness the full potential of OER to advance equitable, high-quality education. This chapter reaffirms OER's role as a critical driver of progress toward SDG 4, contributing to the creation of sustainable educational ecosystems that empower learners worldwide.

As education systems evolve to meet the demands of the twenty-first century, the promise of OER lies in its ability to adapt and scale, offering practical, cost-effective solutions for global educational challenges. While challenges remain, the collaborative efforts of policymakers, educators, and researchers can ensure that OER continues to promote inclusion, innovation, and lifelong learning for all.

References

Abelson, H. (2008). The creation of OpenCourseWare at MIT. *Journal of Science Education and Technology, 17*, 164–174.

Ahlf, M., McNeil, S., & Nguyen, P. (2024). Open educational resources: Time, resources and sustainability in an ephemeral digital world. In *Educational research and the question (s) of time* (pp. 345–369). Springer Nature Singapore.

Ajiga, D., Okeleke, P. A., Folorunsho, S. O., & Ezeigweneme, C. (2024). Methodologies for developing scalable software frameworks that support growing business needs.

Arnavut, A., Bicen, H., & Nuri, C. (2019). Students' approaches to massive open online courses: The case of Khan Academy. *BRAIN. Broad Research in Artificial Intelligence and Neuroscience, 10*(1), 82–90.

Baker, W., & Dutton, J. E. (2017). Enabling positive social capital in organizations. In *Exploring positive relationships at work* (pp. 325–346). Psychology Press.

Bammel, J. (2014). The impact of copyright on the enjoyment of right to science and culture. *Publishing Research Quarterly, 30*(4), 335–343.

Barger, R. P. (2020). Democratization of education through massive open online courses in Asia. *IAFOR Journal of Education, 8*(2), 29–46.

Butcher, N., & Moore, A. (2015). Understanding open educational resources.

COL. (2017). Open educational resources: Global report 2017.

Coleman, T. (2021). Digital divide in UK education during COVID-19 pandemic: Literature review.

Connell, M., & Connell, J. (2020). Critical evaluation of quality criteria and quality instruments in OER repositories for the encouragement of effective teacher engagement.

Costa, A. C. F., de Mello Santos, V. H., & de Oliveira, O. J. (2022). Towards the revolution and democratization of education: A framework to overcome challenges and explore opportunities through Industry 4.0. *Informatics in Education, 21*(1), 1–32.

Curtis, A. (2021). What do we mean by under-resourced context? In *Research on teaching and learning English in under-resourced contexts* (pp. 14–28). Routledge.

Darwish, H. (2019). Open educational resources (OER) Edupreneurship business models for different stakeholders. *Education and Information Technologies, 24*(6), 3855–3886.

Fatayer, M. (2024). Designing inclusive learning experience through open educational practices. In *Designing learning experiences for inclusivity and diversity: Advice for learning designers.*

Farrell, O., O'Regan, M., Whyte, A., Aceto, S., Brown, M., & Brunton, J. (2022). Strategic support for OER value proposition.

Fleming, J. I., Wilson, S. E., Hart, S. A., Therrien, W. J., & Cook, B. G. (2021). Open accessibility in education research: Enhancing the credibility, equity, impact, and efficiency of research. *Educational Psychologist, 56*(2), 110–121.

Ganapathi, J. (2019). User-generated content's impact on the sustainability of open educational resources. *Open Praxis, 11*(2), 211–225.

Griffiths, R., Mislevy, J., & Wang, S. (2022). Encouraging impacts of an open education resource degree initiative on college students' progress to degree. *Higher Education, 84*(5), 1089–1106.

Griggs, R. A., & Jackson, S. L. (2017). Studying open versus traditional textbook effects on students' course performance: Confounds abound. *Teaching of Psychology, 44*(4), 306–312.

Gurung, R. A. (2017). Predicting learning: Comparing an open educational resource and standard textbooks. *Scholarship of Teaching and Learning in Psychology, 3*(3), 233.

Hewlett Foundation. (2013). *Hewlett Foundation 2013 Annual Report.* https://www.hewlett.org/wp-content/uploads/2016/09/2013AnnualReport.pdf.

Herrera-Cubides, J. F., Gaona-García, P. A., Montenegro-Marín, C. E., & Sánchez-Alonso, S. (2022). Improving OER descriptions to enhance their availability, reuse, and enrichment. *Education and Information Technologies*, 1–29.

Holl, O., Cunha, F. S., Streuli, D., & Laborie, T. (2024). EduQuest: lecture texts and questions for higher education. In *Proceedings of the 17th International conference on educational data mining* (pp. 849–856).

Hoque, M. N., & Abdullah, M. F. (2021). The world's oldest university and its financing experience: A study on al-Qarawiyyin university (859–990). *Journal of Nusantara Studies (JONUS), 6*(1), 24–41.

Huang, R., Tlili, A., Chang, T. W., Zhang, X., Nascimbeni, F., & Burgos, D. (2020). Disrupted classes, undisrupted learning during COVID-19 outbreak in China: Application of open educational practices and resources. *Smart Learning Environments, 7*, 1–15.

Huzooree, G., & Doargajudhur, M. (2023). Promoting Learners' engagement and motivation in a technology-mediated learning environment during COVID-19 and beyond. In *Handbook of research on creating motivational online environments for students* (pp. 161–187). IGI Global.

Imberman, S., & Fiddler, A. (2019). Share and share alike: Using Creative Commons licenses to create OER. *ACM Inroads, 10*(2), 16–21.

Khayyat, M., & Bannister, F. (2015). Open data licensing: More than meets the eye. *Information Polity, 20*(4), 231–252.

King, M., Pegrum, M., & Forsey, M. (2018). MOOCs and OER in the Global South: Problems and potential. *The International Review of Research in Open and Distributed Learning, 19*(5).

Kitchin, R., Collins, S., & Frost, D. (2015). Funding models for open access digital data repositories. *Online Information Review, 39*(5), 664–681.

Kukulska-Hulme, A., Dawadi, S., Giri, R. A., Gaved, M., Khan, R., Bedri, A., et al. (2024). Reaching out to marginalised populations in under-resourced countries. *Available at SSRN 4913316*.

Jones, S. (2018). A solution to OER publication resistance: Using blockchain technology to protect scholar copyright. *The International Journal of Open Educational Resources, 1*(1).

Lambert, S. (2019). The Siyavula Case: Digital, collaborative text-book authoring to address educational disadvantage and resource shortage in South African schools. *International Electronic Journal of Elementary Education, 11*(3), 279–290.

Luo, T., Hostetler, K., Freeman, C., & Stefaniak, J. (2020). The power of open: Benefits, barriers, and strategies for integration of open educational resources. *Open Learning: THe Journal of Open, Distance and e-Learning, 35*(2), 140–158.

Marczewska, M., Weresa, M. A., & Lachowicz, M. (2024). Towards creativity and innovation in universities: Study on Central and Eastern Europe. *Journal of the Knowledge Economy, 15*(1), 1363–1385.

Maristela, J. (2024). EdChains: Connecting data and information to problem solvers of global challenges.

Meng, S. (2023). Enhancing teaching and learning: Aligning instructional practices with education quality standards. *Research and Advances in Education, 2*(7), 17–31.

McShane, M. Q. (2017). Open educational resources. *Education next, 17*(1), 18–24.

Mishra, S. (2017). Open educational resources: Removing barriers from within. *Distance Education, 38*(3), 369–380.

Menzli, L. J., Smirani, L. K., Boulahia, J. A., & Hadjouni, M. (2022). Investigation of open educational resources adoption in higher education using Rogers' diffusion of innovation theory. *Heliyon, 8*(7).

Mncube, V., Olawale, B. E., & Bitso, C. (2021). Democratisation of knowledge through open educational resources: Opportunities and challenges at a South African University. In *Global Education Network 2021, South Africa, Conference Proceedings* (pp. 207–220). Global Education Network.

Mullens, A. M., & Hoffman, B. (2023). The affordability solution: A systematic review of open educational resources. *Educational Psychology Review, 35*(3), 72.

Munir, H., Linåker, J., Wnuk, K., Runeson, P., & Regnell, B. (2018). Open innovation using open source tools: A case study at Sony Mobile. *Empirical Software Engineering, 23*, 186–223.

Muthanga, B. N., Ogalo, J. O., Were, S. M., & Maku, P. G. (2023). A review on utilization of OER in teaching and research in academic institutions. *Open Access Library Journal, 10*(3), 1–14.

Naab, T. K., & Sehl, A. (2017). Studies of user-generated content: A systematic review. *Journalism, 18*(10), 1256–1273.
Ndebele, N. C., Masuku, M. M., & Mlambo, V. H. (2023). Funding open educational resources in higher education: A South African public policy perspective. *Social Sciences, 12*(1), 49.
Nilashi, M., Abumalloh, R. A., Zibarzani, M., Samad, S., Zogaan, W. A., Ismail, M. Y., et al. (2022). What factors influence students satisfaction in massive open online courses? Findings from user-generated content using educational data mining. *Education and Information Technologies, 27*(7), 9401–9435.
Norris, M. E., Swartz, M., & Kuhlmeier, V. A. (2023). The importance of copyright and shared norms for credit in open educational resources. In *Frontiers in education* (Vol. 7, p. 1069388). Frontiers Media SA.
Nykyporets, S., Stepanova, I., & Hadaichuk, N. (2023). The use of open educational resources in Ukraine: Unleashing the potential for knowledge democratization and lifelong learning. *Journal of Innovations and Sustainability, 7*(1), 07–07.
OECD. (2007). *Giving knowledge for free: The emergence of open educational resources.* OECD Publishing.
Onopriienko, K., Vasylieva, T., Bourekkadi, S., Rehak, R., & Artyukhova, N. (2023). Effectiveness of open education in the lifelong learning system.
Oriakhogba, D., & Ndlovu, L. (2024). Copyright, access to knowledge, and SDG 4 in Southern Africa. In *The elgar companion to intellectual property and the sustainable development goals* (pp. 142–174). Edward Elgar Publishing.
Otero, I., Pantazatos, D., & Docampo, D. (2019). Good practices adopting copyright for creating and reusing oer in ngdle environment. In *INTED2019 Proceedings* (pp. 5504–5513). IATED.
Perifanou, M., & Economides, A. A. (2023). Measuring quality, popularity, demand and usage of repositories of open educational resources (ROER): A study on thirteen popular ROER. *Open Learning: THe Journal of Open, Distance and e-Learning, 38*(4), 315–330.
Perryman, L. A., & Coughlan, T. (2014). When two worlds don't collide: Can social curation address the marginalisation of open educational practices and resources from outside academia? *Journal of Interactive Media in Education, 2014*(2).
Ramirez-Montoya, M. S. (2020). MOOCs and OER: Developments and contributions for open education and open science. *Radical Solutions and Open Science: An Open Approach to Boost Higher Education*, 159–175.
Rodrigo, C., & García-Serrano, A. (2015). Social accessibility action to improve quality of OER user-generated video-classes and associated resources. *Procedia Computer Science, 67*, 19–27.
Rodríguez-Abitia, G., Martínez-Pérez, S., Ramirez-Montoya, M. S., & Lopez-Caudana, E. (2020). Digital gap in universities and challenges for quality education: A diagnostic study in Mexico and Spain. *Sustainability, 12*(21), 9069.
Rohayati, M. I., Najdi, Y., & Williamson, J. C. (2016). Philanthropic fundraising of higher education institutions: A review of the Malaysian and Australian perspectives. *Sustainability, 8*(6), 541.
Schnackenberg, H. L. (2019). Are we part of the problem?: Teacher preparation programs, educational technology, and the mis-education of future teachers. In *Preparing the higher education space for Gen Z* (pp. 110–123). IGI Global.
Schuwer, R., & Baas, M. (2023). Reuse of OER, a process model approach. *Distributed learning ecosystems: Concepts, resources, and repositories* (pp. 117–137). Springer Fachmedien Wiesbaden.
Stafford, D., & Flatley, R. (2018). Openstax. *The Charleston Advisor, 20*(1), 48–51.
Soncin, M., Agasisti, T., Frattini, F., Patrucco, A., & Pero, M. (2022). The costs, quality, and scalability of blended learning in postgraduate management education. *Journal of Management Education, 46*(6), 1052–1085.
Swain, B. K., & Pathak, R. K. (2024). Benefits and challenges of using OER in higher education: A pragmatic review. *Discover Education, 3*(1), 81.
Terras, M. (2015). Opening access to collections: The making and using of open digitised cultural content. *Online Information Review, 39*(5), 733–752.

Tlili, A., & Burgos, D. (2022). Unleashing the power of Open Educational Practices (OEP) through Artificial Intelligence (AI): Where to begin?. *Interactive Learning Environments*, 1–8.

Tlili, A., Zhang, J., Papamitsiou, Z., Manske, S., Huang, R., Kinshuk, & Hoppe, H. U. (2021). Towards utilising emerging technologies to address the challenges of using Open Educational Resources: a vision of the future. *Educational Technology Research and Development*, *69*, 515–532.

Tonegawa, Y. (2022). Education in SDGs: What is inclusive and equitable quality education?. In *Sustainable development disciplines for humanity: Breaking down the 5Ps—People, planet, prosperity, peace, and partnerships* (pp. 55–70). Springer Nature Singapore.

UNESCO. (2002). Forum on the impact of open courseware for higher education in developing countries. *Digital Library*.

UNESCO. (2005). *Second global forum on International Quality Assurance, Accreditation and the Recognition of Qualifications*. ED/2005/RP/H/1. https://unesdoc.unesco.org/ark:/48223/pf0000141768

UNESCO. (2019). *Towards implementation of Open Educational Resources Recommendation in West and Central African countries*. Available at: https://www.unesco.org/en/articles/towards-implementation-open-educationalresources-recommendation-west-and-central-african-countries.

Veletsianos, G. (2020). *Learning online: The student experience*. JHU Press.

Wiley, D., Bliss, T. J., & McEwen, M. (2014). Open educational resources: A review of the literature. In *Handbook of research on educational communications and technology* (pp. 781–789).

Williams, K., & Werth, E. (2021). Student selection of content licenses in OER-enabled pedagogy: An exploratory study. *Journal of Copyright in Education & Librarianship*, *5*(1).

Zaid, Y., Olatise, O., & Alabi, A. (2024). OER accessibility for students with visual disabilities in higher education in the era of open knowledge. *Mousaion, 42*(1).

Zhang, X. (2023). Teachers' self-directed professional development in under-resourced contexts: How do open educational resources matter? *Education and Information Technologies, 28*(2), 1849–1863.

Zhou, X., Chen, L. H., & Chen, C. L. (2019). Collaborative learning by teaching: A pedagogy between learner-centered and learner-driven. *Sustainability, 11*(4), 1174.

Open Access in Education: Unlocking Quality Learning for All to Achieve SDG 4

Mitali Chugh and K. H. Keoy Alan

Abstract

Ensuring inclusive and equitable quality education and Lifelong Learning is one of the United Nations Agenda 2030 Sustainable Development Goals 4 (SDG 4). However, in higher education, a lack of equity in resource allocation and distribution restrains the achievement of this goal. There is a viable solution in open access (OA) that increases access, fairness, and innovation, and hence plays a pivotal role in the achievement of SDG 4. This chapter explores the contribution of OA to the democratization of knowledge and consequently narrows down the education gap. OA, which gives the public access to research articles and other materials for free, particularly those that are not behind paywalls, benefits higher learning institutions by filling gaps between institutions and regions. It facilitates the distribution of books such as text and research papers, as well as Open Educational Resources (OER), especially intended for low-income and under-resourced communities. Through implementing equity and inclusion, OA helps minorities and brings quality education all around the world. OA also enables cooperation in research as it enables scholars to provide results and methods as approaches to international issues such as environmental change and health. Also, opponents argue that OA helps people learn throughout their lives, and provides materials for gaining skills and becoming better, and this is especially important now as the world changes every day. However, since the inception of OA, some factors have hindered it, including funding sustainability, quality, and technology. These problems can only be solved by investing in infrastructural changes, new business models, and promoting a new culture of

M. Chugh (✉)
School of Computer Science, UPES, Dehradun, India
e-mail: mitalichugh11@gmail.com

K. H. Keoy Alan
UCSI Graduate Business School, UCSI University, Selangor, Malaysia

T. Choudhury et al. (eds.), *Open Access as a Knowledge Platform for Implementing the United Nations Sustainable Development Goals*, World Sustainability Series,
https://doi.org/10.1007/978-3-031-99605-4_6

open scientific practices. Therefore, it is to emphasize that OA has the potential to support SDG 4 and contribute to the changes in higher education concerning its accessibility, innovation, and unbiased reputation. Implementing and coming up with solutions to the challenges faced by OA can greatly enhance the quality of education and lead to a more sustainable and inclusive world.

Keywords

Open access (OA) • Equitable education • Sustainable development goal 4 (SDG 4) • Knowledge democratization • Higher education equity

1 Introduction

Quality Education captured under the United Nations Sustainable Development Goal 4 (SDG 4) is a critical component of the United Nations 2030 Agenda for Sustainable Development (Connor, 2015). The implementation of education for all is a United Nations-recognized universal human right and a key enabler of sustainable development. It retains a pivotal position in facilitating individuals, promoting social participation, and encouraging economic development. SDG 4 aims to address persistent inequalities in educational access and quality, advocating for a holistic approach that considers the needs of marginalized and disadvantaged groups. This goal includes a variety of endeavors, such as calls for free, equitable, and quality primary and secondary education, which directly highlights the importance of inclusive educational frameworks. Affordable vocational training, as well as promotion of higher education for all (UNESCO, 2021). Furthermore, it emphasizes the value of continuous education, recalling that education should not end with formal schooling and should signify lifelong learning.

Higher education institutions are strategic organizational entities whose roles are multifaceted as knowledge generators, disseminators, and users (Kivunja, 2014). This chapter presents details of SDG 4 with a focus on the central role of higher education in the general aim and scope of this SDG. However, disparities in access to quality educational resources—especially in developing and underfunded regions—continue to obstruct the equitable achievement of SDG 4. This chapter aims to address this issue by exploring how OA can bridge these gaps and support inclusive, lifelong learning opportunities in higher education.

The chapter presents details of SDG 4 at its inception, focusing on the central role of higher education in the general aim and scope of this sustainable development goal. Hence, opening access to and increasing the quality of higher education through OA is critical to generating a brighter and more sustainable world.

Higher education institutions also play a vital role in fostering inclusivity and diversity. They have the responsibility to create an environment that is welcoming and accessible to all students, particularly those from marginalized backgrounds (Miller et al., 2017). In the case of accessibility to learning resources, one of the best chances that has been opened by OA in Higher Education is the enhancement of access to such resources. This creates a seamless connection to the discussion on how OA increases access in the dissemination of textbooks, research papers, and open educational resources (OERs), where the cost of user access is nil, hence reducing the costs imposed on students and other institutions. This is especially beneficial to the students and instructors in low-income and developing countries where access to quality learning resources is quite challenging. OA helps in eradicating some of the financial and geographical barriers to learning and hence is in line with one of the targets of SDG 4, which is to promote equity in education. In addition, OA works to improve equity and inclusion in the higher learning environment. This democratization of knowledge is an important step towards the realization of the global society's goal of education for all, as subscribed to by SDG 4.

The chapter also looked at the relationship between OA and innovation as well as research collaboration. OA platforms help to make available research findings, data, and methods for use by researchers across the globe, enhancing what different researchers have done and putting into practice a culture of collaborative research. Such an approach is especially relevant to the current global issues like climate change, health crises, and inequality, as they do not fit into a solitary discipline or a nation's framework. Since OA avails research in its unrestrained form, it fosters novel ideas for the development of solutions to SDG challenges.

Other than affecting research, OA also improves access to lifelong learning, which is an element of SDG 4. OA facilitates people to get qualitative resources for educational purposes, career upgrades, extending new professions, and other aims, as well as purely informational objectives. This accessibility is crucial for promoting ongoing professional education and guarantees that people have all the resources to adapt to new conditions in society and the economy.

Despite the critical role that higher education plays in achieving SDG 4, numerous challenges persist in the pursuit of these objectives. One of the most significant barriers is inequitable access to educational resources and opportunities. Many students, particularly those in low-income regions, face significant obstacles in accessing high-quality educational materials, research publications, and learning opportunities due to financial constraints and institutional limitations (Wilkinson, 2020). To address these challenges, the chapter discusses potential sources of funding for OA, emphasizing the role of institutional members, government funding, and funding from a consortium of institutions, among others, as solutions to the above challenges to make OA sustainable in the long run.

Policing the quality and standards of OA publications is a critical requirement for the credibility of the system and the exclusion of misleading or low-quality information. Another major challenge is the technological and infrastructure limitations, as more institutions are faced with big challenges, especially those in developing regions. The chapter concludes that all institutions must be empowered to take advantage of OA, and therefore, advocating for funding of digital infrastructure as well as offering all the technological support is essential. Moreover, addressing the cultural aspects is vital, as many professions have kept traditional academic or publishing practices close to heart. To overcome this trend, culture change must be initiated through the communities of researchers, who must see value in OA.

This is also to do with culture and institutions, as many professions have kept traditional academic or publishing practices close to heart. To overcome this trend, culture change must be initiated through the communities of researchers, who must see value in OA. In this context, OA emerges as a transformative solution that can significantly enhance higher education and contribute to achieving SDG 4. OA refers to the free and unrestricted online access to scholarly research, educational materials, and other academic resources (Suber, 2012). By eliminating the financial and institutional barriers associated with traditional publishing, OA democratizes knowledge and empowers individuals to engage with academic content without facing prohibitive costs. Furthermore, the chapter will highlight the role of Open Educational Resources (OER) within the OA framework. OER encompasses a wide range of freely accessible teaching, learning, and research materials, which can be adapted and reused by educators and learners alike (Wiley et al., 2014).

OA is synonymous with the improvement of learning and innovation in computer science because it offers free access to state-of-the-art literature, open-source software, and knowledge resources. This chapter elaborates on how OA platforms contribute to the conception and development of fresh tools and technologies that underpin the growth of the field (Piwowar et al., 2018). The section offers an outlook on OA, especially about higher learning institutions, considering various reflections on future trends in OA. In this context, IT will go steadily forward, and new types of possibilities for fostering OA and succeeding SDG 4 will appear. In this context, it is anticipated that IT will go steadily forward, and new types of possibilities for fostering OA and succeeding SDG 4 will appear (refer Fig. 1).

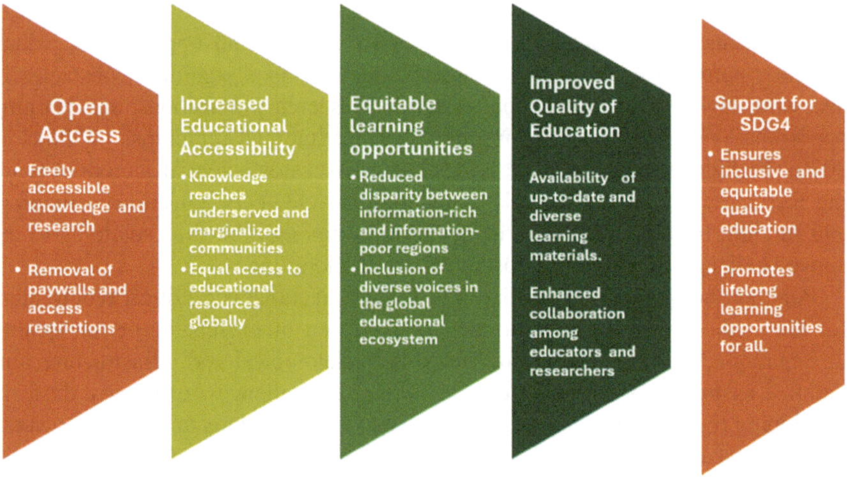

Fig. 1 Open access (OA) contributions to SDG 4: quality education for all

This chapter is organized as follows: Sect. 2 explores the historical development of OA and its relationship with SDG 4. Section 3 examines key OA models and platforms relevant to higher education. Section 4 discusses the barriers and challenges of OA implementation, while Sect. 5 highlights its benefits and contributions to equitable learning. Section 6 presents case studies from global and Indian contexts. Section 7 outlines strategies to overcome implementation challenges, and Sect. 8 offers future perspectives. The chapter concludes with key insights and recommendations for stakeholders in higher education.

2 Understanding Open Access in Higher Education

2.1 Definition and Importance of OA

Open Access (OA) has changed the academic model, with scholars gaining free online access to scholarly resources from anywhere in the world. This increases equal opportunity for everyone in higher education. This work defines and outlines the concept of Open Access and its models, as well as the repositories and platforms that have been deployed in higher learning institutions to support and enhance OA practices.

Open access can be defined as unrestricted access to scholarly works, allowing individuals to download, read, or reuse academic works without financial, legal, and technological barriers (Piwowar et al., 2018). Conventional databases are paid for and were formerly available only for educational institutions, but failed to provide general open access. Open Access solves this problem by making research freely available, enhancing understanding and collaboration around the world (refer Table 1).

OA is most important in higher education since universities and other research institutions are the providers of most academic output, and OA enables their works to gain a greater appreciation and reach out to those who may benefit from them most (Zhu, 2017).

Through the adoption of OA in higher learning institutions, the institutions enhance a culture of openness, responsibility, and teamwork. Shifts toward OA are therefore in tandem with part of the spirit of SDG 4, which calls for quality education for all. OA contributes to widening access to materials; fosters and improves knowledge sharing; and disseminates knowledge, especially regarding institutions in contexts of low resources (Tennant, 2020).

2.2 Types of Open-Access Models

There are three primary types of OA models—Gold, Green, and Hybrid OA—that offer distinct pathways to making academic content freely available.

Table 1 Global Open Access Milestones (2016–2025)

Year	Africa	Asia	Europe	North America	South America	Oceania
2016	Dakar Declaration on OA Publishing	China launches Open Access Plan	Horizon 2020 OA Mandate	#GoOpen OER Initiative	SciELO 20 years with new OA guidelines	ARC strengthens OA policy
2017	–	India's UGC recommends OA	Plan S launched	Tri-Agency OA Policy takes effect	–	NZ supports OER in schools
2018	–	–	Plan S reaffirmed	–	Argentina's National Repository grows	–
2019	AOSP launches coordinated OA initiative	–	–	–	CLACSO launches new OA repositories	NZ supports OER for higher ed
2020	–	Japan integrates Plan S	–	–	–	–
2021	South African NRF mandates OA	–	UK adopts Plan S policies	–	CLACSO expands OA programs	Open Access Australasia formed
2022	AOSP strategic plan to enhance Open Science	China mandates OA for research	Open Research Europe platform launched	OSTP mandates immediate OA for US-funded research	FAPESP policy on OA for funded research	ARC mandates immediate OA
2023	African Academy of Sciences launches OA publishing platform	India promotes a national OER policy	EU launches Digital Europe Programme	Canada's Tri-Agency mandates OA	CLACSO expands OA access	NZ launches Open Research policy

(continued)

Table 1 (continued)

Year	Africa	Asia	Europe	North America	South America	Oceania
2024 (OpenAI Moments)	AI adoption for agriculture and education	–	–	GPT-5, AI in Education & Healthcare, Climate, Ethics	–	–
2025	African Union launches continental OA policy harmonization framework for member states	Indonesia's Ministry of Education announces OA mandate for all publicly funded research	European Commission introduces AI-integrated OA metrics to assess research impact	U.S. Department of Education pilots OA textbooks initiative for community colleges	Chile launches national OA repository in collaboration with UNESCO	Australian universities consortium funds OA infrastructure expansion across regional institutions

2.2.1 Gold Open Access

This model involves publishing articles in OA journals where the content is freely accessible. Often, authors or their institutions pay Article Processing Charges (APCs) to cover publication costs. Gold OA journals, such as *PLOS ONE* and *BMC Biology*, remove access barriers for readers but may require institutional support for APCs. Higher education institutions that fund APCs contribute to a global ecosystem of open, accessible research while strengthening their visibility in academia (Bolick et al., 2024).

2.2.2 Green Open Access

Also called "self-archiving," Green OA allows authors to publish in traditional subscription-based journals but to archive a version of their work in an institutional repository. Although publishers may impose embargo periods on Green OA content, this model is financially advantageous as it does not require APCs, making it accessible even to institutions with limited funding. Higher education institutions with dedicated repositories for Green OA enable their faculty and researchers to disseminate their work without incurring publication costs (Pinfield, 2024).

2.2.3 Hybrid Open Access

In the Hybrid OA model, authors can pay APCs to make their articles openly accessible in a subscription journal, providing a flexible but costly approach to OA. This option allows authors to publish in high-impact journals while complying with OA mandates, though concerns around the "double-dipping" issue (charging both subscription fees and APCs) have been raised (Shen, 2020). For higher education institutions, Hybrid OA offers strategic options for specific high-value research outputs while managing overall publication costs.

These OA models provide higher education institutions with several routes to open research outputs. The decision on which model is to be employed largely depends on the institutional and academic budget constraints, scientific focus, and needs of each discipline; while many continue to subscribe to traditional journals, an increasing number are gravitating towards dedicated OA journals and/or hybrid models.

3 OA Repositories and Platforms in Higher Education

For OA, all higher educational institutions have adopted the creation of repositories and digital resources that include scholarly outputs that are shared with the global academic community, and which meet OA policies and mandates.

3.1 Institutional Repositories

Scholarly works of faculty from universities, research institutions, and or students are usually compiled to form digital repositories within their institutions. There are eScholarship platforms at the University of California, DSpace@MIT, and DASH at Harvard University. An institutional repository is home to a variety of products, including preprints, postprints, theses, datasets, and other materials; thus addressing the question of the availability of academic content and the preservation of content for long-term usage (Severin et al., 2020).

3.2 Disciplinary Repositories

These are discipline-based and enable researchers to contribute as well as gather materials in their respective specialty areas. For physics, there is arXiv, for biological sciences, there is bioRxiv, and for the social sciences, there is SSRN. It adds that disciplinary repositories raise participation levels since writers can post preprints, obtain peer responses, and follow advancements in their disciplines. The higher education-based platforms enable collaboration and increase the visibility of academic works worldwide (Björk, 2021).

3.3 Open Educational Resources (OER) Platforms

Open Educational Resources encompass OpenStax, OER Commons, and MERLOT. These platforms allow instructors to upload textbooks, curricula, and multimedia resources for free and help to bring down the cost of instructional materials. Open educational resources aim to provide equal access to learning materials, thus aligning closely with the goals of SDG 4 (Huijboom et al., 2023).

3.4 National and Regional OA Platforms

Some geographic areas and even countries have developed their own national OA platform to promote OA practice on a larger scale. This is reflected in OpenAIRE in Europe and SciELO in Latin America. They foster the partnership and make the region's research more visible to the rest of the world. They are especially useful for highlighting the contributions of scholars from the local context and for enhancing networks between organizations of higher learning across the globe (Tennant, 2020).

These repositories and platforms assist in the advancement of OA in higher education, as well as meet the requirements of funders, provide useful resources for sharing knowledge, and encourage more creativity. It is more than the storage of academic content, enabling higher education institutions and other scholarly institutions to contribute more positively to a global pool of available knowledge.

4 Literature Review

The chapter seeks to establish OA and its significance in meeting SDG 4, which centers on quality education for all. The chapter also traces the evolution of OA and its existing policies and concerns the radical models of Gold, Green, Hybrid, and pedagogies of how they can impact higher education. Moreover, it focuses on the ability of OA to increase access to educational resources, engage the different fields, and embrace the broader community, and on the known existing challenges, including limited funds for funding the OA schemes, technical issues, and lack of willingness to adopt new changes. Despite its strengths in presenting detailed analyses and highlighting innovative tools like Open Educational Resources (OER), the review notes limitations in empirical data, regional applicability, and diverse case studies. By addressing these gaps and focusing on actionable strategies, OA can play a transformative role in bridging knowledge disparities and advancing quality education for all, thus contributing significantly to the realization of SDG 4 (Refer Table 2).

5 Opportunities of OA for Higher Education and SDG 4

Open Educational Resources (OER) and their evolution into Massive Open Online Courses (MOOCs) have emerged as significant contributors to the advancement of SDG4: Quality Education. Before the advent of the OER policy in 2013, UNESCO conceived of OER in 2004, and then, subsequently, through the OER Paris Declaration of 2012, awareness has spread rapidly in terms of possibilities to solve global educational issues. That said, it must be recognized that although OER and MOOCs do not enact these crises, each has a transformative function in broadening access, increasing equity, and improving quality. OER specifically provides to address licensing restrictions so that material can be shared across institutions, and

Table 2 Reviewing OA: its role in enhancing SDG4

Key themes/ findings	Key authors/ references	Strengths	Limitations/ gaps	Relevance to SDG 4
Overview of OA and its importance for SDG 4	Chankseliani and McCowan (2021), Boeren (2019), Saini et al. (2023), Unterhalter (2019)	Provides a comprehensive background on OA	Limited focus on higher education contexts	Establishes a baseline for SDG 4
Importance of equitable access to education	Ertz et al. (2022), Adipat and Chotikapanich (2022), Baños et al. (2019), Cano et al., (2020), Tonegawa (2023)	Highlights the direct link between OA and educational equity	May not cover regional differences	Critical for achieving the SDG 4 goals
Key milestones in OA development	Luo et al. (2020), Majekolagbe (2024), Severin et al., (2018), Tang (2021)	Provides historical context	Focuses primarily on developed countries	Understanding OA evolution aids SDG 4
Impact of mandates and policies on OA	Atenas et al. (2020), Huang et al., (2020), Martin and Godonoga (2020)	Identifies influential policies	Limited analysis of regional variations	Policies affect the implementation of SDG 4
Changes in Institutional Practices for OA Education	Bygstad et al. (2022), Chugh et al., (2023), Miranda et al., (2021), Rawashdeh et al. (2021)	Explores practical implications	Examples may not be universally applicable	Directly relates to educational equity
Comparison of Gold, Green, and Hybrid models of OA	Huang et al., (2020), Sheikh and Richardson (2023), Verma and Sonkar (2021), Young and Brandes (2020)	Detailed model comparison	Lacks case studies that compare the models	Models Impact SDG 4 Implementation
Effectiveness of different OA models in higher education	Alamri et al. (2021), Chankseliani and McCowan (2021), Ghanem (2020)	Provides practical insights	Need for more empirical data	Aligns model effectiveness with SDG 4
Role of Institutional Repositories in Higher Education	Marín et al. (2022), Neumann et al., (2022), Otto (2021)	Identifies key platforms	Limited evaluation of the impact on research	Facilitates knowledge sharing for SDG 4

(continued)

Table 2 (continued)

Key themes/ findings	Key authors/ references	Strengths	Limitations/ gaps	Relevance to SDG 4
Importance of OER and its accessibility	Luo et al. (2020), Rets et al., (2023), Tillinghast (2020), Zhang et al., (2020)	Highlights innovative educational tools	Lack of user engagement studies	Supports equitable access
Barriers to OA, such as APCs and funding issues	Budzinski et al. (2020), Klebel and Ross-Hellauer (2023), Nicholas et al., (2024), Vervoort et al., (2021), Williams et al., (2023)	Addresses real-world barriers	Needs more focus on solutions	Directly affects SDG 4 progress
Issues regarding the acceptance of OA	(Kılıçkaya & Kic-Drgas, 2021; A. Kumar et al., 2021; Tao et al., 2022; Zhang et al., 2020)	Explores institutional culture	Limited case studies of successful transitions	Important for policy and strategy
Challenges in OA repository management	Ramirez-Montoya (2020), Tlili et al., (2021)	Identifies significant barriers	More data is needed on technical solutions	Crucial for effective implementation
OA's Role in widening access to education	Haleem et al. (2022), Lambert (2020)	Shows positive impacts of OA	Needs case studies to illustrate benefits	Essential for achieving SDG 4
Role of OA in improving educational outcomes	Khan et al. (2022), Saini et al. (2023)	Supports inclusivity	Requires more qualitative analysis	Directly supports SDG 4 objectives
Role of OA in facilitating interdisciplinary research	(Banyas et al., 2023; Nishikawa & Murakami, 2019)	Promotes collaborative work	May overlook challenges in collaboration	Enhances research contributing to SDG 4
Successful OA integration strategies	Chan (2023), Miao et al., (2023)	Practical insights from real cases	Limited diversity in case studies	Provides models for SDG 4 implementation

(continued)

Table 2 (continued)

Key themes/ findings	Key authors/ references	Strengths	Limitations/ gaps	Relevance to SDG 4
Analysis of successful OA mandates	Alqahtani and Rajkhan (2020), Fensie et al., (2024)	Highlight effective policies	Limited focus on unsuccessful policies	Informs strategic planning for SDG 4
Policy implications for stakeholders of OA implementation	Arancio et al. (2022), Khan et al. (2022), Saini et al., (2023), Shabalala and Ngcwangu (2021)	Provides actionable insights	Limited analysis of potential challenges	Essential for sustainable education

offers flexibility for use in local settings. Most especially in developing countries, they enhance cheap and universal learning solutions, improving equity in education, and promoting international cooperation. They relate well with the current digital economy, thereby enabling the learners to do their work in diverse interconnecting systems. As the global community maintains its support and makes strategic investments in OER, these open educational resources are thus poised for a unique role in developing effective and sustainable education systems for the future.

5.1 Improving Access to Educational Resources

OA, on the other hand, offers free access to plenty of academic information such as textbooks, research papers, and journals. This aspect makes a difference from other conventional approaches to publishing, where access is granted to those willing to pay for costly subscriptions. Such democratization of knowledge makes a place for students, educators, and researchers across the globe, particularly in the low-income and developing parts of the world, to benefit from the high-quality educational content. For example, OA contributes to offering recent ideas to researchers who otherwise cannot participate in modern advances emerging in the relevant fields. This broader accessibility improves learning experiences and makes more offers equal between all institutions since all will be able to access a wealth of academic information.

With the help of the OER, students gain not only access to knowledge but also repurposing rights and the ability to engage with everybody and content of all formats. More than solely opened licensed Web content, OER enables the user to download, revise, and redistribute resources (Herrera-Cubides et al., 2022). These resources conform to the participatory approaches embraced by Indigenous people by providing forums for the documentation of their knowledge systems. Also, OER has facilitated the sharing of knowledge in courses developed in the south by the south, for instance, the African Virtual University, in a language understandable by the Brazilian learners (Oviedo & Krimphove, 2022). This is an innovative approach from the dominant model, where flows are from developed countries in the Global North to developing countries in the Global South.

5.2 Enhancing the Availability of Open Educational Resources (OER) for Students and Educators

OER is promoting access to knowledge and its reuse and adaptation in various forms. These resources enable the learners and educators since the learning materials are available in an open-source format that can be easily downloaded, edited, and shared by anyone, unlike expensive, restrictive licensing (Matthew Bingley, 2024). The web icon flexibility provides equal access and encourages people's involvement, which is in concord with the cultural and learning models of various organizations. For example, OER resonates with the participatory principles of Indigenous communities, offering opportunities to **preserve and share their knowledge systems**. Moreover, OER supports **South-South knowledge exchanges**, exemplified by Brazilian learners accessing Portuguese-language OER developed by the African Virtual University, signaling a shift from the traditional unidirectional knowledge flow from the Global North to the Global South (Armijo et al., 2024).

From the educational quality enhancement perspective, OER responds to significant deficits in teaching content and presentation. Compared to other commercial resources and materials, OERs can be easily customized to reflect cultural and environmental differences, and other conditions, as well as teaching–learning needs. According to the National Knowledge Commission (NKC) of India, OER will help address the problem of poor teaching quality, insufficient and poor teaching infrastructure, and limited availability of educational resources, making it more inclusive and accessible to learners (P. Kumar & Wiseman, 2021).

Global OER and **Massive Open Online Course (MOOC)** initiatives, including Siyavula and AVU in South Africa, the Indian NKC OER initiative, and several Asian university projects, underscore the growing adoption of open resources. Such initiatives, combined with the development of the new AI-based adaptive learning systems, are already broadening the opportunities for the scale and individualization of Open Educational Resources as worldwide educators and learners (Supriya et al., 2024).

5.3 Promoting Equity and Inclusion

Perhaps the biggest strength of the OER is the fact that the resources can be shared between institutions as well as across regions. For a long time, conventional teaching methods have faced issues with disparity in the availability of quality resources, especially in developing nations or in institutions that lack adequate funds. OER also provides a sustainable version by providing all and varied free access to quality resources irrespective of their geographical location or their institutional affiliations (Tennant et al., 2016).

By encouraging the creation of open-to-use content, OER eliminates reliance on expensive proprietary content. For example, in regions where schools and universities never get enough funds to purchase textbooks and other necessary learning materials, OER acts as their supplement, which they can access for free and at the same time can be adapted to their local settings. This customization makes sure that, in addition to ensuring that the educational content is available, it is indeed useful for learners those having issues with culture, environment, and even instructional requirements (Butcher, 2015).

The African Virtual University (AVU) case provides a good example of how OER can respond to geographic disparities. In this way, having translated the educational materials into several languages, AVU has provided institutions in Africa with free access to high-quality materials. Siyavula in South Africa, for instance, offers open textbooks in such subjects as mathematics and science, so that the students who are in the regions of relative deprivation experience the same high-quality materials as those in the regions of affluence. (McGreal, 2017).

Further, the expansion of computer usage in the classrooms has greatly increased the reach of OER. Resources, which include OER Commons and MERLOT, give teachers and institutions a rich library of learning materials that can be accessed at any time. It has also enhanced the features of search and put AI into these platforms, and has also enhanced the ease of searching for materials that fit the needs of a user (Bozkurt, 2023).

While OER cannot single-handedly eliminate all disparities in education, it represents a significant step towards leveling the playing field. By providing free and adaptable resources, OER enables institutions in resource-scarce regions to deliver quality education comparable to that of well-funded institutions, thereby promoting equity in global education.

OER also has a utility in facilitating learners with disabilities affected by their impairment. There may be significant barriers in accessibility to conventional material that exists for this group, which OER can be easily tailored with facilities like text-to-speech conversion, increased font size, and other media formats for learners with Visual or hearing impairment. OER meets the needs of universally designed education materials that fit all kinds of users (Wiley et al., 2014).

Therefore, it is evident that OER offers a means through which inequities that surround marginalized/underrepresented groups can be redressed. Free, easily modifiable, and easily accessible, these tools eliminate the prejudice inherent in schooling and guarantee that education is a recognized right for every citizen, regardless of his or her social class, place of residence, or disability.

6 Fostering Innovation and Research Collaboration

OER and MOOC have now become innovative solutions that enhance collaboration and enrich education. They have opened doors letting institutions, researchers, and educators make and share knowledge without limitations on the quality of content available. These applications facilitate collaborative investigations in areas of climate, health, AI, and others by providing public datasets, tutorial materials, and tools for working on research (Raffaghelli et al., 2023). For example, the Open Research Knowledge Graph (ORKG) and the Open Science Framework (OSF) enable researchers to deposit and access deposited academic knowledge, facilitating interdisciplinary interactions and global shareability of research.

MOOCs, particularly, are leaning toward transforming learning by incorporating features such as low cost, OA, and Flexibility. Schools like Athabasca University and the Commonwealth of Learning use MOOCs to provide teachers with training and technology-enhanced learning to increase their capacity to offer quality education in developing countries (Mishra & Panda, 2020). They also focus on issues such as climate change, with resources such as the Climate Knowledge Brokers Group: a platform that provides for OA of key data sets and coordinates cross-continental research.

However, criticisms have been leveled against MOOCs as being highly dependent on video content, completion rates, and privileged learning. These have in many cases only been developed to be used mainly by center and postgraduate students rather than the subject vulnerable groups that are constrained from conventional schooling. Solutions to these problems involve enhancing local language support and identifying, implementing, and evaluating the effectiveness of digital literacy approaches for non-traditional learners (Ossiannilsson, 2021). However, MOOCs and OER are the cornerstones that enable access to education for everybody, and innovation through facilitating the development of new models of cooperation for researchers and educators to test new teaching strategies.

They have also fueled the growth of teaching and learning practices, and these platforms empower instructors to modify the content presented to learners to suit their environment, thus improving the learners' interest. For instance, the environmental science modules can be made to incorporate the local example to enrich the material presented to the learners. Here, self-organized and often small-scale education initiatives are supported by MOOCs, as these bring new concepts like gamification, peer learning, or adaptive learning systems that are based on artificial intelligence in the orientation of different learning styles, affecting educational results positively.

In addition, the integration of the OER in the future with testing and prior learning portfolio assessment will help to legitimize the OER and align them with the existing educational systems. The open resources are actively promoted, and certifications also already exist. For example, within the framework of the OER programs, learners receive credits for their achievements, thanks to which usage of open resources in education is gradually becoming firmly established, and unlimited participation in it is encouraged (Thornhill-Miller et al., 2023). Through their ability to foster collaboration, drive educational innovation, and address global challenges, OER and MOOCs represent a vital step toward inclusive, accessible, and transformative education.

7 Enhancing Lifelong Learning Opportunities

The OER and OA are crucial in promoting lifelong learning through the provision of uncontrolled access to quality learning materials. These resources enable individuals to expand his or her knowledge and skills for a lifetime career, to enhance skills in various areas from technology to health care. For example, Coursera or edX is offering more specific courses that would respond to the new necessary competencies in the modern workplace (Likovič & Rojko, 2022).

OER also serves the need for continuous professional development (CPD) as it provides the learning content flexible enough to be used in various contexts. The information, professional certificates, and training programs can be obtained without being physically present at a workplace and without paying monetary fees. Projects like the Commonwealth of Learning's MOOC for Technology-Enabled Learning show how organizations for CPD utilize OER by training educators in underdeveloped places with programs specialized for their needs.

In particular, the choice of learning path and support of professional development make OERs an important avenue in achieving universally inclusive and lifelong learning. Apart from providing employment advantages to individual participants, these resources also benefit in developing a globally competitive and versatile workforce.

8 Challenges in Implementing OA in Higher Education

Despite its transformative potential, OA faces several significant challenges related to funding, quality assurance, technological barriers, and institutional resistance. Addressing these issues is crucial to achieving equitable and sustainable access to knowledge.

8.1 Funding and Sustainability Issues

A major barrier to OA adoption lies in financial constraints. OA publishing often relies on article processing charges (APCs), which can burden researchers, particularly those from low-income countries and underfunded institutions (Bosman et al., 2021). While some initiatives, such as Plan S, aim to promote equitable OA publishing models, the reliance on APCs exacerbates disparities. Similarly, the maintenance of OA repositories demands sustained funding to cover technical infrastructure, staffing, and long-term digital preservation, which many institutions find challenging to secure (Pinfield, 2024). Public and private funding models must evolve to ensure that OA initiatives remain viable without imposing excessive financial burdens.

8.2 Quality and Credibility Concerns

Ensuring the academic rigor and reliability of OA materials is another pressing challenge. Critics often question the credibility of OA journals, citing concerns about predatory publishers that exploit the APC model to prioritize profit over quality(Siler, 2020). Additionally, inconsistencies in peer review processes and editorial standards can undermine trust in OA publications (Tennant & Ross-Hellauer, 2020). Strengthening peer review mechanisms and developing robust quality control frameworks are essential to enhance the credibility of OA content.

8.3 Technological and Infrastructure Barriers

The success of OA depends on reliable digital infrastructure, which remains a challenge for many low-resource settings. Limited access to high-speed internet, inadequate technological support, and insufficient training in digital tools hinder the widespread adoption of OA initiatives in developing regions (Faturoti, 2022). Institutions often struggle with the cost and complexity of implementing and maintaining repositories, further widening the digital divide. Investments in technological capacity and inclusive policies are critical to overcoming these barriers.

8.4 Institutional and Cultural Resistance

Resistance to OA often stems from entrenched institutional and cultural norms. Many academic communities remain skeptical of OA's value, fearing a decline in prestige or quality associated with traditional subscription-based publishing models(Hamzah et al., 2024). Additionally, researchers may hesitate to share their work openly due to concerns about copyright, misuse, or loss of intellectual property (Ploman & Hamilton, 2024). Promoting cultural shifts toward open sharing and providing clear guidelines on copyright and licensing can help reduce these concerns. Institutional leadership also plays a vital role in encouraging OA adoption by integrating it into research policies and academic reward systems.

9 Strategies to Overcome Challenges

OA requires a combination of sustainable funding sources, quality assurance, more technology, and organizational culture transformations to address the various concerns it has. These objectives seek to achieve the sustainable nature of

OA alongside support for equity, accessibility, and preventing cases of academic dishonesty in the management of OA.

9.1 Developing Sustainable Funding Models

Sustainability is central to OA's success. Institutions, governments, and consortia should collaborate to explore alternative funding sources. Institutional budgets and government grants can be allocated specifically to OA initiatives, reducing reliance on article processing charges (APCs) (Pinfield, 2024). Additionally, transformative agreements, such as "read-and-publish" models, can provide financial support for OA publishing while lowering costs for researchers(Borrego et al., 2021). Non-APC-based models, such as diamond OA (funded entirely by institutions or grants), offer a promising pathway to more equitable publishing systems.

9.2 Ensuring Quality and Academic Integrity

Sustaining the integrity of OA must involve strict peer-reviewed processes or systems as well as other standard scholarly procedures. The journal and repositories should follow clear policies and shouldn't close themselves up in making decisions about the submissions, this should involve qualified peer reviewers. Letting out distinct regulations measuring OA publications and partnering with well-known indexing service providers can uplift the OA content recognition factor. It is equally important that the researchers are informed of how they should avoid such journals (Flores-Vivar & García-Peñalvo, 2023).

9.3 Building Robust Infrastructure

OA initiatives require a good deal of investment to put in place decent digital platforms. Several governments, non-profit organizations, and international organizations need to ensure they develop robust, affordable internet connections, cloud storage, and management of repositories for the less fortunate regions. Institutional repositories are required to be trained on matters concerning repository management, copyright, and data-sharing tools that will enable them to develop sustainable measures on OA.

9.4 Promoting Institutional Buy-In and Cultural Change

The cultural resistance to OA can be reduced when institutions promote the policy of sharing research. It is also suggested that universities and research institutions incorporate the pursuit of OA into their strategic plans, whereby these institutions give creators of content motivators, including career trajectories and grant-making, for submitting their work to OA platforms (Zhu, 2017). It is by inviting academic communities into participatory workshops, seminars, and dialogue, more espousing the virtues as well as the contentious discussions about OA, that changes of attitude are catalyzed collectively. Intentional cooperation with other countries AND the dissemination of successful overseas policies can also promote the acceptance of OA.

By addressing these challenges strategically, the OA movement can achieve its potential to democratize access to knowledge and support the global academic community in tackling pressing societal issues.

10 OA in Computer Science: Enhancing Learning and Innovation

Computer science has been revolutionized by OA, which provides access to important data, resources, and tools. This openness not only contributes positively to learning and innovation processes but also enriches intercultural cooperation and advances innovations in this active section.

10.1 Access to Cutting-Edge Research and Tools

OA enables students, educators, and researchers to access cutting-edge resources, such as open-source software, algorithms, and datasets, which are critical for computer science education and research. Platforms like GitHub and SourceForge host millions of open-source projects, providing invaluable tools for learning and experimentation (Cosmo & Zacchiroli, 2017). Similarly, repositories like Kaggle and OpenML make publicly available datasets that empower researchers to tackle complex problems in machine learning, artificial intelligence, and data analytics (Tandon et al., 2025).

Secondly, OA guarantees the current discoveries, and the latest technologies found in the computer science field are easily retrievable. After analyzing the organic results, scientists discovered that research papers shared on platforms like arXiv and HAL Archives are useful for getting early signs of advancements in AI, quantum computing, and blockchain technologies. This accessibility enables researchers and developers to base their next study on recent research without the added costs of having to subscribe to journals (Willinsky, 2018).

10.2 Supporting Collaborative Development

OA enables highly significant cooperation among researchers and computer scientists in institutions and even across countries. Another outstanding feature is that big platforms such as OpenAI, TensorFlow, and Apache are collaboratively designed by numerous contributors across the world in aspects of perfecting algorithms, creating frameworks, and experimenting with software tools. For instance, the Apache Hadoop project, which was originally open source, has transformed the traditional industry into something more profound, such as big data analytics and cloud computing (Jain, 2017).

Likewise, the OpenCV computer vision library has evolved over the years into an adopted source with contributing developers around the world. This 'open source' approach also helps advance tool making while at the same time, maintaining a culture that supports ongoing enhancements and modifications of the technology for use in areas ranging from health diagnostics to self-driving cars.

10.3 Educational Resources and Curriculum Development

OA has significantly impacted computer science education by providing access to high-quality, up-to-date open educational resources (OER). Platforms like MIT OpenCourseWare and edX host free courses on computer science topics, including programming, algorithms, and artificial intelligence. These resources make advanced computer science education accessible to learners worldwide, regardless of their financial circumstances (Ashadevi & MuthamilSelvi, 2017).

OA also helps curriculum development by allowing instructors to adapt and share course materials tailored to specific institutional needs. Educators can develop courseware that can be of specific use in other institutions. Many materials, like the resources ACM's Digital Library or the open sources of textbooks, provide comprehensive material to which an educator can add context or alter the content according to the contributions of a particular field. This idea makes sure that students gain modern-day knowledge and skills to match changing industry dynamics. These changes mean students are well-equipped to meet current or emerging industry challenges.

10.4 Promoting Innovation and Research

The place of OA in raising the pace of innovation in the field of computer science is immeasurable. Collaborative work and shared knowledge make it easier to create 'fail fast, fail forward' environments that unlock new areas for growth, such as AI, as well as cybersecurity. For instance, popular deep learning libraries PyTorch and TensorFlow, which are open-source, have eased the entrance to funding the innovative AI tools and models (Nguyen et al., 2019).

Case studies illustrate the transformative potential of OA in computer science. The Linux operating system, developed collaboratively through open-source contributions, serves as the backbone for numerous modern technologies, from servers to smartphones. Another example is the RISC-V project, an open-source hardware instruction set architecture that has driven innovation in computer hardware design, offering a customizable and cost-effective alternative to proprietary architectures.

11 Case Studies

OA has impacted the higher education sector by making it possible to offer equal access to resources, embarking on research, and associating globally. The current section provides cases from universities and institutions that have utilized OA to contribute to SDG 4 of inclusive quality education and lifelong learning.

The information provided in these case studies concerns various activities, including establishing OA repositories as well as organizing cross-border activities to tackle obstacles and disseminate ideas for effectiveness. The insights provide helpful recommendations for the institutions intending to implement and expand OA policies and practices, which proves its potential for revolutionizing education and research processes.

11.1 Successful OA Initiatives in Higher Education

Expanding on the foundations that global and national declarations offer, universities and institutions across the globe have effectively adopted OA to improve the educational and research processes. Established universities such as the Massachusetts Institute of Technology (MIT) have developed OA programs with applications like MIT OpenCourseWare, the initiative offering free access to an array of quality education information. In the same regard, through the OpenUCT project, the University of Cape Town has continued to make local research and teaching material available to the global community.

OA has been one of the most impactful approaches towards breaking the knowledge gap and enhancing equal access to scholarly materials throughout the world. The most famous is the Budapest Open Access Initiative (BOAI), started in 2002, and created the basis for the OA policies worldwide, declaring free access for peer-reviewed articles. Likewise, Plan S, launched by a coalition, requires that the papers arising from research funded by public grants be published in journals and platforms meeting the OA requirements. However, in Malaysia, one of the universities has made a groundbreaking attempt to implement an OA policy in their university system through the MyUniNet OA Initiative, which allows universities to publish and access research outputs. Overseen by the Malaysian Ministry of Higher Education, this venture unites institutional repositories in Malaysia and extends research dissemination worldwide. These international and continental processes leave no doubt that the effective development of open knowledge environments requires collaboration at the global and continental levels.

India's National Knowledge Commission (NKC) was set up on June 13, 2005, as an advisory body to help India evolve into a knowledge-driven society. The NKC's Working Groups on OA and OERs specifically and convincingly called for making the research literature funded by the public openly accessible. Furthermore, they called for establishing open courseware repositories to allow the posting of quality educational materials across different populations.

The NKC also pointed out that the intellectual products developed by State-funded institutions, including research and learning resources for lifelong education, should be retrievable through Local and Institutional Repositories. Additionally, the NKC recommended that the government set up a national knowledge infrastructure for priority areas, including water knowledge, energy knowledge, environmental knowledge, education knowledge, as well as health knowledge. These reforms seek to corroborate the parochial-funded research through subject matter review by independent experts and to increase the dissemination of knowledge more equitably.

UGC in India came out with the 'UGC (Submission of Metadata and Full Text of Doctoral Theses in Electronic Format) Regulations, 2005' in the year 2005 that laid down the establishment of the Indian National Theses Database (INTED) and compulsory electronic submission of the thesis for the award of PhD. These regulations' objectives are to facilitate the increased availability of doctoral research and improve the bibliographic management of theses and dissertations.

Though a certain number of universities supported by UGC have implemented OA repositories, there was a tendency of resistance because of the rise in fear of being over-exposed to academic content. Nevertheless, such repositories are now putting pressure on universities to produce better-quality research. UGC's initiatives also include initiatives of modernizing Information and Communications Technologies through schemes like the proposed national education grid to host universities' information services and electronic theses repositories. These actions, therefore, seek to increase the uptake of high-quality doctoral research.

The following initiatives showcase India's efforts to bridge the information divide between the "haves" and "have-nots" by leveraging OA in education, research, and cultural preservation.

11.2 Digital Library Initiatives

India has implemented several digital library projects to preserve and disseminate knowledge:

- **Digital Library of India (IISc)**: Digitizing millions of books with support from IIIT Hyderabad and CDAC.
- **Traditional Knowledge Digital Library (TKDL)**: A repository of India's traditional medicinal knowledge, supported by AYUSH and CSIR.
- **Muktabodha Digital Library**: Focused on preserving Indian scriptures and supported by global institutions like EFEO and SYDA Foundation.

These projects employ advanced digital archiving software like Greenstone and DSpace to ensure efficient long-term preservation and access.

11.3 Open Access Journals

Leading OA journal initiatives in India include:

- **E-Journals @ INSA**: A platform by the Indian National Science Academy for science journals.
- **Indian Academy of Sciences**: Partnering with SpringerLink to host high-quality research publications.
- **MEDIND@nic**: A repository of Indian medical research, managed by NIC and ICMR.

11.4 Open Courseware Initiatives

Significant advancements in OA courseware have been made through:

- **e-Gyankosh (IGNOU)**: A national repository for open educational resources.
- **National Programme on Technology Enhanced Learning (NPTEL)**: Collaborative open courseware from IITs and IISc.
- **NCERT Online Textbooks**: Providing free access to essential educational materials for school students.

11.5 Institutional Repositories

Indian institutions have embraced repositories to host research and theses:

- **ETD@IISc**: A digital thesis repository by IISc Bangalore.
- **DSpace@NCL**: Managed by the National Chemical Laboratory to archive research outputs.
- **DSpace@INFLIBNET**: Hosting scholarly content from UGC-recognized universities.

11.6 Metadata Harvesting Initiatives

Metadata harvesting projects facilitate seamless access to distributed repositories:

- **CASSIR (IISc)**: Aggregates metadata from multiple repositories in India.
- **Open J-Gate**: A gateway for OA journals managed by Informatics India Limited.

Through the present OA initiatives in India, we are directly receiving positive impacts of SDG 4 concerns of education in fostering education quality and free research papers. Programs like NPTEL and e-Gyankosh help students get free-of-cost quality study materials, and repositories like ETD@IISc help research scholars to get global access to their work.

Successful implementations highlight the need for robust digital infrastructure, institutional commitment, and sustainable funding. Open collaboration among academic institutions and governmental support remain key drivers for effective OA integration in higher education.

These initiatives have directly supported SDG 4, which seeks to ensure inclusive and equitable quality education and promote lifelong learning opportunities for all. By making high-quality educational resources and research outputs freely accessible, these programs help bridge educational disparities and foster innovation across diverse regions. For example, OA repositories have allowed underserved communities to access cutting-edge research, supporting their educational and developmental aspirations.

Key lessons from these initiatives include the importance of institutional buy-in, robust digital infrastructure, and active engagement with academic communities. Successful implementation requires collaborative efforts, sustainable funding models, and adherence to high-quality academic standards. These insights serve as a blueprint for other institutions seeking to adopt open-access practices effectively.

12 Future Perspectives

12.1 Emerging Trends in OA and Higher Education

The OA movement is now at a stage of rapid evolution due to innovations that are set to revolutionize higher learning. Among them, the main one is the use of Artificial Intelligence (AI) and machine learning technologies to improve open-access content. These technologies can assist in the creation of metadata, improve search functionality, and deliver recommendations on academic content to researchers/learners, enhancing personalized learning and knowledge sharing, all of which are highly beneficial (Severin et al., 2020). Moreover, blockchain has been pointed out as a rising solution to enhance the credibility and protection of research articles, especially guaranteeing that authors and researchers receive credit and remuneration for their work (Dwivedi et al., 2022).

Another trend is the rise of interest in Open Educational Resources (OERs)—the educational materials available for free and further modification and sharing. With the help of combining OER with OA, education could be provided even more freely for students coming from different backgrounds and of different abilities. Such a syncretism of OA and OERs is capable of transforming educational practices by extending cooperative and open learning activities enhanced by the accessibility to up-to-date, shared, and improved resources worldwide through contributions from teachers and scholars (McGreal, 2017).

Consequently, predictions indicate that OA will extend into more specialized spheres of study as governments and other institutions extend their requirements for open-access journal publishing and repository deposits (Saimah et al., 2022). The OA can be expected to spread to textbooks, multimedia products, datasets, as well as curricula. Since more of such resources are becoming both available and accessible, they will force OA to change education practices for the better, open, and participative to increase the literate population and improve the knowledge economy system.

12.2 Opportunities for Further Research and Development

Despite such advancements OA has made in higher education, there still exist some distinct niches that, if explored further, can go on to advance the domain further. One area that has received the least attention when it comes to OA publishing is the existing and viable business models. The current dependencies on APCs affect researchers in many ways, with researchers from low-resource environments being most impaired; Thus, new sustainable financial models for OA publishing that do not impose more costs on authors are required (Severin et al., 2020). More effort can be directed towards searching for institutional or government funding sources to help solve this problem.

The second potential is an enhancement of the supporting community infrastructure for OA, particularly in LAMI countries. Several institutions in developed countries have set up OA repositories, but more infrastructure has to be created in the Global South so that learners and researchers around the world have access to OA resources (Asadi et al., 2019). Thus, more international cooperation could be made to enhance OA structures used in these areas, including equipment, tools, and training.

Additionally, there is room for further research on the impact of OA on teaching and learning practices. As more educational institutions adopt OA materials and OERs, studying the effects of these resources on student learning outcomes, teacher-student engagement, and the quality of education could provide valuable insights into how OA can further transform the educational landscape (Weller et al., 2018).

Finally, as OA continues to grow, there is an emerging need for research on ethical considerations in OA publishing, including concerns around data privacy, intellectual property rights, and the potential for misuse of open data. Exploring these ethical dimensions will ensure that the OA movement remains grounded in principles of equity and fairness, protecting the rights and interests of all stakeholders involved.

In conclusion, the future of OA in higher education holds tremendous potential for transforming access to knowledge, creating new learning opportunities, and fostering an inclusive, global educational environment. The evolving trends and ongoing research will play a pivotal role in ensuring that OA continues to contribute to the achievement of SDG 4 by making quality education universally accessible (Neumann et al., 2022).

13 Conclusion

The chapter highlights how OA can significantly mark a shift in the future of education and research in higher education institutions. The main realized trends include global interconnectedness, increased dissemination of knowledge, as well as equity in achieving understanding. However, some barriers like inadequate infrastructure, end-user resistance to change, and funding are required for the true potential of OA to be realized. The following strategies enhance success in overcoming these barriers: Strong and sound policies, excellence in infrastructure, and international cooperation.

OA to knowledge and information is an inspiring idea that addresses social and economic inclusions of inequalities in the access to knowledge and information "haves" and "have-nots." It makes more sense to say that OA is easier to attain within policy systems, institutions, information infrastructure, manpower, and financial resources that are already intact. For example, specific capacity-building programs on digital preservation, digital literacy, and OA have elicited positive results in India. Substantial advancement has been observed in the last decade, such as the appearance of OA and digital literacy projects. Hundreds of workshops and training sessions have disseminated knowledge among thousands of library and IT specialists regarding how to construct OA repositories employing free software. Moreover, many library schools in India have incorporated open-source digital archiving software in their curriculum, due to which OA is being adopted enormously. The OA movement has also received support from national and international conferences.

Higher education institutions, policymakers, and researchers must recognize the critical role OA plays in achieving SDG 4—ensuring inclusive and equitable quality education for all. Embracing OA is not only an opportunity to foster greater knowledge sharing but also a responsibility to make research accessible to diverse communities. Stakeholders must create supportive environments for OA, including developing policies that encourage open publishing and establishing repositories to facilitate access to knowledge.

In the case of OA, the sustainability of the solutions is a question that should be answered. In some cases, institutional OA portals may not be functional because of internal reasons, thus, there is a need for fortifying institutional frameworks and investing in continuous manpower capacity building. It is therefore recommended that institutional repository coordinators educate the members of the community on self-archiving techniques as well as the need to adopt OA.

As a result of OA, higher education can be expanded by providing improved and transformed educational materials and emphasizing lifelong learning. Since more institutions and countries embrace OA, the world is set to get a better and more open knowledge environment, which creates a more inclusive and equitable global knowledge ecosystem. Through OA, higher education will be part of a shift where quality education and research will be available to all, hence the achievement of SDG 4 'Quality Education' and coming up with a knowledge society that is beneficial for all. OA to scholarly and development-related literature not only ensures global visibility and recognition but also enables digital inclusion, particularly for underprivileged communities, thereby bridging social divides.

References

Adipat, S., & Chotikapanich, R. (2022). Sustainable Development Goal 4: An Education Goal to Achieve Equitable Quality Education. *Academic Journal of Interdisciplinary Studies, 11*(6), 174–183. https://doi.org/10.36941/ajis-2022-0159

Alamri, H. A., Watson, S., & Watson, W. (2021). Learning technology models that support personalization within blended learning environments in higher education. *TechTrends, 65*(1).

Alqahtani, A. Y., & Rajkhan, A. A. (2020). E-learning critical success factors during the covid-19 pandemic: A comprehensive analysis of e-learning managerial perspectives. *Education Sciences, 10*(9), 1–16. https://doi.org/10.3390/educsci10090216

Arancio, J., Morales Tirado, M., & Pearce, J. (2022). Equitable Research Capacity Towards the Sustainable Development Goals: The Case for Open Science Hardware. *Journal of Science Policy & Governance, 21*(02). https://doi.org/10.38126/jspg210202

Armijo, L. E., Fraundorfer, M., & Rhodes, S. D. (2024). *South American policy regionalism: Drivers and barriers to international problem solving*. Taylor & Francis.

Asadi, S., Abdullah, R., Yah, Y., & Nazir, S. (2019). Understanding institutional repository in higher learning institutions: A systematic literature review and directions for future research. *IEEE Access, 7*, 35242–35263. https://doi.org/10.1109/ACCESS.2019.2897729

Ashadevi, B., & MuthamilSelvi, P. (2017). Open educational resources in computer science. *Opportunities and Challenges, 14*(5), 42–49.

Atenas, J., Havemann, L., Neumann, J., & Stefanelli, C. (2020). *Open education policies: Guidelines for co-creation.*

Baños, J.-E., Orefice, C., Bianchi, F., & Costantini, S. (2019). Ten Targets about SDG 4 : ensuring inclusive and equitable quality education and promoting lifelong learning opportunities for all. In *Good health, quality education, sustainable communities, human rights : The scientific contribution of Italian UNESCO Chairs and partners to SDGs 2030* (Vol. 5870).

Banyas, K., College, W., Cloutier, M., & Farry, C. (2023). Leading an open revolution: Promoting awareness of open resources through an interdisciplinary learning community. *Collaborative Librarianship, 13*(2).

Bingley, M. (2024). *Adopting and adapting: How faculty reuse, revise and remix open educational resources.*

Björk, B. C. (2021). Publishing speed and acceptance rates of open access megajournals. *Online Information Review, 45*(2), 270–277.

Boeren, E. (2019). Understanding Sustainable Development Goal (SDG) 4 on "quality education" from micro, meso and macro perspectives. *International Review of Education, 65*(2), 277–294. https://doi.org/10.1007/s11159-019-09772-7

Bolick, J., Emmett, A., Greenberg, M. L., Peterson, A. T., Bolick, J., Greenberg, M. L., Peterson, A. T., & Rosenblum, B. (2024). Planet of the APCs : A decade of progress and setbacks in open access. *Journal of Librarianship and Scholarly Communication, 12*, 0–27.

Borrego, Á., Anglada, L., & Abadal, E. (2021). Transformative agreements: Do they pave the way to open access? *Learned Publishing, 34*(2), 216–232. https://doi.org/10.1002/leap.1347

Bosman, J., de Jonge, H., Kramer, B., & Sondervan, J. (2021). Advancing open access in the Netherlands after 2020: From quantity to quality. *Insights: The UKSG Journal, 34*. https://doi.org/10.1629/UKSG.545

Bozkurt, A. (2023). Generative AI, synthetic contents, open educational resources (OER), and open educational practices (OEP): A new front in the openness landscape. *Open Praxis, 15*(3), 178–184. https://doi.org/10.55982/openpraxis.15.3.579

Budzinski, O., Grebel, T., Wolling, J., & Zhang, X. (2020). Drivers of article processing charges in open access. *Scientometrics, 124*(3), 2185–2206. https://doi.org/10.1007/s11192-020-03578-3

Butcher, N. (2015). A basic guide to open educational resources (OER). In *A Basic guide to open educational resources (OER)*. https://doi.org/10.56059/11599/36
Bygstad, B., Øvrelid, E., Ludvigsen, S., & Dæhlen, M. (2022). From dual digitalization to digital learning space: Exploring the digital transformation of higher education. *Computers and Education, 182*(January). https://doi.org/10.1016/j.compedu.2022.104463
Cano, N. A., Velasco, J. O., & Franco, I. B. (2020). *Science for Sustainable societies actioning the global goals for local impact* (Issue January). https://doi.org/10.1007/978-981-32-9927-6
Chan, C. K. Y. (2023). A comprehensive AI policy education framework for university teaching and learning. *International Journal of Educational Technology in Higher Education, 20*(1). https://doi.org/10.1186/s41239-023-00408-3
Chankseliani, M., & McCowan, T. (2021). Higher education and the sustainable development goals. *Higher Education, 81*(1), 1–8. https://doi.org/10.1007/s10734-020-00652-w
Chugh, M., Upadhyay, R., & Chugh, N. (2023). An empirical investigation of critical factors affecting acceptance of E-Learning platforms: A learner's perspective. *SN Computer Science, 4*(3), 1–15. https://doi.org/10.1007/s42979-022-01558-3
Connor, R. (2015). *The United Nations world water development report 2015: water for a sustainable world*. UNESCO publishing.
Di Cosmo, R, & Zacchiroli, S. (2017). Software Heritage: Why and how to preserve software source code. In *14th International Conference on Digital Preservation, Srp*, 1–10. https://doi.org/10.1145/nnnnnnn.nnnnnnn
Dubey, R., Dutot, V., Felix, R., Goyal, D. P., Gustafsson, A., Hinsch, C., Jebabli, I., et al. (2022). Metaverse beyond the hype: Multidisciplinary perspectives on emerging challenges, opportunities, and agenda for research, practice and policy. *International Journal of Information Management, 66*(July), 102542. https://doi.org/10.1016/j.ijinfomgt.2022.102542
Ertz, M., Kupaei, H. R., Rad, D., Redeş, A. R., Roman, A., Ignat, S., Lile, R., Demeter, E., Eger Au, A., Dughi, T., Balaş, E. B., Maier, R., Kiss, C., Torkos, H., & Rad, G. (2022). Pathways to inclusive and equitable quality early childhood education for achieving SDGG goal-a scoping review. *Frontiers in Psychology*, 1–18.
Faturoti, B. (2022). Online learning during COVID19 and beyond: A human right based approach to internet access in Africa. *International Review of Law, Computers and Technology, 36*(1), 68–90. https://doi.org/10.1080/13600869.2022.2030027
Fensie, A., Pierre, T. S., Jain, J., & Sezen-Barrie, A. (2024). Engaged learning during distraction: a case study of successful working moms in distance education. In *Journal of Computing in Higher Education, 36*(2). https://doi.org/10.1007/s12528-023-09359-0
Flores-Vivar, J. M., & García-Peñalvo, F. J. (2023). Reflections on the ethics, potential, and challenges of artificial intelligence in the framework of quality education (SDG4). *Comunicar, 30*(74), 35–44. https://doi.org/10.3916/C74-2023-03
Ghanem, S. (2020). E-learning in higher education to achieve SDG 4: Benefits and challenges. In *2020 2nd International Sustainability and Resilience Conference: Technology and Innovation in Building Designs*. https://doi.org/10.1109/IEEECONF51154.2020.9319981
Haleem, A., Javaid, M., Qadri, M. A., & Suman, R. (2022). Understanding the role of digital technologies in education: A review. *Sustainable Operations and Computers, 3*(February), 275–285. https://doi.org/10.1016/j.susoc.2022.05.004
Hamzah, B., Mahmod, R., Hafiza, N., Samad, A., Zainuddin, A. J., Haron, H., Asiah, W., Muhamad, W., Azahari, M., & Yusof, M. (2024). The dark side of academic dissemination. *Predatory Publishing, 44*(4), 1383–1391.
Herrera-Cubides, J. F., Gaona-García, P. A., Montenegro-Marín, C. E., & Sánchez-Alonso, S. (2022). Improving OER descriptions to enhance their availability, reuse, and enrichment. *Education and Information Technologies*, 1–29.
Huang, C. K., Neylon, C., Hosking, R., Montgomery, L., Wilson, K. S., Ozaygen, A., & Brookes-Kenworthy, C. (2020). Evaluating the impact of open access policies on research institutions. *eLife, 9*, 1–13. https://doi.org/10.7554/ELIFE.57067
Huijboom, F., van Meeuwen, P., Rusman, E., & Vermeulen, M. (2023). Differences and similarities in the development of professional learning communities: A cross-case longitudinal study.

Learning, Culture and Social Interaction, 42(July), Article 100740. https://doi.org/10.1016/j.lcsi.2023.100740

Jain, V. K. (2017). *Big data and hadoop.*

Khan, P. A., Johl, S. K., Akhtar, S., Asif, M., Salameh, A. A., & Kanesan, T. (2022). Open innovation of institutional investors and higher education system in creating open approach for SDG-4 quality education: A conceptual review. *Journal of Open Innovation: Technology, Market, and Complexity, 8*(1). https://doi.org/10.3390/joitmc8010049

Kılıçkaya, F., & Kic-Drgas, J. (2021). Issues of context and design in OER (open educational resources). *Educational Technology Research and Development, 69*(1), 401–405. https://doi.org/10.1007/s11423-020-09852-8

Kivunja, C. (2014). Innovative pedagogies in higher education to become effective teachers of 21st century skills: Unpacking the learning and innovations skills domain of the new learning paradigm. *International Journal of Higher Education, 3*(4), 37–48. https://doi.org/10.5430/ijhe.v3n4p37

Klebel, T., & Ross-Hellauer, T. (2023). The APC-barrier and its effect on stratification in open access publishing. *Quantitative Science Studies, 4*(1), 22–43. https://doi.org/10.1162/qss_a_00245

Kumar, A., Baishya, D., & Deka, M. (2021). Open Educational Resources (OER) issues and problems experienced by social scientists of select higher educational institutions in India. *Library Philosophy and Practice, 2021*(June), 1–20.

Kumar, P., & Wiseman, A. W. (2021). *Teacher quality and education policy in India: Understanding the relationship between teacher education, teacher effectiveness, and student outcomes.* Routledge.

Lambert, S. R. (2020). Do MOOCs contribute to student equity and social inclusion? A systematic review 2014–18. *Computers and Education, 145*(September 2019), 103693. https://doi.org/10.1016/j.compedu.2019.103693

Likovič, A., & Rojko, K. (2022). E-Learning and a case study of Coursera and edX online platforms. *Research in Social Change, 14*(1), 94–120. https://doi.org/10.2478/rsc-2022-0008

Luo, T., Hostetler, K., Freeman, C., & Stefaniak, J. (2020). The power of open: Benefits, barriers, and strategies for integration of open educational resources. *Open Learning, 35*(2), 140–158. https://doi.org/10.1080/02680513.2019.1677222

Majekolagbe, F. O. (2024). Copyright and quality education for all (SDG4). *European Intellectual Property Review, 1*, 1–26.

Marín, V. I., Zawacki-Richter, O., Aydin, C. H., Bedenlier, S., Bond, M., Bozkurt, A., Conrad, D., Jung, I., Kondakci, Y., Prinsloo, P., Roberts, J., Veletsianos, G., Xiao, J., & Zhang, J. (2022). Faculty perceptions, awareness and use of open educational resources for teaching and learning in higher education: a cross-comparative analysis. *Research and Practice in Technology Enhanced Learning, 17*(1). https://doi.org/10.1186/s41039-022-00185-z

Martin, M., & Godonoga, A. (2020). SDG 4-policies for flexible learning pathways in higher education taking stock of good practices internationally (Issue March). https://doi.org/10.13140/RG.2.2.31907.81449

McGreal, R. (2017). Special report on the role of open educational resources in supporting the sustainable development goal 4: Quality education challenges and opportunities. *International Review of Research in Open and Distributed Learning, 18*(7), 292–305. https://doi.org/10.19173/irrodl.v18i7.3541

Miao, F., Holmes, W., Huang, R., Zhang, H., & Unesco. (2023). *AI and education : Guidance for policymakers.* https://discovery.ucl.ac.uk/id/eprint/10130180/1/MiaoandHolmes-2021-AIandeducationguidanceforpolicy-makers.pdf

Miller, T., Howe, P., & Sonenberg, L. (2017). *Explainable AI: Beware of inmates running the Asylum Or: How I Learnt to stop worrying and love the social and behavioural sciences.* http://arxiv.org/abs/1712.00547

Miranda, J., Navarrete, C., Noguez, J., Molina-Espinosa, J. M., Ramírez-Montoya, M. S., Navarro-Tuch, S. A., Bustamante-Bello, M. R., Rosas-Fernández, J. B., & Molina, A. (2021). The core

components of education 4.0 in higher education: Three case studies in engineering education. *Computers and Electrical Engineering, 93*. https://doi.org/10.1016/j.compeleceng.2021.107278
Mishra, S., & Panda, S. (2020). *Technology-enabled learning: Policy, pedagogy and practice*.
Neumann, J., Schön, S., Bedenlier, S., Ebner, M., Edelsbrunner, S., Krüger, N., Esposito, G. L., Marín, V. I., Orr, D., Peters, L. N., Reimer, R. T. D., & Zawacki-Richter, O. (2022). Approaches to monitor and evaluate OER policies in higher education—Tracing developments in Germany, Austria, and Switzerland. *Asian Journal of Distance Education, 17*(1), 125–147. https://asianjde.com/ojs/index.php/AsianJDE/article/view/619
Nguyen, G., Dlugolinsky, S., Bobák, M., Tran, V., López García, Á., Heredia, I., Malík, P., & Hluchý, L. (2019). Machine learning and deep learning frameworks and libraries for large-scale data mining: A survey. *Artificial Intelligence Review, 52*(1), 77–124. https://doi.org/10.1007/s10462-018-09679-z
Nicholas, D., Revez, J., Abrizah, A., Rodríguez-Bravo, B., Boukacem-Zeghmouri, C., Clark, D., Xu, J., Swigon, M., Watkinson, A., Jamali, H. R., & Herman, E. (2024). Purchase and publish: Early career researchers and open access publishing costs. *Learned Publishing*. https://doi.org/10.1002/leap.1617
Nishikawa, K., & Murakami, A. (2019). Does Open Access foster Interdisciplinary Citation? Decomposing Open Access Citation Advantage. *Sustainability (Switzerland), 11*(1), 1–14. http://scioteca.caf.com/bitstream/handle/123456789/1091/RED2017-Eng-8ene.pdf?sequence=12&isAllowed=y%0A10.1016/j.regsciurbeco.2008.06.005%0A, https://www.researchgate.net/publication/305320484_SISTEM_PEMBETUNGAN_TERPUSAT_STRATEGI_MELESTARI
Ossiannilsson, E. (2021). *MOOCS for lifelong learning, equity, and liberation*.
Otto, D. (2021). How to promote the use of open educational resources (OER) in higher education. A Parley with OER experienced teachers. *Open Praxis, 13*(4), 354–364. https://doi.org/10.55982/openpraxis.13.4.264
Oviedo, L. E. Z., & Krimphove, J. (2022). Virtual exchange contributions to the development of intercultural competence: A Brazilian higher education institutions' perspective. *Actualidades Investigativas En Educación, 22*(1), 312–336.
Pinfield, S. (2024). Achieving global open access: The need for scientific, epistemic and participatory openness. In *achieving global open access: The need for scientific, epistemic and participatory openness*. https://doi.org/10.4324/9781032679259
Piwowar, H., Priem, J., Larivière, V., Alperin, J. P., Matthias, L., Norlander, B., Farley, A., West, J., & Haustein, S. (2018). The state of OA: A large-scale analysis of the prevalence and impact of Open Access articles. *PeerJ, 2018*(2), 1–23. https://doi.org/10.7717/peerj.4375
Ploman, E. W., & Hamilton, L. C. (2024). *Copyright: Intellectual property in the information age*. Taylor & Francis.
Raffaghelli, J. E., Foschi, L. C., Crudele, F., Doria, B., Grion, V., & Cecchinato, G. (2023). *The ENCORE approach: Pedagogy of an AI-driven system to integrate OER in higher education & VET* (In ENCORE).
Ramirez-Montoya, M. S. (2020). Challenges for open education with educational innovation: A systematic literature review. *Sustainability (Switzerland), 12*(17). https://doi.org/10.3390/su12177053
Rawashdeh, A. Z. Al, Mohammed, E. Y., Arab, A. R. Al, Alara, M., & Al-Rawashdeh, B. (2021). Advantages and disadvantages of using E-learning in university education. *Electronic Journal of E-Learning, 19*(2), 107–117. https://academic-publishing.org/index.php/ejel/article/view/2168
Rets, I., Coughlan, T., Stickler, U., & Astruc, L. (2023). Accessibility of open educational resources: How well are they suited for English learners? *Open Learning, 38*(1), 38–57. https://doi.org/10.1080/02680513.2020.1769585
Saimah, B., Gul, S., Bashir, S., Nisa, N. T., & Ganaie, S. A. (2022). Evolution of institutional repositories: Managing institutional research output to remove the gap of academic elitism. *Journal of Librarianship and Information Science, 54*(3), 518–531.

Saini, M., Sengupta, E., Singh, M., Singh, H., & Singh, J. (2023). Sustainable development goal for quality education (SDG 4): A study on SDG 4 to extract the pattern of association among the indicators of SDG 4 employing a genetic algorithm. *Education and Information Technologies, 28*(2). https://doi.org/10.1007/s10639-022-11265-4

Severin, A., Egger, M., Eve, M. P., & Hürlimann, D. (2018). Discipline-specific open access publishing practices and barriers to change: an evidence-based review. *F1000 Research, 7*, 1925. https://doi.org/10.12688/f1000research.17328.1

Severin, A., Egger, M., Eve, M. P., & Hürlimann, D. (2020). Discipline-specific open access publishing practices and barriers to change: An evidence-based review. *F1000 Research, 7*, 1925.

Shabalala, L. P., & Ngcwangu, S. (2021). Accelerating the implementation of SDG 4: Stakeholder perceptions towards initiation of sustainable community engagement projects by higher education institutions. *International Journal of Sustainability in Higher Education, 22*(7), 1573–1591.

Sheikh, A., & Richardson, J. (2023). Open access movement in the scholarly world: Pathways for libraries in developing countries. *Journal of Information Science.*

Shen, C. (2020). *A Study of Three Alternative Open Access Publishing Models.*

Siler, K. (2020). Demarcating spectrums of predatory publishing: Economic and institutional sources of academic legitimacy. *Journal of the Association for Information Science and Technology, 71*(11), 1386–1401.

Supriya, Y., Bhulakshmi, D., Bhattacharya, S., Gadekallu, T. R., Vyas, P., Kaluri, R., Sumathy, S., Koppu, S., Brown, D. J., & Mahmud, M. (2024). Industry 5.0 in smart education: Concepts, applications, challenges, opportunities, and future directions. *IEEE Access, 12*, 81938–81967. https://doi.org/10.1109/ACCESS.2024.3401473

Tandon, A., Awasthi, A., & Pattnayak, K. C. (2025). Efficacy of machine learning in simulating precipitation and its extremes over the capital cities in North Indian states. *Scientific Reports, 15*(1), 10345.

Tang, H. (2021). Implementing open educational resources in digital education. *Educational Technology Research and Development, 69*(1), 389–392. https://doi.org/10.1007/s11423-020-09879-x

Tao, D., Fu, P., Wang, Y., Zhang, T., & Qu, X. (2022). Key characteristics in designing massive open online courses (MOOCs) for user acceptance: An application of the extended technology acceptance model. *Interactive Learning Environments, 30*(5), 882–895.

Tennant, J. P., Waldner, F., Jacques, D. C., Masuzzo, P., Collister, L. B., & Hartgerink, C. H. J. (2016). The academic, economic and societal impacts of open access: An evidence-based review. *F1000Research, 5*. https://doi.org/10.12688/f1000research.8460.1

Tennant, J. P. (2020). Web of science and scopus are not global databases of knowledge. *European Science Editing, 46*, 1–3. https://doi.org/10.3897/ese.2020.e51987

Tennant, J. P., & Ross-Hellauer, T. (2020). The limitations to our understanding of peer review. *Research Integrity and Peer Review, 5*(1), 1–14. https://doi.org/10.1186/s41073-020-00092-1

Thornhill-Miller, B., Camarda, A., Mercier, M., Burkhardt, J. M., Morisseau, T., Bourgeois-Bougrine, S., Vinchon, F., El Hayek, S., Augereau-Landais, M., Mourey, F., Feybesse, C., Sundquist, D., & Lubart, T. (2023). Creativity, critical thinking, communication, and collaboration: Assessment, certification, and promotion of 21st century skills for the future of work and education. *Journal of Intelligence, 11*(3). https://doi.org/10.3390/jintelligence11030054

Tillinghast, B. (2020). Developing an open educational resource and exploring OER-enabled pedagogy in higher education. *IAFOR Journal of Education, 8*(2), 159–174.

Tlili, A., Zhang, J., Papamitsiou, Z., Manske, S., Huang, R., Kinshuk, & Hoppe, H. U. (2021). Towards utilising emerging technologies to address the challenges of using Open Educational Resources: a vision of the future. *Educational Technology Research and Development*, *69*(2), 515–532. https://doi.org/10.1007/s11423-021-09993-4

Tonegawa, Y. (2023). Education in SDGs: What is inclusive and equitable quality education? In *Sustainable development goals series: Vol. Part F2748.* Springer. https://doi.org/10.1007/978-981-19-4859-6_4

UNESCO. (2021). *The United Nations world water development report 2021: Valuing water.*

Unterhalter, E. (2019). The many meanings of quality education: Politics of targets and indicators in SDG4. *Global Policy, 10*(January), 39–51. https://doi.org/10.1111/1758-5899.12591

Verma, A. K., & Sonkar, S. K. (2021). Growth of Open Access Scholarly communication in BRICS countries. *Library Philosophy and Practice.*

Vervoort, D., Ma, X., & Bookholane, H. (2021). Equitable open access publishing: Changing the financial power dynamics in academia. *Global Health Science and Practice, 9*(4), 733–736. https://doi.org/10.9745/GHSP-D-21-00145

Weller, M., Jordan, K., Devries, I., & Rolfe, V. (2018). Mapping the open education landscape: Citation network analysis of historical open and distance education research. *Open Praxis, 10*(2), 109–126.

Wiley, D., Bliss, T. J., & McEwen, M. (2014). *Open educational resources: A review of the literature. Handbook of research on educational communications and technology.*

Wilkinson, J. (2020). Educational leadership as practice. In *Oxford research encyclopedia of education.*

Williams, J. W., Taylor, A., Tolley, K. A., Provete, D. B., Correia, R., Guedes, T. B., Farooq, H., Li, Q., Pinheiro, H. T., Liz, A. V., Luna, L. W., Matthews, T. J., Palmeirim, A. F., Puglielli, G., Rivadeneira, M. M., Robin, V. V., Schrader, J., Shestakova, T. A., Tukiainen, H., et al. (2023). Shifts to open access with high article processing charges hinder research equity and careers. *Journal of Biogeography, 50*(9), 1485–1489. https://doi.org/10.1111/jbi.14697

Willinsky, J. (2018). Scholarly Associations and the economic viability of open access publishing. *Open Journal System Demonstration Journal, 196*, 3-Chas.

Young, J. S., & Brandes, P. M. (2020). Green and gold open access citation and interdisciplinary advantage: A bibliometric study of two science journals. *Journal of Academic Librarianship, 46*(2). https://doi.org/10.1016/j.acalib.2019.102105

Zhang, X., Tlili, A., Nascimbeni, F., Burgos, D., Huang, R., Chang, T. W., Jemni, M., & Khribi, M. K. (2020). Accessibility within open educational resources and practices for disabled learners: a systematic literature review. *Smart Learning Environments, 7*(1), 1–19. https://doi.org/10.1186/s40561-019-0113-2

Zhu, Y. (2017). Who support open access publishing? Gender, discipline, seniority and other factors associated with academics' OA practice. *Scientometrics, 111*(2), 557–579. https://doi.org/10.1007/s11192-017-2316-z

Bridging Gender Gaps in Digital Literacy Through Industry-Led Interventions: A Strategic Pathway for Achieving SDG 5

Priti Sharma, Rupa Khanna Malhotra, Himani Upreti, and Tanupriya Choudhary

Abstract

In this modern era, digital literacy is a vital competency that allows each and every individual to access services like health, education, employment, and civic participation etc. in an electronic manner with an ease of 24*7 access. It creates a climate that promotes the overall development of local communities and individuals. Despite this, the gender gap persists and inappropriately excludes girls and women, particularly in developing countries. Women are 21% less likely to be online in comparison to male across the globe; this gap rises to 52% in developed countries (*Bridging the Digital Literacy Gender Gap in Developing Countries*, 2022). Industrial organizations continuously trying for gender equality solutions by leveraging their resources alongside their innovation capabilities and extended networks. The gap becomes bridgeable through multiple government-supported programs including training initiatives and electronic inclusion programs and industry partnerships. This paper studies the digital gender divide's contextualization process through the relationship between industry involvement and digital literacy and gender factors to drive transformative organizational change. The partition of digital access reveals its structural and sociocultural as well as economic characteristics while its elimination remains vital for achieving Sustainable Development Goals (SDGs) 5 (Gender Equality) and SDG 4 (Quality Education). The main role in digital inclusion played by governments and civil society but this chapter brings attention to how the industrial sector fights against digital divides based on gender. The digital literacy rates of women in multiple regions improved because

P. Sharma (✉) · R. K. Malhotra · H. Upreti
Department of Commerce, Graphic Era Deemed to be University, Dehradun, Uttarakhand, India
e-mail: pritisharma@geu.ac.in

T. Choudhary
School of Computer Sciences, UPES, Dehradun, Uttarakhand, India

T. Choudhury et al. (eds.), *Open Access as a Knowledge Platform for Implementing the United Nations Sustainable Development Goals*, World Sustainability Series,
https://doi.org/10.1007/978-3-031-99605-4_7

of industry leadership although multiple barriers continue to exist (*Bridging the Gender Digital Divide*, 2023). The implementation faces barriers because of restricted enlargement and public opposition to change as well as inadequate integration between programming and existing government guidelines. Furthermore, the findings emphasises the significance of collaboration between industries, governments, and civil Society in achieving sustainable progress and bridging the gender gap.

Keywords

Sustainable development goals (SDG) • SDG 4 • SDG 5 • Gender-gap • Gender equality • Quality education

1 Introduction

In the twenty-first century, digital literacy has become increasingly important. It is not a luxury but a necessity. The ability to access, manage, understand, integrate, and create information safely and accurately is known as digital literacy. Digital literacy is a pathway to education, employment and actively participation in interconnected world. After a lot of advancement in each and every field digital divide still persists adversely impacting girls and women. These gaps will have far reaching consequences limiting women access to the world of knowledge. Digital literacy gender Gap is entrenched in sociocultural, economic, and infrastructural challenges. Traditional gender roles keep women from accessing education and technology in many places, while systemic issues such as poverty and the absence of digital infrastructure increases the gap. In the digital age, as the world shifts online, these divides not only prevent women from reaching their full potential as individuals, but hold back wider social and economic advancement. This gap is not just a matter of access but also involves the social and economic advancements that arises from it (Valli, 2025). Women often face unique challenges and obstacles when it comes to utilizing the digital technologies (Kaur & Sharma, 2023). Quality education is necessary to bridge this gap and improve society in this era.

Quality education is essential for everyone, as it cultivates the mind, fosters creativity, and establishes an environment that promotes self-respect. It opens the door to various opportunities in life and provides avenues for self-development and society's growth. Keeping the belief that education is a fundamental right and should be given to all to promote social and economic development, Sustainable Development Goal 4—Quality Education has been framed. However, SDG 4 alone cannot change this entire world. Therefore, in order to achieve SDG 4 we must achieve SDG 5—Gender equality. Without providing equal opportunities to all and empowering women and girls, it is not possible for a nation to develop and achieve the full potential of SDG 4. The objective is not just to offer equal education opportunities or eradicate gender discrimination, but also to ensure equal access to essential resources and opportunities for a society as a whole. This may require multi-stakeholder collaborations and industry-led initiatives as strategies to bridge the gender divide. Educational initiatives, capacity-building programs, and promoting gender-responsive technology design are among the activities that aim to improve women's digital literacy and participation in the digital sphere (Roy et al., 2024). As per the sustainable development Goals reports issued by united nations, parity in women's participation in management positions at current rates will require another 176 years to bridge the gap (UNSD, 2024). There is an urgent need for strong and ongoing commitments to eradicating harmful practices, reforming biased social norms, and repealing discriminatory laws. Increasing the representation of women in leadership and decision-making positions and appropriately increasing funding for gender equality on priorities are high at the national, regional, and international levels (Summary for policymakers, 2023).

The companies have recognized that addressing these gaps, is not just a social responsibility, but is in fact a business opportunity for growth and innovation. As engines of innovation and economic development, industries have a distinct role to play in tackling gender inequality in digital literacy. Through corporate social responsibility and public private partnerships and focused programmes, business can use its resources, technological know-how and global connectivity to inclusively involve and benefit women—and narrow the digital divide (Malik, 2024). Global technology giants like Microsoft, Google, Intel etc. have already started pioneering programs like "DigiGirlz", "Women Techmakers", "She Will Connect" etc. which are targeted to provide women and girls vital digital skills. They not only demonstrate the capacity of industry-led actions to be change-enabling, they also represent the significance of collective action partnerships in the fight for equality in the digital era. Industry was certainly actively involved on facilitating means to empower women and bridge gap as digital divide through its range of coding to digital literacy programs including the affordable devices offering. This article discusses how such initiatives are indeed addressing the gender gap in digital literacy, the hurdles they encounter, and the opportunities they present.

2 Literature Review

Digital divide is typically defined as the gap between those who possess access to and are able to utilize information and communication technologies (ICTs) and those who lack such access (Hackl, 2018). Gender digital divide, a case of the above phenomenon, is the deeply entrenched disparities in accessibility, utilization, and benefits accrued from digital technology between women and men. As stated in the definition by UNESCO in 2019 access, affordability, digital literacy, safety, and representation in the digital realm are constituting a phenomenon. The term digital divide represents much more than a lack of technology; it speaks to a deeper socio-cultural system and the economic stratification which exhibits prevailing inequalities. Economy and variety of mobility restrictions in Global South countries like India greatly limit women's access.

Recent studies show the effectiveness of industry-led initiatives to address the gaps in infrastructure and scalability. Through various public–private partnerships, the social responsibility (CSR) of the private sector is able to greatly drive digital inclusion for women through the design of products. Frameworks such as the Sustainable Development Goals clearly emphasize the importance of digital inclusion for achieving Gender Equality, Quality Education, and Decent Work. The ITU together with UN Women has founded EQUALS Global Partnership where industry partnerships are identified as key drivers crucial in closing the gender gap in technology (Harputlu & Güryay, 2024).

In various tech empowerment literature, the Intel initiative "She Will Connect" and Google's "Internet Saathi" are often highlighted as best practice cases. Intel's program, which started in community digital centers (the Centers of Excellence) in India and Sub-Saharan Africa, emphasized peer instruction, training, and supervision. Women trained through these programs were three times more likely to get selected for online jobs, and were twice as inclined to use e-governance services (Broaband India Forum, 2021).

World bank assessed the Internet Saathi program which claimed to have trained over 30 million women across approximately 300,000 villages in the country. Foundational findings indicated that there was a significant increase in digital self-confidence and entrepreneurial activities alongside increased educational pursuits at the inordinate level across the nation. That program's train-the-trainer model gets commended often for cultural appreciation and sustainability in social contexts somehow. Scholars sternly warn of trying to balance out adequate CSR perception as 'solutions to problems' which has marked progress in a great number of other areas (Google's Internet Saathi Program, n.d.).

Thomas (2022) critiques the "NGO-ization" of corporate responsibility arguing that these initiatives are shallow, short-term, and often aimed at brand enhancement rather than bringing about real change. Chetty in 2018 strongly emphasize the need for feminist frameworks of digital justice centered on women's agency and participation in online spaces (Chetty et al., 2018).

Rural India's women's mobile internet usage has shot up markedly by over 20 percentage points in certain districts thanks largely to Internet Saathi. Peer-training models are often touted as exemplary approach for delivery of culturally sensitive skills in various training settings effectively nowadays. Intel's She Will Connect program trained over 250,000 Indian women and evaluation reports indicate improved access to various government e-services nationwide. (Intel® She Will Connect: Closing Gender Gaps in Technology, n.d.)

Meta's GOAL program mentored over 5000 tribal girls alongside Ministry of Tribal Affairs helping them launch online ventures or become trainers. Evaluations reveal glaring shortcomings despite notable accomplishments including woefully limited scalability outside pilot regions and overreliance on local NGOs with disparate capacities. Scarce program continuity after initial funding dries up is also a major issue.

Use of resources and networks for women empowerment and backwards area needs attention from industries to reduce the digital literacy gender based gaps. These programs frequently concentrate on boosting digital literacy, granting access to technology, and creating welcoming conditions that encourage women's involvement in the digital economy and could add value to digital literacy programs, which can take gender issues into account, in uplifting underprivileged communities. Enhancing digital competencies may indirectly benefit women and other marginalised genders by increasing work possibilities and community participation, according to the research (Avilés et al., 2018). The gender gap in digital literacy may be addressed by implementing suggestions for inclusive and successful program design. (Prasastiningtyas et al., 2024).

Suleiman (2023) in his study in Nigeria suggested that Infrastructure development is also required to attain the target of digital awareness. The paper suggested that industry collaboration is essential to bridge the gap. There should be targeted campaigns and industry awareness programs to bridge the gaps. By putting females and empowering women economically industry led initiatives can effect digital literacy significantly but the approach needs cooperation and support—two things that are essential for conquering obstacles in the market place.

Despite efforts to close the digital divide through regulation, many women still do not have consistent access to digital devices and the internet, which limits their use of online resources and digital literacy programs (Jeffrie et al., 2023). Study encourages teamwork to build a more digitally inclusive workplace, which may involve industry-led programs to improve women's digital literacy and offer tools to close the gender gap in digital occupations as women cannot be left behind society that is moving at a faster pace towards digital inclusions (Anjana, 2023).

Literature highlights critical gaps despite encouraging narratives and data. Most available studies are short-term and lack tracking of outcomes like economic mobility or digital independence over time. Few studies break down data by factors like caste or disability thereby missing key barriers to women's full participation digitally. Limited scrutiny exists around companies adhering strictly to inclusive design standards or gender-sensitive policies during product development processes lately (Table 1).

Table 1 Gender gap in mobile and internet access in India

Aspect	Men (%)	Women (%)	Gender gap (%)
Own mobile phone	84.0	68.0	16.0
Use internet weekly	62.0	41.0	21.0
Use smartphone	59.0	36.0	23.0
Digital financial transaction	45.0	24.0	21.0

Source GSMA report (2023)

3 Digital Divide and Sustainable Development

Bridging gender gap is pivotal to achieve United Nations Sustainable development goal by 2030 because as the use of information technology is increasing within the society, the digital divide between different age groups are also increasing (Povetkina et al., 2023). Mainly, following SDG's are related with this issue, SDG 4 which ensures inclusive and equitable quality education and promote learning, SDG 5 focuses on achieving gender equality and empowering girls and women, SDG 8 Along with employment opportunities, it provides sustainable and inclusive economic growth.

Digital literacy allows women to access e-commerce, government services, online learning, health information, and voice their opinions on the internet. It also brings economic, social and environmental Sustainability in the economy. Research has shown that widening digital inclusion has the capability to drive economic growth and gender equality (Looker, 2008). Whenever women came forward and contribute for the growth of economy through participation in remote work or freelancing due to family responsibilities or working as an entrepreneur, it boosts the overall economy of the country. Digitally literate women can actively become working women and improve household income as well as status of the family. Women run business through digital platforms are more flexible and versatile. In 2015, TATA trust in collaboration with Google launch a programme known as Internet Saathi in India. The idea was to take help of rural women activities organizers in promoting digital literacy within their rural communities. These women need to act as a link between the trust and rural community thereby providing the required data regarding the internet usage. This programme has given aspirations to the contributors to flourish.

Digital literacy empowers women accessing education health governance and social services thereby making societies fairly just. Women surfing online are rather more aware of childcare nutrition and reproductive health services readily. Women participate heavily in government online through various advocacy groups and complex democratic processes with access readily available nowadays. Women empowered digitally can operate vehicles swiftly and work diligently on community-owned sustainable eco-projects nationwide with great enthusiasm. Women today can utilize the internet to access novel tools that foster renewable agriculture, digital technologies, and eco-friendly resources. Sites showcase women embracing rather obscure eco-friendly products and various related services online nowadays Women organize climate consciousness online via internet platforms rather effectively (Fig. 1).

4 Role of Industries in Bridging the Gap

Industries, as key drivers of technological innovation, are uniquely positioned to address gender disparities in digital literacy. Corporates have increasingly been concerned with empowering women through skills training, mentorship programs,

Fig. 1 Digital divide and sustainability (please cite this in text)

and resource facilitation. Although NGO activities and public policy have provided a scaffold for access to the digital world, initiatives spearheaded by industry have multiple benefits to offer. Industries have the financial ability and human capacity to successfully carry out large-scale programs for digital inclusion that leverage the latest technological infrastructure at great scale. Industries tend to reach huge numbers of beneficiaries, without the burden of resources which many non-profit/government initiatives face like in the Internet Saathi program which was jointly undertaken by Google and TATA Trust, to train more than 30 million rural populations across India.

IT and telecom industries are areas of technological innovation which happen mostly deep in edtech areas with very great amount of technology innovation going on in rapid pace (Hothi et.al., 2019). New tools like gamified content and AI based learning platforms designed to cater to women with low literacy can be introduced pretty well today—digital education becomes quite more engaging fairly soon and accessibility goes up in some way. With adequate CSR and ESG plans for their enterprises, businesses can invest in making a long-term impact. Industry-led initiatives often have a strong economic impact (surprisingly enough) within their infrastructure time and again—Amazon and Flipkart and Paytm have launched several programs to help women entrepreneurs acquire digital skills and market their products wildly online. The work undertaken is presumably aligned with SDG 8 (good work and economic growth), therefore ensuring adequate livelihoods across the globe. These companies also promoting girls and women in making their careers and providing them a secured and safe work culture. Industry can create direct links with employment and entrepreneurial prospects. Industry involvement is able to create fairly robust digital ecosystems through very extensive support and detailed toolkits, together with elaborate mentoring programs, in order to ensure that digital literacy develops as a long-term journey, rather than as a one-off intervention, so that women feel empowered to succeed in a rapidly evolving digital economy. Many companies and Government of India started up the initiatives to promote and build entrepreneurship skills among the females so that they can come forward and make their ways without any hesitation.

5 Leading Industry–Digital Literacy Initiatives Examples

In 2015, Internet Saathi was launched as a partnership between Google and Tata Trusts to empower rural women digitally by providing them with basic digital literacy skills. The initiative uses local women ("Saathis") and trains them to use smartphones and the internet. The women then train other women in their village community. Major achievements recorded till the year 2022 for this programme are that it extended to over 300,000 villages in 20 Indian states and Educated more than 30 million rural women. While growing women's use of the internet from 1 in 10 to 3 in 10 in program areas according to the report of KPMG, 2020. It helps in fueled entrepreneurship and education awareness among the participants. Research conducted by the Indian Council for Research on International Economic Relations (ICRIER) revealed that educated women were more likely to undertake distance learning, utilize health information, and establish home-based businesses.

Intel's She Will Connect initiative was launched in 2013 to bridge the digital gender gap in sub-Saharan Africa and Asia. The program combines digital literacy training with an online learning platform and mentorship networks. Major achievements of this progamme are that this programme reached more than 5 million women worldwide and entered into partnerships with state governments and NGOs. It also incorporated digital security modules to fight cyber attacks based on gender. In India, Intel also sponsored the My Digital Identity campaign to facilitate rural women getting digital IDs and e-governance services.

Microsoft has introduced many programs globally to advance young women's technological skills such as DigiGirlz to Empowers girls to become STEM professionals by training them in coding sessions, mentorship, and leadership and YouthSpark to offer free online access to digital skill training, coding, and career counselling for young people, with a specific focus on girls from low-income families. In India, Microsoft has collaborated with state governments and organizations such as NASSCOM Foundation to facilitate skilling initiatives in tier-2 and tier-3 cities.

The Connected Women program partners with mobile operators to bridge the gap in mobile internet and mobile money usage. The GSMA has partnered with more than 40 telecommunications operators. This programme reached over 55 million women with targeted mobile services. It also helps to close the mobile internet penetration gender gap in countries like Bangladesh, Nigeria, and India. Mass dissemination of women-oriented mobile literacy initiatives through IVR (Interactive Voice Response) and SMS also became very successful.

Mastercard's Girls4Tech is a STEM education program that encourages girls to seek careers in data science, AI, and cybersecurity. It provides girls aged 8–16 years with hands-on workshops and has partnered with NGOs and schools globally. Impact of this programme till 2023 is that it reached 5.5 million girls in 60 countries and it translated into 16 languages. It also Comprises modules of digital ethics and privacy.

GOAL is a mentorship program launched by Meta (formerly Facebook) in partnership with the Ministry of Tribal Affairs. It empowers urban women with digital training as mentors to tribal girls between the age group of 18–30 years for digital mentorship and soft skill development. More than 5000 tribal girls from 20 states headed were covered under the program. They learned digital marketing, business acumen, content creation, and leadership. Approximately 25% of mentees started online businesses or online advocacy campaigns. The program also imparted to mentees skills on how to use Facebook, WhatsApp Business, and Instagram as instruments of entrepreneurship and social change.

Reliance Jio, in association with Facebook, launched Digital Udaan, a mass digital literacy initiative for first-time internet users, mainly women in rural and semi-urban India. Training is provided on a weekly basis at Jio stores or village community centers. This program is to educated over 25 million users, with a significant majority of women. It functions in 13 regional languages across 10,000+ stores. The Content includes mobile internet usage, security, online payment, and standard applications. It Increases confidence among women to shop online, use UPI, and access government schemes online.

Ideosync Media in association with a NGO started an initiative "Free/Dem" that provides a platform to underdeveloped slum areas girls and women to get equipped with digital literacy and bridge the gender gap. This program was started in Slum areas of National Capital Region (NCR) (Prakashshravani, 2024). They provide public library with the help of digital pathshalas which allows women to explore the functionality of smart phones and internet. Once girls will complete sixteen hours course offered by them, they are given platform in media to raise voice for their concerns on paid basis.

6 Features of Industry-Led Effective Initiative

From the initiatives discussed above, Some important features of these programmes are reported such as Programs like Internet Saathi and Intel's bottom-up training model suggest that localisation of content and learning from peers is more effective, especially among women in rural or conservative settings. Forming partnerships with governments and NGOs is crucial in scaling programs, especially in reaching last-mile communities. For example, GSMA works with telecom regulators to enable services to be affordable and accessible in extended networks.

Many programs now include cyber safety, netiquette, and digital rights education to counter online harassment and disinformation, issues which are highly relevant for women users. With the high penetration of mobile phones in the developing world, most programs target mobile-based learning platforms (SMS, apps, IVR) so that digital literacy is within reach even without the presence of computers. Work and Enterprise Emphasis Some initiatives go beyond basic literacy to empower women to have access to freelancing, the gig economy, or microenterprises. For instance, Internet Saathi members have set up YouTube channels, online tailoring shops, and digital kiosks.

7 Challenges and Opportunities in Industry Driven Digital Literacy Programme

There are lot of challenges in front of Industries for implementing digital literacy programmes. Although industries are promoting all these programmes as a part of corporate social responsibility also but still they are facing a lot of challenges and issues. Some of the challenges and their proposed solutions are being compiled as follows (Table 2).

Table 2 Analysis of challenges and opportunities in industry-led digital literacy initiatives

Challenge	Reason	Proposed solution
Focus	Programmes generally focused on the residents of urban areas while neglecting rural and semi-urban areas	Companies can increase investment in rural digital infrastructure and partner with local NGOs for better reach
Cultural resistance	Social and cultural barriers discourage women's participation in training and development programs	Need to sensitize community stakeholders regarding the importance of these programs
Policy mismanagement	Lack of integration with the country's development policy	Policymakers need to align their educational and economic goals with collaboration
Limited scalability	Many Initiatives relies on CSR funds	Companies need to collaborate with government to provide more funds
Language barriers	Lack of local language content and native language interfaces generally limit reach in multilingual countries like India	Needs to provide more training for local languages to the trainers
Connectivity gaps	Inadequate digital infrastructure	Government need to work on connectivity issues in the far-flung areas
Cultural	Women's participation is subject to family permission and social authorization	Community needs to be sensitized

8 Policy Implications

The finding supports several policy implications for bridging the gap which could be helpful in achieving long term sustainable development goals. Industry may support government in programmes like PMGDISHA, Prime Minister Gramin Digital Saksharta Abhiyaan a digital literacy initiative by Ministry of Electronics and Information Technology in India where aim is to make 6 crore people in rural India digitally literate targeting one member from each household. Industries can collaborate with the government to provide necessary support required for the successful implementation of the programme.

Corporate Social Responsibility (CSR), mandated under Section 135 of the Companies Act, 2013, requires companies meeting certain financial thresholds (net worth ≥ ₹500 crore, turnover ≥ ₹1000 crore, or net profit ≥ ₹5 crore) to spend at least 2% of their average net profits on socially beneficial activities. The law list a number of broad areas to focus upon (e. g. education, gender equality and rural development), however, it does not clearly list digital inclusion/digital initiatives for women. India has historically had a gender digital divide—only 33. 6% of women aged 15–49 reported using the internet (NFHS-5)—a huge case for digital gender equity to be explicitly included in Schedule VII of Companies Act, as an approved CSR activity in the framework of Schedule VII.

The language barrier negatively impacts Digital Literacy Rate among Rural Women. Indian women in rural areas do not have formal schooling hence require a very drastic decline in their digital literacy rates. Over 70% of rural females say that they preferred native and regional language-based digital training, while 8% claimed that they struggled with English-based content somehow. Such data shows the need for linguistic inclusion in digital literacy programs during their often flawed development processes. Assistant management must lead the team for digital initiatives by localizing them through cultural adaptation, taking the needs and culture of the target population into account. The digital training required for rural women is audio & visual media via videos and mobile apps with text to speech feature. Internet Saathi program provided audio-visual and language training to 30 million rural women. It presented Hindi digital training through videos and interactive vernacular training in person to them. Women get better understanding and better memory capacities through localization of the program content, which can therefore increase their confidence and trust in the technology thus overcome their fear of technology. The companies achieve cultural and technical inclusion by native languages or dialects of target population, that make learning of digital skills much easier for rural women.

If digital skills are well integrated in the school and vocational education frameworks, they can be extremely beneficial for long term and sustainable digital inclusion of women quite effectively over a long period of time. Implementing gender-sensitive digital literacy programs in rural tribal schools at early age helps to fill up a huge digital gender divide slowly. Digital modules need to go quite far beyond basic computer literacy and also include training in e-governance tools and responsible use of internet pretty much nowadays. Tailoring content to address specific issues girls face like extreme restrictions on their mobility and online harassment are done so that digital education can be more effective. Industrial Training Institutes and vocational centers are the core locations where you will find young women looking for gainful employment pretty much everywhere nowadays. Advanced digital modules like ecommerce operations and fintech literacy can be well integrated in ITI curriculum, helping them to get employable in India's rapidly changing digital landscape. The key stakeholders like IT companies and fintech firms as well as digital platforms should work together very close with educational boards on developing the curricula effectively. Providing internships and digital toolkits along with mentorship ensures the exposure in real life pretty effectively for them basically. The public/private collaboration is obviously building capacities pretty quickly and making them fairly inclusive in economic participation somehow.

Some industries conduct digital literacy programmes to train women but there are no mechanisms for assessing the impact of such programs over a relatively long period. Most of the digital literacy campaigns reported outputs like number of women trained but did not measure the long term consequences of the projects like economic empowerment or considerable behavioural change. A fundamental component of every project that works towards digital inclusion in isolated communities across India needs to be a Monitoring and Evaluation framework that covers accurate measure of Key Performance Indicators such as percentage of income change, device usage rate and skill retention metrics. For example, according to the Global Society for Marketing & Analysis (GSMA/2023) fewer than 40% of the surveyed women did not use digital tools three months after being trained in short-term programs such as online literacy training in South Asia. This data highlights the need for such types of systematic follow-up and evaluation. Even if M&E is standardized it offers transparency and accountability and also helps to refine the curriculum at a more finely tuned level over a relatively short period through various delivery channels. Companies can draw on aggregated impact data to derive annual Digital Inclusion Scorecards that help policymakers and CSR funders to better target resources. It will help all those engaged in these digital engagements evolve into truly transformational engagements—with substantial impact towards advancing gender equality and the country's economic development.

9 Conclusion

The role of women must be in central in the economy and attention should be on their education and participation in the political and economic sectors. At the same time, it must be ensured that women do not have to contribute financially to the economy just because they have a house to look after and other unpaid tasks to undertake. Men must equally contribute towards this change because their role is also important in upliftment of women. Business has become a significant driver in bridging the gender digital divide through targeted, scalable, and frequently innovative interventions. From multinational technology companies to national telecom companies, business is increasingly linking digital inclusion to core business and CSR strategies. The challenges notwithstanding, the involvement of the private sector, combined with inclusivity, local partnership, and continued investment, can have a strong impact on increasing digital opportunities for women and girls.

Closing the digital gender gap is at the forefront of inclusive, equitable, and sustainable twenty-first-century development. As technology increasingly shapes societies, from education and employment to governance and management of natural resources, women's access to and control over digital technologies determine not only their own empowerment but also the development of communities and nations as a whole.

In the form of training courses and app-based learning modules or collaborations with local organizations, the private sector has shown the potential to supplement government efforts and drive digital inclusion. But the true contribution of such initiatives is experienced when these are seen in the context of the overall initiative and achieving of the Sustainable Development Goals (SDGs). As noted, women's digital literacy drives action under SDG 4 (Quality Education), SDG 5 (Gender Equality), SDG 8 (Decent Work), SDG 9 (Innovation and Infrastructure), SDG 10 (Reduced Inequalities), and SDG 13 (Climate Action). By bridging the gender gap and providing digital education, many issues shall be addressed altogether.

But significant barriers remain. Gendered barriers, including socio-cultural beliefs, affordability, safety online, and lack of representation of women in the technology sector, remain an obstacle. In addition, while many industry-led programs show promise, their size and longevity vary widely, even in rural and disadvantaged areas. In the future, there is a need for a multi-stakeholder setting. Policymakers must create enabling environments by investing in digital infrastructure and gender-responsive policy-making. The business community must provide inclusive design principles, equal access, and long-term capacity building. Civil society organizations must serve as bridges, creating context and trust at the grassroots level. These stakeholders, in concert, can create a robust ecosystem where digital empowerment is a right and not a privilege. They can focus on bridging up the gap in collaboration with government to increase the digital literacy in the country. Moreover, implementation of policies for betterment of women is not the only step which needs to be taken but there should be continuous evaluation of program and policies so that steps could be taken in advance for better results.

In short, closing the gender digital divide is not technology—it is equity, opportunity, and resilience. It is ensuring that all women, regardless of where they reside or their socio-economic status, have the resources to learn, earn, lead, and participate in a sustainable future. Enabling women through digital literacy is one of the strongest investments we can make—for them, and for the world.

References

Anjana, L. (2023). *Gender digital divide in India* (pp. 105–118). Productivity Press. https://doi.org/10.4324/9781003302346-6

Avilés, J. M., Mayne, G., Aneja, U., & Sorgner, A. (2018). *Bridging the gender digital gap*. Research Papers in Economics. https://econpapers.repec.org/RePEc:zbw:ifwedp:201860.

Antonet, O., Galasso, L., Zen, V. D., & Cianfrini, L. (2022). Impact report. In L. Galasso, M. Ciccolini, G. D. Colle, S. Hardy, G. Marrone, N. Perricone, R. Ristucci, M. Rovazzi, L. Simone, W. Simone & N. Wright, (Eds.), *Meta centre*. CA. https://metacentre.ca/wp-content/uploads/2022/04/202122-META-FOUNDATION-IMPACT-REPORT.pdf

Bridging the digital literacy gender gap in developing countries. (2022). https://www.global-solutions-initiative.org/publication/bridging-the-digital-literacy-gender-gap-in-developing-countries/

Bridging the gender digital divide. (2023, May). *Unicef*. https://data.unicef.org/resources/ictgenderdivide/.

Broadband India Forum. (30 C.E.). Building an inclusive digital society for rural India. In *ICRIER*.

Chetty, K., Aneja, U., Mishra, V., Gcora, N., & Josie, J. (2018). Bridging the digital divide in the G20: Skills for the new age. *Economics,12*(1). https://doi.org/10.5018/economics-ejournal.ja.2018-24

Girl4Tech | MasterCard. (n.d.). *Girls4Tech*. https://www.girls4tech.com/

Explore key topics | Tata Trusts. (n.d.). https://www.tatatrusts.org/topics/digital-literacy.

Google's Internet Saathi program. (n.d.). *World Bank*. https://documents.worldbank.org/en/publication/documents-reports/documentdetail/732651613724922041/google-s-internet-saathi-program.

GSMA. (2023). *The mobile economy 2023*. https://www.gsma.com/solutions-and-impact/connectivity-for-good/mobile-economy/wp-content/uploads/2023/03/270223-The-Mobile-Economy-2023.pdf

Hackl, A. (2018). Mobility equity in a globalized world: Reducing inequalities in the sustainable development agenda. *World Development, 112*, 150–162. https://doi.org/10.1016/j.worlddev.2018.08.005

Harputlu, L., & Güryay, B. (2024). Gender differences in digital literacy: An analysis. Batı Anadolu Eğitim Bilimleri Dergisi. https://doi.org/10.51460/baebd.1562102

Hothi, N., Awasthi, A., & Bisht, S. (2019). A perspective interrogation of students" epistemological beliefs and anticipations of studying physics as an introductory course in computer science engineering. *International Journal of Innovative Technology and Exploring Engineering (IJITEE), 8*(12), 2100–2108.

ICUBE 2022 Internet in India. (2022). *In Annexure on GDPR* (Vol. 2, pp. 3–17). https://www.iamai.in/sites/default/files/research/Internet%20in%20India%202022_Print%20version.pdf

In Nigeria: Challenges, policies, and collaborative solutions for sustainable peace. *International Journal of Peace, Education and Development,11*(2). https://doi.org/10.30954/2454-9525.02.2023.2

Intel® She will Connect: Closing gender gaps in technology. (n.d.). *Intel*. https://www.intel.com/content/www/us/en/corporate-responsibility/social-impact-and-educational-initiatives/she-will-connect.html.

Internet saathi. (n.d.). *Tata Trusts*. https://www.tatatrusts.org/our-work/digital-transformation/digital-literacy/internet-saathi

Intel. (2022). *Corporate responsibility report 2021–22*. https://csrreportbuilder.intel.com/pdfbuilder/pdfs/CSR-2021-22-Full-Report.pdf.

Jeffrie, N., Bahia, K., Carboni, I., Lindsey, D., Sibthorpe, C., & Zagdanski, J. (2023). *The mobile gender gap report 2023*. https://www.gsma.com/r/wp-content/uploads/2023/07/The-Mobile-Gender-Gap-Report-2023.pdf

Kaur, M., & Sharma, J. (2023). *The role of digital literacy to promote the gender equality*. https://doi.org/10.59231/sari7642

KPMG Assurance and Consulting Services LLP, & Mishra, N. (2020). *India's rural economy: Key to economic revival and sustainable and equitable growth.* https://assets.kpmg.com/content/dam/kpmg/in/pdf/2020/12/india-s-rural-economy.pdf

Looker, E. D. (2008). *Gender and information technology* (pp. 779–788). Springer. https://doi.org/10.1007/978-0-387-73315-9_45

Malik, C. D. S. (2024). Technology and women empowerment: Bridging the gender gap in the digital era. *International Journal for Research Publication and Seminar, 15*(4), 132–126. https://doi.org/10.36676/jrps.v15.i4.19

Prakashshravani. (2024, October 10). The power of partnerships: 6 collaborative initiatives to bridge digital gender gaps. Ellenomics.https://ellenomics.in/2024/07/28/the-power-of-partnerships-5-initiatives-to-bridge-digital-gender-gaps/

Prasastiningtyas, W., Ubaidillah, A. Z., Aprianti, I., & Nurfadilah, L. (2024). Improving literacy in the digital age: Unleashing critical thinking potential. *Toplama, 2*(1), 18–27. https://doi.org/10.61397/tla.v2i1.218.

Press Release. (2024, July 8). *Microsoft invites teens to Digigirlz Day.* Tech News TT. https://technewstt.com/pr-digigirlz-2022/

Povetkina, N. A., Zueva, A. S., & Kopina, A. A. (2023). Reducing the digital divide as a mechanism to ensure sustainable economic and social development. In *Advances in science, technology & innovation/advances in science, technology & innovation* (pp. 27–31). https://doi.org/10.1007/978-3-031-34256-1_5

Roy, P., Raut, A., Samantaray, S., & Rana, A. (2024). Women and digital networking: Digital skill gap and its challenges. In *2024 International Conference on Advancements in Smart, Secure and Intelligent Computing (ASSIC).*

Suleiman, M. S. (2023). Bridging the digital gender divide for peace and development in Nigeria: Challenges, policies, and collaborative solutions for sustainable peace. *International Journal of Peace, Education and Development, 11*(20). https://doi.org/10.30954/2454-9525.02.2023.2

Summary for policymakers. (2023). In Cambridge University Press eBooks (pp. 3–34). https://doi.org/10.1017/9781009325844.001.

Thomas, Z. (2022). Mujeres Guerreras: Negotiating women & 2019; Empowerment in Colombia. *Journal of Contemporary Ethnography, 52*(4), 514–536. https://doi.org/10.1177/08912416221126519 (Original work published 2023).

UNSD. (2024). *Sustaianle development Goals report.*

Valli, V. A. (2025). Women and technology: Bridging the digital divide. *Indian Scientific Journal of Research in Engineering and Management, 09*(01), 1–9. https://doi.org/10.55041/ijsrem40877

Vol. 5 No. 6 (2024): Dinasti International Journal of Management Science (July–August 2024) | Dinasti International Journal of Management Science. (n.d.). https://dinastipub.org/DIJMS/issue/view/125

Energy Transitions: Balancing Growth, Sustainability, and Technology

Shivani Bisht, Shrijan Kamboj, Shipra Agarwal, Malika Pahwa, and Tanya Rawal

Abstract

The world is undergoing a profound transition from dependence on fossil fuels to sustainable and renewable energy sources, including solar, wind, hydropower, and biofuels. This transition is not less important in light of several mega environmental challenges and efforts to achieve net-zero carbon emissions by 2050. Given the challenge, the energy industry needs to contribute to this with immense decreases in emissions, innovatory technological advances, and unparalleled amounts of funding and international collaboration. The paper shows certain parts of the complex interaction of forces that push this supply shift and emphatically move from fossil fuel reliance to renewables. It gives a comprehensive overview of the current energy production and consumption landscape, pointing out the large environmental and economic costs tied to continued dependence on fossil fuels. This analysis reviews key renewable energy technologies such as solar, wind, and hydroelectric power, considering with care their potential for meeting rising global energy demands. The paper identifies some critical challenges and opportunities toward a sustainable energy future through strict review of the policy framework and technological innovations. These findings put in relief that an integrated approach combining technological innovation, strong policy support, and public engagement is needed for successful transition toward a new energy paradigm. This transition therefore offers an opportunity to rethink and create the whole energy landscape afresh, embedding the principles of resilience and improvement of living standards while protecting the environment for future generations. It enriches the ongoing discourse on

S. Bisht · S. Kamboj · S. Agarwal · M. Pahwa (✉) · T. Rawal
Department of Commerce, Graphic Era Deemed to be University, Dehradun, India
e-mail: malikapahwa17@gmail.com

T. Choudhury et al. (eds.), *Open Access as a Knowledge Platform for Implementing the United Nations Sustainable Development Goals*, World Sustainability Series,
https://doi.org/10.1007/978-3-031-99605-4_8

conserving energy and making it sustainable by showing the way ahead toward a resilient and sustainable energy system.

Keywords

Energy • Economic growth • Policy • Renewable • Technology • Fossils

1 Introduction

The phrase "energy transition" refers to the global shift from traditional fossil fuels such as coal, oil, and natural gas to more sustainable and renewable energy sources like solar, wind, hydropower, and biofuels. Renewable energy technologies can deliver vital services such greater healthcare, better education, and more affordable connectivity for the 759 million who do not have access to power worldwide (Correia, 2019).

The shift to a world with zero carbon emissions, with major obstacle facing human society is the pursuit of global carbon neutralism. In this regard, the energy industry is essential to solving this problem. An effective, sustainable, competitive, and safe energy system is necessary to achieve net zero emission targets and keep global warming to 1.5 °C by 2050. Although the IEA analysis described only one potential route to reach the net zero goal, it emphasized a more general idea: the next thirty years will call for a substantial decrease in emissions, requiring unprecedented amounts of funding and international cooperation (Mensah, 2019).

Many businesses have faced severe financial difficulties as a result of the electrical sector's transition, which has also caused their conventional business models to be disrupted. But this change has also made it easier for new market players to emerge and for the growth of various tactics and business structures, propelled by greater innovation and improved customer involvement. "The greatest challenge of our times," as it is called, calls for nothing less than a total overhaul of the energy infrastructure that powers our economy. By the end of the twentieth century, the average daily energy intake worldwide was approximately 50,000 kcal, or incorporating conventional and traditional energy sources, to about 76 GJ per year. About Coal, oil, and natural gas-based systems accounted for 80% of this energy consumption. Together, nuclear energy and hydroelectric power made up 6 and 2% of the non-organic portion of the world's energy mix. The majority of the biomass in the remaining 125 came from organic vegetable sources. When trash utilized to produce energy is excluded, this 12% mostly consisted of firewood and food for humans and working animals (now a small source of energy). It only continues to be a major source of energy in developing nations (Alhaddi, 2015; Tseng et al., 2020). The World Energy Council and Kearney co-published a research titled "The Global Energy Transition," which defines an energy transition as a substantial change in a country's energy composition involving an improved energy efficiency, a larger proportion of renewable energy, and the steady decline of reliance on fossil fuels (Alhaddi, 2015) Making the switch to renewable energy sources is essential for achieving sustained prosperity and enhancing climate change resilience. As nations grow plans for COVID-19 rehabilitation, it is essential that every country participate in an energy transition that optimizes chances to greatly improve human well-being and sustainability of the environment. Nevertheless, the energy transition process is not consistent or uniformly applicable, necessitating customized strategies for various situations (Yu, 2021). By switching to renewable energy sources and improving energy efficiency, researchers from the German Institute of Applied Ecology proposed that economic growth may be maintained while lowering energy usage. In 1982, the book "Energy Transition Growth Prosperity without Oil and Uranium" was released by the institute presented the idea of "Energiewende." This piece emphasized the necessity of changing the dominating energy sources, including nuclear and oil, as well as renewables, creating the foundation for what would eventually be acknowledged as the notion of the shift to renewable energy (Kumar, 2020). One characteristic of sustainable energy is its lower greenhouse gas emissions per person. Developing countries have accused rich nations of using excessive amounts of energy, which contributes to environmental damage strong demand-driven consumption. On the other hand, developed countries have highlighted the environmental deterioration in developing nations due to rising energy consumption linked to population expansion (Laccourreye & Maisonneuve, 2019).

2 Objective of the Study

This paper examines the shift from fossil fuels to renewable and sustainable energy sources, including solar power, wind energy, hydroelectric power, and biofuels. It shall focus mainly on the huge environmental challenges arising due to climate change and the demand to reach zero carbon emissions no later than 2050. The paper also aims at a general overview of present production and consumption models of energy, pointing to massive environmental and economic costs resulting from high dependence on fossil fuel. It also goes over the most promising renewable energy technologies in terms of how well they can provide the world's energy needs. It discusses the obstacles and possibilities for technology advancement and policy in the direction of a sustainable energy future. In this sense, the significance of an integrated strategy that incorporates technological innovation, robust legislative backing, and proactive public involvement in accomplishing energy transition cannot be overstated.

3 Methodology

A critical analysis of the switch from fossil fuels to renewable energy sources was conducted. Keywords associated with "energy transition," "renewable energy," "sustainability," "economic growth," and "technological innovation" were used to collect literature from a variety of websites and databases, including IEEE Xplore, Google Scholar, and Scopus. While the technology evaluation examined the problem of cutting-edge technologies like artificial intelligence (AI), the Internet of Things (IoT), and smart grids, the policy study focused on the frameworks and regulatory policies that facilitate energy transitions (Awasthi et al., 2024). In the end, sustainability analysis focuses on the impacts that energy transitions have on the environment, economy, and society by using the Triple Bottom Line framework in the determination of the balance among environmental stewardship, social equity, and economic development.

4 Global Energy Demand and Economic Growth

Energy is the primary driver of economic progress and a crucial material foundation for the evolution of human society. When examining the relationship between economic progress and energy from an economics standpoint, there are two primary points to consider: first, economic development depends on energy. In other words, energy and economic progress are inextricably linked. Second, specific conditions of economic development determine the extent and magnitude of energy opening and usage. Large-scale energy development and use can be encouraged by economic growth.

In order to promote economic growth, many developing nations have seen a sharp rise in imports, exports, and energy consumption over the past 50 years. As national economies expand, so does the demand for energy, which has always been one of the most important resources for governments and communities. Additionally, the industrialization process and the pace of urbanization are increasing the need for energy in nations, and the intense use of energy benefits both the nations' overall production and welfare levels. Rising living standards, industrial sector advancements, and the world's growing population all contribute to rising energy demand and consumption.

Though their per capita consumption is much lower than that of industrialized nations, developing nations currently consume more than half of the world's energy. Their demand, which accounts for a growing portion of the world's energy demand, has nearly doubled over the past 15 years and is expected to increase by another 30% over the next 15 years. Creating policies and initiatives that can maximize the advantages of energy in fostering healthy economic and social development while lowering any associated costs at the local and international levels is the ultimate goal of the work on energy for development. Because of this, energy for development is crucial for everyone, including the wealthiest in industrialized nations, the poorest in emerging nations, and everyone in between. As a result, it is becoming more and more central to global energy strategy.

Because of its widespread use of fossil fuels like coal, petroleum, and natural gas, which leads to substantial emissions of greenhouse gases and carbon dioxide, the industrial sector is a key contributor to hazardous environmental problems like global warming. Using wind energy to provide electricity, integrating energy-efficient technologies, and implementing solar power for industrial processes are a few examples of sustainable energy for industry. Geothermal and biomass energy implementation can support environmentally friendly industrial processes. This is accomplished by minimizing the environmental effect across multiple sectors and decreasing dependency on non-renewable resources. Responsible decision-making by industry owners is essential to paving the road for a healthier future because it establishes a standard that can benefit society overall and guarantee company excellence. Explore our website to find out more about how we help raise the industrial sector's sustainability standards globally. New energy resources like unconventional sources like shale oil and renewable energy sources like the sun, wind, and others will be developed as a result of the increasing demand for fossil fuels worldwide and their rising costs.

5 Sustainability in Energy System

With the United Nations Conference on the Human Environment in the 1970s, which highlighted the necessity of putting in place sufficient measures to solve environmental concerns, environmental awareness significantly increased (McNamara, 1972). It establishes a framework for societal engagement to interact with the environment that aims the conservation of resources for future generations. As both a development paradigm and a conceptual framework, it supports raising living standards without threatening the planet's ecosystem or adding to environmental problems like air and water pollution or deforestation. If these issues are not addressed, there may be dire repercussions, such as species extinction and climate change.

The rising trend of sustainability has welcomed the innovation of the Triple Bottom Line framework, also referred to as TBL. The TBL framework incorporates the three crucial dimensions of sustainability: environmental stewardship, social equity, and economic development. The TBL approach towards sustainability was first introduced in the Brundtland report back in 1987. The economic pillar of the TBL refers to the impact that an organization's business activities have on the greater economic system. It emphasizes the ability of the economy as a key component of sustainability to endure and adapt over time to ensure the capacity to support future generations. This dimension also links the growth of an organization with the development in the economy as a whole since it focuses on what a business does to contribute toward development. In the last decade or so, from the perspective of the Triple Bottom Line, there has evolved some recent literature that has emphasized an environmental pillar as one path to achieve sustainability. Most consequently, various studies have developed systems where environmental and social outcomes were given priority over economic growth because they regarded these benefits as major drivers on a path toward sustainable development. Implementing fair and socially responsible corporate practices that have a positive effect on labor, human capital, and the larger community is the focus of the Triple Bottom Line's (TBL) social pillar. The underlying principles is that these practices are intended to create value for society and contribute to community welfare. In the context of energy transition for workers unconventional energy sectors fostering the creation of new employment opportunities within the renewable energy industry and addressing issues related to energy access and affordability for vulnerable populations (Markard, 2018). With the growing population, electricity demand is soaring. Energy sources are divided into conventional (fossil fuels) and renewable (solar, hydro, nuclear, etc.). Fossil fuels power plants remain central to global energy, but their environmental impact is increasingly evident.

Renewable energy projects have contributed to mitigating environmental impacts, notably through the reduction of carbon dioxide emissions and raising community awareness about climate change. However, notable positive impacts were observed in terms of improved living standards, the creation of social cohesion and overall community development (Harichandan et al., 2022). Additionally, the research highlighted the complexity of installing renewable energy projects which are complex to implement and highly sensitive to local environmental conditions. Moreover, the forecasting, execution and planning of these projects require more comprehensive attention and expertise compared to conventional energy projects.

5.1 The First Pillar: Environmental Sustainability

Environmental sustainability is one of the founding aspects of the Triple Bottom Line that intends to balance economic activities at par with ecological preservation. This main tenet promotes the protection of Earth's natural systems by businesses, balancing them between ecological preservation and economic development. By prioritizing sustainability, organizations commit to the health and resilience of the environment for long-term viability into the next generation.

5.1.1 Understanding Environmental Sustainability

Environmental sustainability is defined as finding ways of life and work that do not harm the planet. It calls for the protection of natural resources, reduction of waste and pollution, and shifting toward cleaner, renewable sources of energy by individuals, businesses, and communities. At its most basic, it's making sure we use the resources on Earth in a manner that satisfies our needs today, without inhibiting the ability of future generations to do the same. It's a commitment to care for the world around us, so our children and grandchildren can have a healthy, thriving planet, too.

5.2 The Second Pillar: Social Sustainability

As of crossing the environmental to human aspect of the Triple Bottom Line, one enters the domain of social sustainability, which indeed is quite compelling. This pillar reflects the fact that business activities are not solely economic entities but also an integral part of the social fabric, with strong impacts on communities, workers, and society in general. Social sustainability will deal with the just, equitable, and inclusive conditions under which individuals and communities will thrive, now and in the future.

5.2.1 What Is Social Sustainability?

Social sustainability refers to all those practices and policies a business adopts in order to make a positive impact on the lives of its employees, customers, and communities it operates within. This would address worker rights, social equity, community development, among other concerns. At the heart of it all, social sustainability deals with investing in people and guaranteeing that there exists a relationship between businesses and the larger society in ways that are both fair and mutually beneficial.

5.3 Third Pillar: Economic Sustainability

Economic sustainability, the last pillar in the Triple Bottom Line, exists at the confluence of profitability and responsibility. This business principle redefines traditional priorities for businesses, shifting them from short-term financial gains to long-term economic stability and resilience. It stresses that only strategies for continued financial viability should be adopted which could allow social and environmental commitments for an eventual contribution to a balanced and inclusive economic ecosystem.

5.3.1 Understanding Economic Sustainability

Economic sustainability is the process of adopting such practices that would ensure a firm's capability for sustainable and equitable growth over time. It goes beyond profit maximization to include responsible financial strategies that are compatible with social and environmental goals.

6 Technological Innovations in Energy

Technological innovations have completely changed the landscape of green energy. The technological betterment made the sources of renewable energy such as solar, wind, hydropower, geo-thermal power, and biomass power to come more into the scene. Solar panels are deployed far and wide today because these have become far more affordable and efficient. The design and efficiency of wind turbines have improved to make wind power a more competitive element in the energy economy

(Azadani, 2023). In addition, energy storage technologies have played a critical role in overcoming the intermittency of these renewable energy sources. Innovation in battery storage with technologies such as lithium-ion batteries and newer ones like flow batteries allows the storage of excess energy generated for periods when the output of these renewable energies is low.

Drone and robotics integration has been a game-changer in the green energy industry, providing unmatched capabilities for monitoring, maintaining, and streamlining operations across fossil fuel and renewable energy infrastructure. Energy management and distribution have been completely transformed by the use of smart grid technology. In order to optimize energy use, minimize waste, and smoothly integrate renewable energy sources into the grid, smart grids enable bidirectional communication between energy providers and consumers. In the field of green energy, artificial intelligence and data analytics have become revolutionary. Large volumes of data are analysed using machine learning algorithms, increasing the effectiveness of renewable energy systems. Predictive models powered by AI optimize energy production and consumption, enhancing system performance and cutting down on operating expenses. One of the main factors propelling the world's shift to green, sustainable energy sources is technology. Unquestionably, it has contributed to the development of renewable energy sources, the optimization of energy systems, and the transformation of numerous industries, including transportation.

Figure 1 highlights the growth of various renewable energy sources in India, including solar, wind, hydropower, and other renewables, from 2014 to 2023. Solar energy exhibits the most remarkable growth, rising from 705.89 in 2014 to 1641.58 in 2023, reflecting significant investments, technological advancements, and policy incentives. Wind energy also demonstrates steady growth, increasing from 831.32 in 2014 to 2325.31 in 2023, showcasing its established role in India's renewable energy mix. Hydropower remains the largest contributor throughout the period, growing from 3888.8 in 2014 to 4240.01 in 2023, though its growth rate appears more gradual, possibly due to limited expansion opportunities or saturation. In support of the main sources, other renewables, such as biomass and small-scale projects, are also growing steadily, rising from 505.6 in 2014 to 781.5 in 2023. Overall, India's attempts to diversify its energy mix, lessen its reliance on fossil fuels, and meet renewable energy targets are demonstrated by the consistent rise in all sources. India's commitment on sustainable energy development in the fight against climate change and its pursuit of energy independence, spearheaded by solar energy, is highlighted by the data.

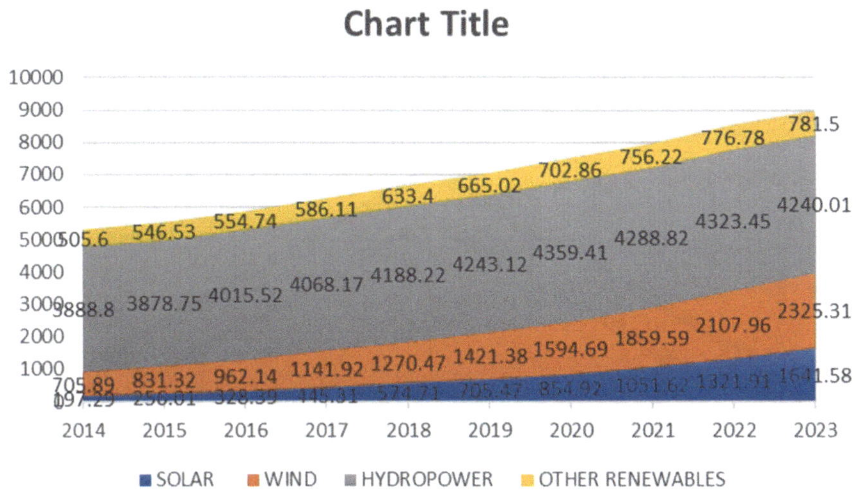

Fig. 1 Growth of various renewable sources in India

7 Innovations in Solar Energy

7.1 Solar Paints and Coatings

But now imagine the paints on the wall lighting up not only the space but the advance kind of paint to lighten up your homes to power. That is what solar paint is—photovoltaic paint, in simple words—billions of special light-sensitive particles convert light energies into electrical forms. This particular paint will cover just about anything that can absorb sunlight, hence changing all kinds of uniquely shaped houses into a generator of renewable energy. The beauty of solar paint for a homeowner lies in its simplicity and low cost. Not only is it easier and much cheaper to install compared with traditional solar panels, but quite exciting, too, for those who want to harness the power of the sun at minimal hassle as shown in Fig. 2.

7.2 Next-Generation Photovoltaics

The term "next-generation photovoltaics" describes those materials beyond the current market leaders—now considered leading edge—in solar technology. While traditional crystalline silicon-based solar panels are considered first-generation, and thin-film technologies such as CIGS and cadmium telluride form the second generation, it is new materials such as organic and perovskite solar cells that are seen as the next wave of innovation. These new materials are far more efficient and less expensive to make and could revolutionize solar energy. Many of these technologies are scaling up from the lab into the real world and will make solar

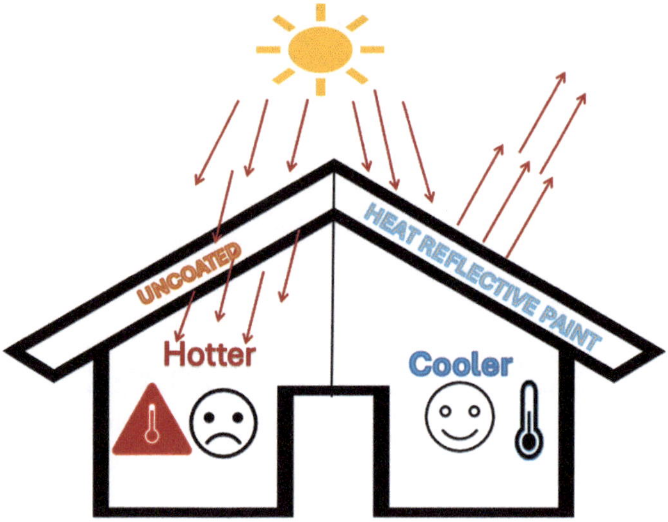

Fig. 2 Reflective solar cooling paint

power accessible and inexpensive like never before. With every development, we get a little closer to a future of even more efficient, widely available solar energy.

7.3 Solar Windows

Solar windows are an ingenious blend of functionality and sustainability. They look just like regular windows, but they come with photovoltaic glazing that converts sunlight into renewable electricity. This technology enables your windows to generate clean energy without compromising the aesthetics of your home. It doesn't need any bulky solar panels because BIPV or solar panel windows are integrated into the very structure of the building itself. These capture solar energy through photovoltaic cells embedded in those glasses for powering home appliances and gadgets. In other words, solar-enabled windows let such materials that let your light in your home enable harnessing the energy of the sun for moving us towards energy-efficient buildings and homes.

8 Innovations in Wind Energy

8.1 Floating Wind Farms

Floating wind farms are in an innovative technology class set to capture energy produced at swaths of ocean going over deeper and wider areas; these turbines sit atop floating platforms—quite different from being attached to a seabed, as land

and water wind farms are doing—thus making new areas of good winds accessible. Conventional wind farms have been typically limited by their geographic position; this innovative technology offers a continuous source of clean energy. The floating wind farm will be very much contributing to the carbon-free future with a sustainable and practically unlimited supply of renewable energy.

8.2 Vertical Axis Wind Turbines (VAWTs)

They have a unique design in wind turbines, having their rotor axis vertical. Because of this, they are ideal for those locations where wind directions frequently change. VAWTs do not need to face the wind compared to traditional horizontal turbines; therefore, they can be used in urban or residential areas. They generate less noise than horizontal turbines, minimizing noise disturbances in populated areas. However, the VAWTs experience more air drag that cuts down their efficiency compared to horizontal-axis turbines. Engineers are developing ways to minimize drag and enhance the efficiency of VAWTs, making them a viable clean energy option in urban environments.

8.3 Kite Systems for Wind Energy

Imagine kites flying hundreds of meters high, harnessing wind energy. This is the concept of wind-powered kite systems. Controlled by automated systems, these kites fly in a figure-eight pattern, unwinding a tether connected to a generator on the ground. As the kite ascends, the tether's pull generates energy, stored in the generator. Once the tether is fully extended, the kite moves to a neutral position, and the generator reels it back in, creating a continuous energy cycle. Operating at altitudes of 200–400 m, these systems capture stronger, consistent winds, offering an innovative and efficient renewable energy source.

9 Innovations in Hydroelectric

9.1 Harnessing the Power of Tides and Rivers

Oceans, covering over two-thirds of Earth's surface, are a vast, untapped energy source. Tidal energy harnesses the natural movements of tides to generate electricity (Khojasteh et al., 2023). Although drawing energy from water is not new, modern engineering has refined it into three methods: tidal range, tidal stream, and wave energy. Traditionally costly and location-restricted, tidal energy is now becoming more viable thanks to advancements in turbine technology, such as axial and crossflow turbines, and new designs like dynamic tidal power and tidal lagoons. These breakthroughs are making tidal energy more environmentally

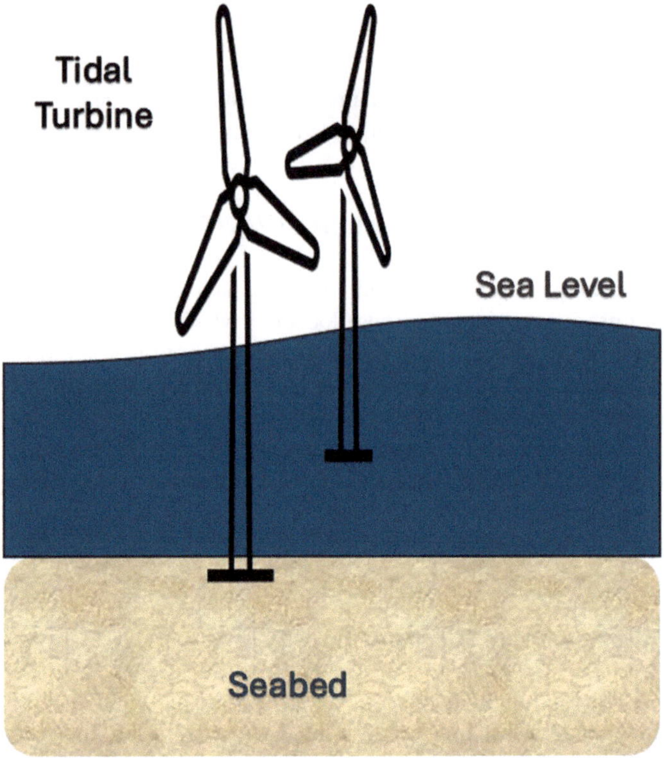

Fig. 3 Tidal powers

friendly and cost-effective, positioning it as a competitive renewable energy source as shown in Fig. 3.

9.2 Run-of-the-River Hydroelectric Systems

While traditional hydroelectric plants require a large dam and a reservoir, run-of-the-river systems generate electricity simply by availing themselves of the natural flow of rivers. If there is a dam at all, it will be a small one, just adequate to ensure good flow; the systems avoid massive water storage. Instead, the current of the river would be availed to generate the energy, with their turbines depending upon the natural movement of the water rather than steep vertical fall.

This is the perfect approach for areas with low water storage requirements; thus, it is more sustainable and less intrusive than conventional hydropower. In a run-of-the-river hydroelectric system, the natural flow of the river is used to minimize environmental disruption while still providing clean, renewable energy.

9.3 Hydroelectric Turbines: The Heart of Waterpower

The hydroelectric turbine is the power conversion core behind hydroelectric power generation. It engineeringly converts kinetic energy contained in flowing water into useful mechanical energy. Indeed, these are very efficient turbines, with more than 90% of the kinetic energy delivered by the water really converted to power the motor driving the generator.

Each turbine is essentially composed of two main parts: one is the wheel or impeller; the other, stationary, is called the stator or distributor.

The stator directs and controls the water flow, channeling its energy to the impeller, which spins and transmits the mechanical energy to a generator. This efficient energy transfer makes hydroelectric turbines an indispensable component of modern renewable energy systems, powering communities with clean and reliable electricity derived from nature's flow.

The adoption of new technologies empowers the functioning of energy systems which shifts the dependence on conventional fuels towards renewable energy sources. Such technologies include systems for energy storage, smart grids, AI, and other modern materials and greatly improve the effectiveness with which renewables are fused into the existing power systems. The integration of renewable sources into the economy is encapsulated with solar and wind power, and hydroelectric power and bioenergy are vital too. The most affordable and easier to implement and construct are the solar PV systems, but their deployment depends on the geography and storage capabilities. Wind energy has favourable expanding and steep raising energy conversion rates, which may be observed through offshore wind farms, even though it demands massive funding for initial implementation alongside grid system infrastructure. Hydroelectric power still remains energy stable and able to grow, with the only limitation of geography because of massive environmental concerns. Organic waste material and bioenergy is easily renewable, however, needs to be watched for extensive usage. This paper approaches the subject from multiple angles, taking into consideration the performance, expenses, and flexibility of these technologies for practical application, analysing all outlining results and deriving a conclusion values with return for investment are constantly needed to face up problems with integrations into energy systems from different global locations.

10 Policy and Regulatory Frameworks

Achieving renewable energy targets by 2030 requires comprehensive, globally unified policies to coordinate efforts, support equitable financing, capacity building and advance technology ensuring all countries and communities benefit from the transition. As the energy markets move toward a low-oil and low-carbon development pathway, globalization will play a significant role. However, there is currently little scholarly research on this subject. Our investigation is motivated by this disparity. The majority of current research has focused on the connection between environmental degradation and globalization, with very little focus on energy use or consumption and even less on clean cooking energy technology. Given the significance of multi-level governance in advancing community energy, the final article selection was categorized based on the government level at which specific instruments were analysed or implemented (a) supranational to national, and (b) regional to local. Some instruments may also be analysed across both levels in the research. Financial support instruments, such as feed in tariffs (FITs), grid services and fiscal incentives were primarily discussed at a supranational to national levels, applying broadly to most instruments.

Exemptions included power purchase agreements (PPAs), energy efficiency programs and climate change instruments which were more frequently examined at regional (e.g., provincial) to local (e.g. municipal or community) levels focusing on their impact on community energy development (di Santo et al., 2022).

To understand the influence of international trade law on the energy transition, it is essential to explore their potential areas of interactions. International trade law governs instruments like tariffs, quotas and subsidies prohibiting certain practices (e.g., discrimination and quantitative restrictions) while limiting others to facilitate free trade. The preambles of both General Agreement to Trade and Tariff and Marrakesh Agreement underscore a commitment to substantially reducing tariffs and trade barriers and eliminating discriminatory treatment in global trade relation. The Indian Renewable Energy Act, 2015 and the Ujjawal Discom Assurance Yojana were enacted to enhance India's approach to the energy trilemma-promoting environmental sustainability, energy security and equitable access via renewable energy initiatives. This act facilitates the establishment of key institutional mechanisms, including the national renewable energy plans, state level renewable energy policies and various authorities like that National Renewable Energy Commission (NREC), National Renewable Energy Advisory Group (NREAG), Renewable Energy Corporation of INDIA (RECI) as well as Renewable Energy and Renewable Electricity Investment Zones.

The deployment of new technology is often hindered by various barriers, challenges and application limitation that need to be resolved for widespread adoption. Economic consideration plays a critical role in process deployment, with renewable energy technologies (RERs) particularly affected by high CAPEX (capital) and OPEX (operating), which elevate energy cost and impede implementation, and this economic challenge has made financing difficult for RERs projects. The primary purpose of creating supportive institutions is to foster coordination and collaborations among entities to effectively plan, implement and promote RERs.

11 Economic Impacts of Energy Transition

Over the once many times, renewable energy has grown in fashion ability, and for good cause. In addition to being more environmentally friendly, it's also less precious than fossil energies. Because they do not need energy, renewable energy sources like wind, solar, and hydro are affordable to run. Still, the energy demanded for reactionary energy power shops—coal, oil painting, and natural gas—can be expensive to gain, transport, and store. Predicting running costs can be gruelling because energy prices can change. In discrepancy, renewable energy sources are less changeable and more predictable because they do not bear energy. But it's inversely important to consider the environmental consequences in addition to the fiscal bones.

The price of fossil energies doesn't regard for the substantial environmental costs associated with them. These charges correspond of hothouse gas emigrations, water pollution, and air pollution. These charges may have a major influence on society's health and frugality. In discrepancy, renewable energy is more charming from a societal perspective due to its significantly lower environmental costs.

Enterprise for renewable energy are propelling profitable growth and job creation as the world moves toward sustainable energy, particularly for youthful people, both manly and womanish. This enterprise, which range from wind granges to solar granges, have a big influence on original employment rates by generating a variety of job possibilities in design, product, installation, and conservation. To succeed in the renewable energy assiduity, women and youthful people bear a combination of soft and specialized capacities. While soft chops like problem-working and cooperation are just as important, specialized chops include engineering, design operation, and knowledge of renewable technologies.

Long-term goods of the renewable energy transition on hiring practices and pool development are substantial. Professionals with moxie in sustainable practices, energy effectiveness, and renewable energy technology are in lesser demand as a result of the change. The elaboration of the assiduity emphasizes the necessity of having a pool that's both flexible and devoted to lifelong literacy in order to keep up with the quick changes in technology and the shifting dynamics of the sector.

In addition to reducing emigrations, switching subventions from fossil energies to renewable energy promotes sustainable profitable growth, job creation, bettered public health, and lesser equivalency—especially for the world's impoverished and utmost vulnerable people.

Indeed, though mileage-scale solar and wind systems are getting more competitive, some clean technologies might still need fiscal backing. These correspond of off-grid options, energy storehouse, and coastal wind. Support for integration costs (similar energy storehouse) will also presumably be needed in order to quicken the increased use of renewable energy sources. turning savings from reforming reactionary energy subventions or perfecting subvention targeting is one possible source of backing. The necessary assessment of energy programs to make sure they're negotiating their pretensions and furnishing value can only be carried out with complete account and openness.

12 Challenges and Opportunities in the Energy Transitions

Similar to other change processes, the energy transition is confronted the energy transition is fraught with challenges driven by uncertainties associated with the decisions required throughout its progression. The global energy transition is redefining energy systems and demands a comprehensive approach to tackle the intertwined challenges of technology, economics and policy (Chipangamate & Nwaila, 2024). It will require not just deploying renewable technologies, but new economic models and smart policy solutions. At the base of such a transition is technological innovation, where progress in key areas—renewable energy, grid integration, energy storage, and smart grids—are basic in accomplishing scaling of the energy systems for sustainability. Furthermore, there is yet a host of technical challenges for the deployment of the already considerably advanced technologies at a larger scale, in particular in developing countries as shown in Fig. 4. This situation could be enhanced through the mitigation or addressing of a number of common constraints in most of the developing countries as far as technology, access to infrastructure, and finance are concerned.

Energy transition involves changes in energy production and consumption, which is a shift in energy sources, infrastructure, and demand patterns. The transition is not just about replacing one energy system with another but also about the reorganization of socio-political structures and processes. The wider perspective on energy transition deepens insights into essential principles such as socio-equity and agency. Social equity is a dynamic concept applied to a wide array of social

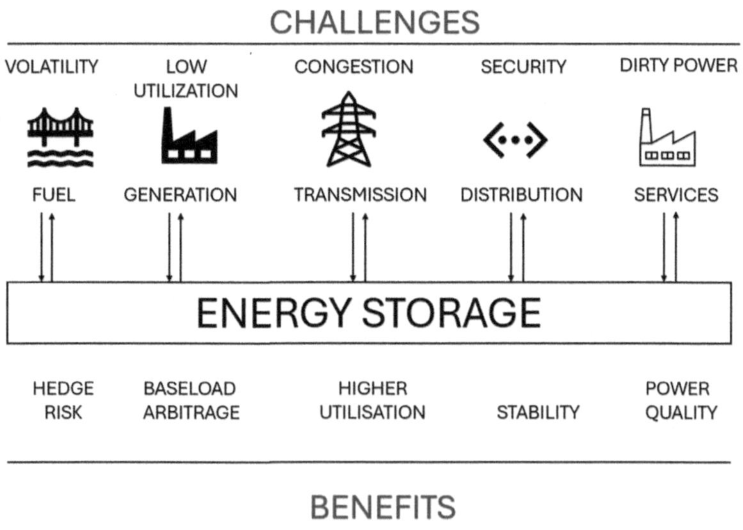

Fig. 4 Challenges and benefits of energy transitions

spheres, from education to environmental sustainability. Social equity is often conceptualized as a socially constructed concept that is negotiated through ongoing debate and contested views about what constitutes the fair, as well as the principal contextual elements necessary to achieve more equitable outcomes. The shift to low-carbon energy sources will inevitably create winners and losers. While beneficiaries enjoy cleaner energy, lower emissions, and new economic opportunities, others may face hardships and not share these benefits. This paper reviews current insights from the academic and policy literature on the potential negative impacts of the energy transition on specific communities and socio-economic groups at the front lines of these changes.

The development of zero-carbon microgrid technologies has been broadened, and their applications have recently come under heightened attention, since these systems will be able to offer sustainable means of satisfying decentralized energy needs. This will entail significant integrations of clean energy sources, extensive energy storage systems, and advanced power electronic devices for realizing zero or near-zero carbon emissions in microgrids (Chen et al., 2024). However, this path will result in a significant increase in initial systems costs. Over the past decades there has been a substantial interest in microgrids globally, despite their status as an emerging niche innovation. As demand for reliable energy rises, microgrid offers a means of increasing self-sufficiency especially in areas lacking central grid access or prone to power outages and natural disasters. As of 2017, the United States held the largest share of global microgrid capacity at 41% of the total 20 GW, followed by China with a share of 30%. For perspective, the total U.S. electricity generation capacity stood at about 1100 GW, with a global capacity of nearly 6800 GW in 2018. As distributed energy resources microgrids, complement the central grid of providing reliable, locally controlled, renewable energy solutions. Various U.S. projects from California to Alabama and New York, are implementing microgrids in collaboration with utilities, fostering energy resilience without fully disconnecting from central grid infrastructure.

13 Energy Transition Around World

The global energy transition refers to the shift from fossil fuel-based energy systems to cleaner, renewable sources such as solar, wind, hydropower, and biomass. This transition is driven by the urgent need to combat climate change, reduce greenhouse gas emissions, and achieve sustainable energy independence. Countries worldwide are implementing policies and investing in technologies to increase the share of renewables in their energy mix. This transformation varies significantly by region, influenced by factors such as natural resource availability, technological advancements, economic priorities, and political commitment.

Uruguay is becoming into a major wind power. In just five years, its wind power has grown from 1 to 35%. Although the nation has historically relied on hydropower, the government was forced to invest in wind energy after droughts

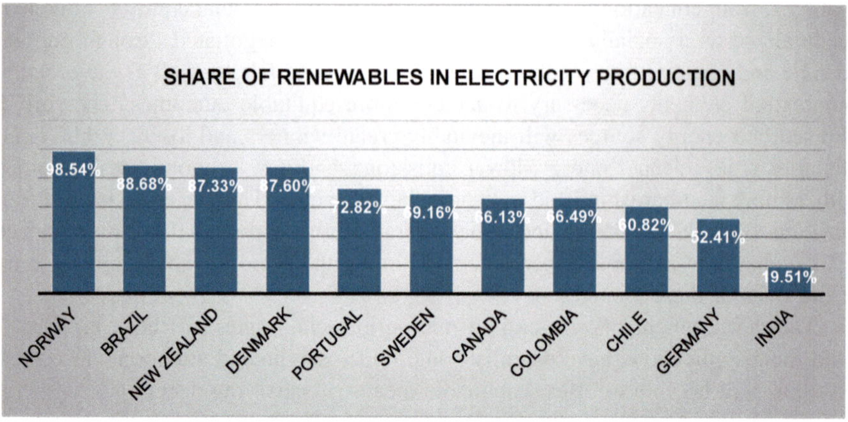

Fig. 5 Share of renewables in energy production by various countries (Ritchie et al., 2024)

in the late 1990s and early 2000s depleted its supplies. From 2008 to 2018, the nation's wind energy output increased dramatically as shown in Fig. 5.

Despite having a smaller GDP than Uruguay and Denmark, Namibia is emerging as a solar hotspot in Africa. Namibia is excellently situated geographically to generate solar energy. The WEF estimates that the nation can receive about 10 h of intense sunlight per day, 300 days a year making it one of the regions with the highest levels of solar radiation that can supply electricity to its citizens; in 2020, 40% of them and their neighbours in Africa did not have access to electricity.

As per the above figure highlights the share of renewable energy in electricity production across various countries. Norway leads with an impressive 98.54% of its electricity generated from renewable sources, primarily due to its abundant hydropower resources. Brazil follows with 88.68%, benefiting from a mix of hydropower and bioenergy. New Zealand and Denmark show significant shares of 87.33% and 87.60%, respectively, with New Zealand relying heavily on hydropower and geothermal energy, and Denmark excelling in wind energy.

Portugal and Sweden also exhibit strong performances, with renewables contributing 72.82% and 69.16%, respectively. Canada, Colombia, and Chile demonstrate renewable shares ranging between 60.82% and 66.49%, reflecting their reliance on hydropower and other sustainable energy sources. Germany, despite being a leader in solar and wind energy deployment, has a lower share of renewables at 52.41%, largely due to its continued use of coal and nuclear power during the transition phase. India, with a renewable share of only 19.51%, lags behind the other countries in the figure, highlighting the challenges it faces in scaling up its renewable energy capacity to meet growing energy demands.

The world's powers have changed. Dominant economies are no longer limited to Western Europe and the United States. Despite producing money at previously unheard-of levels, the world's three superpowers—China, India, and the United States—are lagging behind in reducing their greenhouse gas emissions. There is optimism as the world continues to descend into a climate calamity. A growing worldwide market for renewables is feasible with aggressive reduction targets and Europe leading the way in a potential green revolution. Sustainable development could lead to both economic advancement and the overall well-being of the world if global energy usage patterns are successful in becoming more ecologically friendly.

The data underscores the varying progress of energy transition efforts globally. While countries like Norway, Brazil, and Denmark are nearing complete reliance on renewable energy for electricity production, others like India still have significant ground to cover. The energy transition is crucial for achieving a sustainable future, and countries must accelerate investments in renewable technologies, infrastructure, and policies to meet global climate goals and reduce dependence on fossil fuels. Collaboration, innovation, and political will remain essential to ensure a just and effective energy transition worldwide.

14 Significance of Shift Towards Renewable Energies

Gaining public support, public engagement, and public awareness are key elements in the processes of adopting clean energy technology. A clean energy public is more likely to support policies designed to encourage its use, making it more likely that governments will provide favourable regulations, subsidies, and incentives aimed at reducing fossil fuel dependency. Moreover, the greater the awareness, the greater the change in behaviour as people become more mindful of the need to use less energy and look for more sustainable options which could include the installation of solar panels, wind energy turbines, and the use of energy efficient appliances.

A public that is more knowledgeable will also boost the demand, investment, and growth in the renewable energy market. Such people and communities that appreciate the benefits of renewables are more willing to invest on projects like rooftop solar panels or community wind turbines. Public awareness campaigns help counter the misinformation and misconceptions about renewable energy. They enable the reliability, cost, and efficiency myths to be busted. Stakeholders have to build trust so that sustainable energy marketing is adopted on a larger scale.

In Addition, community engagement and support is vital to the wind farms efficiency. Frequently, there is a gap where local renewable energy projects could be successful but are not, due to push back from the locals as a result of territorial and ecological apprehensions. Constructive conversation can mitigate fears, that helps to gain support to peripheral opposition. Additionally, there is often a notable increase in need in the renewables industry with heightened public awareness. The need and motivation for innovation, technological advancement and jobs in the clean energy industry increases as more realize the benefits of clean energy.

15 Obstacles in Adopting Renewable Energy Technologies

Adopting renewable energy technology comes with its challenges, especially for developing countries. They are often limited by tight budgets and capital. Many developing countries lack sufficient financing resources both internally and through foreigh investment. As a result, renewables become harder to access and increase in price. Furthermore, fossil fuels have short term government subsidies which make it easier to use them. In addition, these countries have other infrastructural and technical problems too. Many developing countries have energy grids that are not only out of date but also insufficient. Because of this, solar and wind power cannot be harnessed optimally. Additionally, the current workforce in these countries is not trained enough to work in renewables. These workers mostly depend on engineers and technicians to operate and maintain these technologies. In addition to this, many of the essential parts that are needed to build these technologies are made in other countries as this increases the cost and reliance on foreign countries.

There are both policies and regulations that pose difficulties. Most developing countries are afflicted by conflicting policies, ineffective regulation enforcement, and bureaucracy that obstructs investment in renewable energy. Also, the lack of strong incentives such as tax reliefs and feed-in tariffs further stagnate the shift toward renewables. Instead of seeking long term solutions, governments concentrate on meeting immediate energy needs which deepens the reliance on fuel.

The areas which bear the greatest impact are social and economic. The general public has low awareness of the advantages of renewable energy and, therefore, communities may resist the idea of switching to renewable energy because they fear losing jobs in conventional energy sectors. Additionally, these projects face opposition due to land usage and environmental impact. Within developing countries, there is a reliance on other countries to provide expertise and materials such as lithium and rare earth metals. Dependency is caused by a small number of countries holding this resource which makes the developing nations vulnerable to price changes and supply chain issues.

Notwithstanding these difficulties, emerging economies can still hasten the uptake of renewable energy. Formulation of coherent policies, improvement of grid systems, and local assembly as well as research activities can lower the reliance on foreign technology. International organizations can assist with financing and technical expertise, while local renewable energy initiatives and education programs can foster local support. Developing countries can successfully navigate impediments towards a more environmentally friendly future by managing these obstacles effectively.

16 International Cooperation and Funding Influence the Progress of Renewable Energy Initiatives on a Global Scale

Funding and cooperation at the international level are crucial for the improvement of renewable energy projects. Many developing countries do not have the funds or the skills to be able to shift to clean energy on their own. Through international relations, developed countries along with other global financial bodies and institutions like the United Nations and the World Bank extend assistance in the form of funds, technology, and capacity-building. Nations receive financial support through grants, low-interest loans, and investment programs that help improve financial status so that countries can adopt bigger scale renewable energy programs that would otherwise not be possible. Additionally, initiatives such as the Paris Agreement help in fostering collaborative approaches by establishing emission reduction goals while promoting climate action across all nations.

Apart from providing funds, international cooperation enhances further development through knowledge sharing and innovation in the renewable energy sector. Countries that have adequate renewable energy technologies can share their relevant research and technical information with countries that are still working on their clean energy infrastructure. This leads to deploying renewable energy solutions at an increased rate which is also more efficient and cost friendly. New technologies along with energy storage, grid integration, and efficiency improvements of renewables can also result from joint research and development programs.

In addition, international co-operate funding and cooperation allows for improvement on common issues such as energy procurement, policy issuance, and market volatility. Global alliances like the International Renewable Energy Agency (IRENA) enable members and non-members to conduct multi-governmental dialogues and establish constructive policies that facilitate fresh investment in renewable energy. Also, multinationals and private sector investors are instrumental in funding clean energy projects while advertising eco-friendly business models and cross-border renewable infrastructure investment. Funding from foreign cooperations, and international aid are integral in providing value toward renewable energy improvement. Nations working together can mitigate monetary and technological obstacles while foster essentially fast-paced groundbreaking processes and equitably allocate a transition toward clean energy. These established partnerships are essential for delivering a sustainable energy future while tackling climate change.

17 GOI Initiatives

17.1 Development of Solar Parks and Ultra-Mega Solar Power Projects

India's push for solar energy is taking shape with the creation of Solar Parks and Ultra-Mega Solar Power Projects, aiming for an impressive 40,000 MW capacity. These parks make it simpler for developers to set up large-scale solar projects since they have all the necessary infrastructure, including roads, water supplies, land, and power evacuation systems. This effort helps accelerate the generation of solar energy by ensuring that all required licenses are in place. This attracts investment and positions renewable energy as a major role in the country's energy landscape.

17.2 CPSU Scheme Phase-II (Government Producer Scheme)

This is a scheme that would enable the government to seize the leading position in the use of solar energy just by setting up grid-connected solar PV projects with domestically manufactured solar cells and modules. The scheme, while providing subsidy support to bridge the cost gap, would strengthen India's self-reliance in solar manufacturing besides promoting clean energy use. Whether used directly by departments or distributed through DISCOMs, the initiative will assure a steady move toward a greener and sustainable energy future.

17.3 Production Linked Incentive Scheme for High-Efficiency Solar PV Modules

It introduces the PLI scheme in India, which encourages high-efficiency solar PV modules for manufacturing, with an objective of eliminating dependence on imports and having a strong solar industry in India. The goal here has been to enable their scale-up and bring fresh, more competitive solar technology. Creating prospects of employment and leading toward global positioning, this is indeed going to put India among the leadership in advanced manufacturing in the world.

17.4 PM-KUSUM Scheme

The PM-KUSUM Scheme aimed at bringing renewable energy to the heart of rural India by promoting solar solutions among farmers. Be it small solar plants on farmland, or solar-powered pumps for irrigation, this would aid both the farmers and the environment. Farmers could use the surplus solar power produced to sell back into the grid, increasing their incomes, and states save on agricultural electricity subsidies. This scheme will help support livelihoods and a cleaner energy future with solar power reaching the most remote areas at an affordable cost.

17.5 Rooftop Solar Programme Phase II

This program makes it very easy for homeowners to go solar. Incentivizing rooftop solar systems through subsidies and rewarding DISCOMs on overachievement, this would lead to a wider diffusion of clean energy. It cuts electricity bills not only for households but also contributes toward a decentralized energy network, easing the load on the national grid. This will be a triple win—families, environment, and the country's renewable energy ambitions.

17.6 Green Energy Corridors (GEC)

While renewable energy production is on the increase, Green Energy Corridors ensure it actually reaches the people who need it. This program is about building transmission infrastructure to carry clean energy from solar and wind farms up to the grid. The program covers ten states of the country to ensure seamless integration of renewable energy into the national electricity supply—a step toward a greener, more reliable energy system.

17.7 Bio-energy Programme

The Bio-energy Programme turns waste into a resource by generating energy from urban, industrial, and agricultural waste. Be it from the waste-to-energy plants, biomass briquettes, and pellets, or even the small-scale biogas plants for rural households, this initiative turns the most common waste into invaluable energy. The initiative thus not only tackles issues of waste management but generates employment and brings forth clean and inexpensive sources of energy, serving communities and industries alike.

17.8 Renewable Energy Research and Technology Development (RE-RTD) Programme

Innovation is at the core of the renewable energy journey in India, and the RE-RTD Programme ensures the country is always ahead. Supporting cutting-edge research in partnership with academic and industry leaders, it enables breakthroughs in energy storage, efficiency, and integration. This focus on research, therefore, essentially drives India's effort to create cost-effective and sustainable solutions for a rapidly growing energy demand.

17.9 Scheme of Human Resource Development

It will take skilled people to make a clean energy future, and this plan invests in building a capable workforce. It will equip the people with knowledge and expertise through training programs, internships, and academic collaborations in renewable energy. More state-of-the-art labs and Renewable Energy Chairs at institutions will make India's talent pool ready to take on leading roles in green energy innovation.

17.10 National Green Hydrogen Mission

The National Green Hydrogen Mission is a big leap toward making India a global hub for the production and export of green hydrogen. With a massive investment of ₹19,744 crore, this mission is powerfully positioned to bring in a revolution in industries like steel, chemicals, and transportation by offering clean, renewable hydrogen as an alternative fuel. This project will significantly contribute to India's climate goals and establish it as a leader in the global green energy movement by lowering reliance on fossil fuels and fostering innovation.

18 Future Developments in Renewable Energy Technologies

New developments in solar and wind energy, as well as improvements in battery technology, are likely to make renewable electricity more reliable and cheaper. New smart grids and energy management powered by artificial intelligence will further streamline energy distribution, helping to minimize waste and maximize value. Moreover, green hydrogen and carbon capture technology could transform energy production by offering cleaner substitutes for fossil fuels. Such developments are likely to hasten the shift towards sustainable energy across the globe, which minimizes the use of traditional energy sources, radically changing the geopolitical energy scenario. As it is forecasted that the cost of renewable technologies will decrease, it is expected that nations will turn toward domestically produced energy, thus improving energy independence while reducing dependency on foreign oil. Adopting these transforms systems not only achieves drastic reductions in carbon emissions but also creates a more resilient and decentralized global energy infrastructure.

19 Energy Justice and Social Implications

Energy justice guarantees that everyone has access to the energy they require in amounts that are reasonable, dependable, and clean, irrespective of their zip code or financial situation. It emanates from the greater struggle for social equity. It strongly relates to environmental justice and underlines disparities in the access and use of energy between communities. Energy justice is particularly significant to those who have been left out of the many benefits of modern energy. This often results from the historical inequalities that have created a division in the current energy world. It seeks to ensure that these communities aren't left behind in the transition toward cleaner energy, working toward righting past wrongs and creating a more equitable, inclusive energy system for future generations.

This Special Issue calls for researchers from both the supply and demand sides to critically examine the social justice and ethical considerations inherent in the historical, current, and future development of low-carbon energy systems. Energy justice works as the conceptual, analytical, and decision-making framework that allows the identification of ethical challenges within energy systems, who the stakeholders to involve are in solving the issues at hand, and further guides a choice of solutions toward a sustainable and just energy transition.

The notion of "leaving no one behind" underlines the need for integrating socio-economic concerns into energy transitions beyond technology and climate issues for the resolution of injustices already at play. While the terms "just," "equitable," and "inclusive" are often used interchangeably, each points to different dimensions of ensuring transitions do not perpetuate disparities. EDI was conceptualized by the International Energy Agency in 2010 to keep track of the development relating to access to modern energies through ranking (Caiafa et al., 2023). EDI monitors access to electric and clean cooking facilities via its four different indicators: per capita consumption of commercial energy, electricity consumption at the residential sector, the share of modern fuel in the residential sector usage of energy, and population that has electricity access. However, since then, the IEA has focused on SDG indicators.

The MTF, developed by the World Bank in collaboration with its partners, measures energy access across several attributes: adequacy, reliability, affordability, and safety. It, therefore, proposes a set of tiered indicators to assess energy access at household, productive activity, and community facility levels. Though it was originally defined in a broad way, the practice of its application has often streamlined its focus on electricity and modern cooking solutions. The MTF has become a standard for measuring energy access and poverty, adaptable to different contexts to reflect the varied challenges of energy.

Usually, methods of measuring energy poverty fall into three approaches: consensual-based, expenditure-based, and direct measurement. These approaches are complementary in understanding energy poverty, focusing on technological, physical, or financial thresholds.

19.1 Consensual-Based Method

It is a self-reporting technique for the measurement of conditions at home that are in relation to thermal comfort and the capability of affording the basic services of energy. This approach identifies the technological threshold that defines whether or not the population has access to modern sources of energy. Indicators include primary ones, such as being unable to heat the home sufficiently or pay utility bills, and secondary ones like housing with damp or rot. While useful for cross-country comparisons, it depends on subjective responses and, therefore, accuracy can be affected because the interpretation of survey questions may vary.

19.2 Expenditure-Based Method

This method compares household income and energy costs to assess how affordable energy is. If energy expenses rise above a certain level, a household is deemed energy-poor. It encompasses both relative and absolute limits, such as median energy load and fixed percentages of income. The approach gives information about affordability and draws attention to the energy burden, or energy costs as a percentage of income. However, fluctuating energy prices and differing economic conditions across countries complicate analysis. Examples include the UK's use of 10%, 2M, M/2, and LIHC indicators.

19.3 Direct Measurement Method

This method evaluates household energy services, such as heating, lighting, and cooling, against predefined standards. It considers theoretical energy needs based on thermal efficiency and acceptable heating levels per World Health Organization (WHO) guidelines (e.g., 18 °C for bedrooms, 20–21 °C for living rooms). Regional variations exist, such as Scotland's higher threshold for vulnerable groups. Challenges include the lack of comprehensive data on energy services and the subjective nature of thermal comfort, influenced by cultural, climatic, and psychological factors.

20 Global Energy Market Trends

The Renewable Energy Market was valued at USD 1092.91 billion in 2023 and is projected to grow to USD 1219.00 billion in 2024 and USD 2277.84 billion by 2032, with a compound annual growth rate (CAGR) of 8.13% over the forecast period (2024–2032). Renewable energy encompasses energy generated from naturally replenished resources such as solar radiation, wind currents, water flow, geothermal heat, and biomass. Unlike finite fossil fuels, which contribute significantly to greenhouse gas emissions and environmental degradation, renewable energy sources harness sustainable natural processes for power generation. These energy systems are pivotal in reducing carbon emissions, mitigating climate change, and facilitating the transition to a sustainable energy framework. By relying on resources that are naturally renewed and minimizing environmental impacts, renewable energy offers a viable and cleaner alternative to traditional energy systems. Fossil fuels have long dominated the global energy mix, their share modestly decreasing from 85% in the 1970s to about 80% today. But times are changing, driven by falling renewable energy costs and increased awareness of the environmental consequences of fossil fuel emissions. The current transition toward renewable energies is marked by a switch from fossil fuel to renewables, further electrification, digitalization, and decentralization of the energy system that fundamentally changes global energy geopolitics.

Where the onus of a decrease in demand for oil falls squarely on hydrocarbon-exporting countries, including GCC members, energy-importing countries come out winners with their newfound energy independence and exports of clean energy. These shifting dynamics will make energy diplomacy pivotal to the conduct of international relations and leveraging strategic gain (Alnaser & Alnaser, 2011).

Sustainability transition scholars have indicated the two phases of the energy transition, which are the 'take-off' phase from the 1990s when the emerging renewable technologies were coming up and the 'breakthrough' phase from the mid-2000s onward characterized by mature renewables starting to replace fossil fuels. This is further supported by the development of niche industries such as electric vehicles and solar energy and destabilization of the fossil fuel regime through taxation and regulation. Recent research also discusses the socio-economic impacts of transitions and policy measures to address those challenges.

Neo-classical and renewable energy geopolitics emphasize natural resources, their locations, and transportation as structuring variables of energy systems and politics. Neo-classical geopolitics focused on fossil fuels—particularly oil and gas—and treated energy and geography as fixed entities, whereas renewable energy geopolitics would include renewables, critical materials, and demand-side technologies like storage (Slaria, 2024).

The concept of GEST evolves, and so does energy materialities, from its core to a wide range of resources, technologies, and global supply chains; this is reflective of how dynamic energy systems are. Energy systems, once hugely dependent on fossil fuels in order to perform their role in socio-economic functions, are being actively redefined in alignment with sustainability goals.

Emphasizing the increasing complication, GEST includes such variables like energy resources, a system of energy conversion, storage technologies, and greenhouse gas emissions, making a strong case for their interlinkages and the transition to more environmentally friendly systems of energy supply. The ongoing global transformation of the energy system is redrawing its geopolitics, and recent literature has highlighted, in particular, the dominance of renewables in a future low-carbon scenario. Yet most of this research skips over the transition phase of GEST, where high- and low-carbon energy systems will be competing. This calls for more interaction between disciplines to understand the complexities of this transition phase.

The ongoing energy transition poses major socio-economic challenges, especially regarding employment and economic development. An shift from fossil fuels to renewables such as solar, wind, and battery storage offer new employment opportunities. However, the transition requires extensive skillset retraining along with new forms of workforce development that provides workers the knowledge needed for employment in green energy. In addition, renewable energy projects that are not concentrated in urban areas, especially those located in the countryside, can boost business activity by providing green jobs and advancing the local economy.

In terms of economic expansion, activity spending in clean energy systems and technological advancement lead to economic growth because new capital is brought in as new ideas are developed. The productivity gains from higher energy efficiency are not only cost savings for a particular firm, but productivity improves in all sectors. While the reallocation of resources may create some short-term costs for incumbent producers of energy, the long term outcome is greater economic sustainability. The shift to renewable energy offers the opportunity to build a stronger economy while achieving environmental goals and still allowing for economic growth.

21 Technological Challenges in Scaling Renewables

A smart grid deploys modern advances in communications and information technology to increase the robustness, performance, and reliability of conventional power systems, as well as services and customer participation, cybersecurity, and integration of renewable energy resources. The renewable generation, mainly solar PV and wind turbines has also grown remarkably, with PV capacity increasing from 3 GW in 2003 to 139 GW in 2013, while wind has gained momentum, rising from 39 to 318 GW in the same period (Debnath et al., 2022).

While renewables provide sustainability, they also bring challenges such as forecasting, energy storage, demand management, voltage control, and system stability. This paper reviews some methods for enhancing voltage control, including low-voltage ride-through capabilities, and the techniques used to improve system stability, focusing on inter-area oscillation damping in systems with PV and wind power plants (Zohuri, 2023).

22 Challenges and Directions to Mitigate Intermittency in Energy Storage Systems (ESSs)

The modern ESSs are playing a crucial role in enhancing the flexibility and reliability of the systems in various sectors of power, including dispatchable power plants, transmission, distribution, and end-user operations. Challenges in their integration lie in the variability of renewable energy sources, among other issues, and require focused research in the following areas:

1. ESS Degradation: Frequent charge/discharge cycles result in ESS degradation, which negatively affects the lifetime of ESS systems. As a result, research has offered a variety of methods for improving charging pattern management in order to boost durability, particularly for BESS.
2. Capacity Optimization in the Electric Transportation Segment: ESSs face problems related to intermittent loads and prolonged operational demands in electric ships and aircraft. Research has been focused on hybrid energy storage solutions for stability in operations and reduced fuel consumption.
3. Energy Efficiency: High energy storage efficiency should be ensured. Since key players like BESS and EVs are at the core of grid support and reduction of emissions, improvement in energy density and conversion efficiency becomes crucial.
4. Geopolitical Stability: Critical materials, like cobalt and lithium, are highly important in the development of ESS. In this regard, geopolitical risks must be addressed to ensure stability in the supply chains of ESS production.
5. Recycling and Reuse: There is growing need to build efficient recycling and reuse programs due to the increased demand for batteries. To optimize the performance of retired batteries and the recovery of their raw materials, life cycle assessments ought to be conducted.
6. Financial Viability: Most people agree that this is one of the most important adoption-related aspects. Finding the ideal capital cost and conducting feasibility studies for hybrid systems that combine ESS with renewable energy sources have been the primary areas of research.
7. Policy and Regulation: The implementation of ESS has been primarily propelled by a robust regulatory environment. To get the most out of ESS, policies should prioritize market involvement, investment incentives, and system dependability.

23 Securing Smart Renewable Energy Systems

These systems require advanced defenses, proactive risk assessments, and stakeholder engagement. Technologies including artificial intelligence (AI), blockchain, encryption, access restrictions, awareness campaigns, and audits are crucial for data protection and resilience. This study report provides an overview of current practices, challenges, and recommendations for enhancing energy system security and resilience against evolving cyberthreats.

24 Energy Storage and Distribution Innovations

Energy storage technologies are emerging as significant solutions to the global energy issues, notwithstanding both technical and non-technical challenges. Finding new mechanisms and processes is essential to resolving energy system issues and fostering industrial expansion through technical development. This article highlights the state of the field now and key directions for the future while examining recent trends and advancements in energy storage.

24.1 Classification of Energy Storage Technologies

Energy storage technologies can be classified on the basis of the form of energy stored in them. The following are some:

1. Electrical Energy Storage Systems (EESS): These store energy in electrical form for grid applications, among other power systems.
2. Electrochemical Energy Storage Systems (ECESS): Utilize chemical reactions, including batteries, for efficient energy storage.
3. Mechanical Energy Storage Systems (MESS): This includes mechanical technologies that can store energy, such as compressed air or flywheels.
4. Thermal Energy Storage Systems (TESS): These systems store energy in the form of heat for industrial and residential purposes.
5. Chemical Energy Storage Systems (CESS): These systems convert energy into chemical bonds and store it for later usage, including hydrogen storage.

Critical challenges are typical for the global energy landscape, including the dwindling resources of fossil fuels, increased energy demand, and environmental impacts. Li-ion batteries have been very important in portable electronics and electric vehicles for many years but are limited by insufficient energy density, safety risks, and scarce materials. These limitations define the urgent need for future sustainability and performance in advanced energy storage solutions.

Until today, WPT has enormous potential for contributing to sustainability through the inefficiencies of the traditional way of transmission or storage. Using resonant inductive coupling and radio waves, WPT enables better energy efficiency with less waste, as such, integrates renewable resources of energy. This is especially useful in remote areas or places where resources are challenged; WPT bypasses resource-intensive infrastructure and democratizes access to energy, therefore allowing education, healthcare, and economic growth.

It is vital to align research with the SDGs since global challenges include poverty, environmental protection, and equitable development. Research will inform policy and nurture innovation while, at the same time, ensure accountability by monitoring the progress of the SDGs. A multidisciplinary approach can strengthen inclusivity and achieve social justice to enable comprehensive solutions for a sustainable and prosperous future.

Microgrids allow for a transition from centralized to decentralized electricity generation, with an emphasis on renewable energy sources like photovoltaic and wind, supplemented by distributed, small-scale conventional units. However, the low inertia in PV systems and decoupling of the mechanical part in wind generation systems due to interfacing power electronics reduce the inertia of microgrids, leading to excessive frequency deviations and high ROCOF during disturbances.

Enhancing inertia requires effective energy storage and utilization. By integrating a supercapacitor with PV and utilizing the intrinsic kinetic energy of wind systems, the scheme optimizes energy sharing and minimizes stress on individual resources. Simulations show a 53.3% improvement in frequency deviation and a 61.9% reduction in ROCOF when PV and wind support frequency simultaneously. The scheme also reduces energy demands on PV and wind systems by 29.5% and 13.7%, respectively. With an equal energy-sharing strategy, energy stress on wind is significantly reduced, enhancing system reliability. Applied to an AC microgrid comprising thermal, PV, and wind systems, the proposed scheme demonstrates superior frequency support under various scenarios, including load changes and grid disturbances, confirming its efficacy and robustness.

25 Future Outlook: Energy Transition Roadmaps

To ensure the success of the field, it is essential to explore, anticipate, and predict future trends. Our analysis focused on publication with the maximum usage frequency of the 180 days by analysing our database for articles with usage frequency of 180 days by analysing our database for articles with usage data from June to December 2021. The use count reflects the number of views an article has received during the last six months. While citation counts often measure long term influence, they do not capture current trends, which is why 'usage count' is more effective in providing real-time insights into the ongoing relevance of publications. Effective resource mobilization for the transition to modern, clean and sustainable energy systems hinges on multistakeholder collaboration across innovation, production, distribution, investment and financial institutions. However, many existing regulatory frameworks, which were designed for large scale conventional power generation and transmission, now impede the integration of renewable energy. These legacy systems hinder the adoption of decentralised solutions and need adjustments to handle the intermittent nature of renewable energy and the rising role of prosumers. Overcoming these barriers requires societal and economic transformation while ensuring long term stability. The intersection of energy transitions and economic growth presents challenges for even the most developed nations. Among 115 countries, less than half could improve their Economic Transition Index (ETI) while simultaneously advancing in economic growth. A major concern is energy affordability, especially in advanced economies, where electricity tariffs have increased by 25% in real terms over past decades. Furthermore, over 40% of developing nations made substantial progress in both environmental sustainability and economic growth. Giving one aspect of the transition priority could also mean exacerbating inequality or standing in the way of the climate goals being met. This strategy pursued will have to balance the simultaneous needs of sustainability, energy security, and economic development. Simultaneously, the energy system is exposed to multiple eventualities created by persistent macroeconomic and geopolitical uncertainty. Successful energy transition sure way will be overcoming these uncertainties, which requires making significant and rapid advances by overcoming barriers that surpass the simple, quick fixes. Resilience in transition could be increased by concomitantly addressing the interrelated demands of affordable access, energy security, economic growth, and environmental sustainability. In record time, onshore wind and solar energy have emerged as the least costly ways of electricity generation worldwide. Some countries rely on renewables to cover more than 80% of their day-to-day energy needs. In the year 2020, more than 260 GW of renewable capacity was added in a single year alone. This sector employed 11.5 million people in 2019, and transitioning to low carbon economies could generate over 200 million jobs by 2030, especially in major emerging markets. Achieving this requires expanding renewable energy support, enhancing power reliability, transitioning from coal, scaling energy efficiency and removing fossil fuel subsidies while ensuring energy access.

26 AI and ML in Improving Operational Efficiency of Renewable Energy Systems

The efficiency of renewable energy systems may be significantly improved with the implementation of AI and machine learning technology, which optimizes numerous energy operations such as generation, distribution, and storage (Tandon et al., 2025). AI enabling predictive maintenance is one of the best examples. To lowers maintenance expenses and downtime, AI analyses sensor data fron wind turbines, solar panels, and other renewable energy assets to guess when equipment failures will happen. AI-driven grid optimization AI also enhances smart energy storage management by cutting battery charge and discharge cycles. AI also extends battery life and reduces operational expenses. Additionally, machine learning models analyse historical weather data, real-time satellite imagery, and IoT sensor inputs for more accurate energy forecasting and predicting solar and wind power generation. In energy trading, AI-powered algorithms enable real-time automated decisioning for cost-effective buying and selling of electricity on the market. These advanced technologies also greatly assist in intelligent malware prevention systems. AI plays a crucial role in balancing loads and responding to demand by distributing energy to eliminate grid straining and mitigate peaks calm system performance. AI-based fault detection systems are able to locate anomalies in energy production systems rapidly and put in place corrective actions, minimizing efficiencies losses planning to reduce costs. In conclusion, Artificial Intelligence Technology (AIT) and Machen Learning (ML) help foster a smarter, more resilient and efficient renewable energy ecosystem which can sustainably and cheaply generate renewable energy.

27 Conclusion

The transition to a sustainable and renewable energy future is both a critical necessity and an immense challenge for global society. With climate change presenting an empirical trouble, reducing reliance on fossil energies and shifting towards cleaner, more sustainable energy sources like solar, wind, and hydropower is essential. still, this transition is not invariant across regions and requires acclimatized results that consider both original and global factors. Despite the significant fiscal challenges and dislocations to traditional business models, this shift has fostered new inventions, request players, and business strategies. Technologies like solar panels, wind turbines, energy storehouse results, and smart grids are transubstantiating the energy geography, making renewable energy sources more effective, affordable, and extensively deployable. likewise, the significance of policy fabrics cannot be exaggerated. Encyclopaedically coordinated programs, similar as feed-in tariffs and renewable energy impulses, can help accelerate the relinquishment of renewable energy. Governments must work together across colourful situations, from supranational to original, to produce a terrain conducive to this transition, icing that communities, diligence, and countries are empowered to share in the energy transition. Eventually, the energy transition must be a multifaceted trouble encompassing technological invention, profitable investment, policy fabrics, and social equity. While obstacles remain, the shift towards a sustainable, zero-carbon future presents an unknown occasion to make flexible husbandry, ameliorate living norms, and cover the terrain (Benoit, 2019). Summary of Key perceptivity the significance of Energy Transition The shift from fossil energies to renewable energy sources is necessary to alleviate climate change and promote sustainable profitable growth. Technological inventions Advancements in renewable energy technologies similar as solar, wind, and storehouse results are making clean energy more accessible and effective. Policy and Regulatory Support Coordinated programs and fabrics at original, public, and global situations are essential to grease the energy transition. Economic and Social Impact The energy transition has significant counteraccusations for profitable development, job creation, and social equity, particularly in developing countries. Walls and Challenges High capital and functional costs, structure limitations, and social acceptance issues present challenges that need to be addressed for wide renewable energy deployment. Strategic Recommendations for Policymakers and Assiduity Leaders Invest in Research and Development Policymakers should allocate coffers towards developing and perfecting renewable energy technologies to reduce costs and increase effectiveness (Sokołowski & Heffron, 2022). Foster Public–Private hookups Governments and private sector realities must unite to address fiscal and technological challenges, participating knowledge and pitfalls. Apply incitement Programs similar as feed-in tariffs, duty impulses, and subventions can stimulate renewable energy relinquishment, particularly in arising husbandry. Promote Education and Awareness Raising public mindfulness of the benefits and eventuality of renewable energy will help gain social acceptance and drive community involvement insure. Just Transition Focus on creating new employment openings in the renewable energy sector to alleviate job losses in traditional energy sectors, supporting workers through training and reskilling programs. Strengthen transnational Collaboration Global challenges bear coordinated action. Countries should work together to set and achieve common renewable energy and carbon impartiality targets.

The global energy transition represents a critical shift from fossil fuels to renewable energy sources, driven by the need to achieve net zero carbon emissions and mitigate climate change. This will require serious investment, innovation, and international cooperation in overcoming obstacles to economic disruption and energy equity. With renewables and efficiency gains, countries can achieve a prosperous and ecologically viable future. However, the transition needs to be contextually different, balancing development needs with climate objectives for both developed and developing countries' benefit. Energy is not only indispensable for economic growth but also for the social development of a nation, as its demand is inextricably related to industrialization, urbanization, and rising living standards. The share of global energy consumption contributed by developing nations is growing increasingly larger, and there is a balancing act here with environmental sustainability. Transitioning to renewable and efficient energy sources would contribute to minimizing environmental impact, reduce reliance on fossil fuels, and support long-term global development. For economic prosperity and tackling environmental challenges, sustainable energy policies are of high priority.

As the growth of environmental awareness took place, especially with the TBL framework, so did the call for sustainable development become an imperative—one that necessitates balancing economic, environmental, and social factors in view. Positive features would include offsetting carbon emissions and improving community welfare through renewable energy projects; challenges regarding project complexity and local conditions. The energy transition presents some creation of employment in the renewable sectors but also spells immense opportunities related to improved access to energy. Further, it requires really careful and knowledgeable planning. It is quite finally clear, therefore, that sustainability can't be replete without this holistic merging of environmental care, social fairness, and resilience in economies.

Technological innovations in this sector are very prominent as they create a field wherein renewable sources are becoming easier and more economically viable day by day. Developments in energy storage, smart grids, and AI have optimized energy systems for far better integration of renewables, besides a remarkable enhancement in operational efficiency. This also involves innovative solutions pertaining to solar, wind, and hydro energy, which is revolutionizing the floating wind farms and tide power industry. These are seemingly setting a pathway toward sustainability and resilience in energy; hence, driving the globe toward green energy. Global policies for coordinated renewable energy targets up to 2030 need appropriate financing at parity and technology developments to ensure that every community has the benefits associated with transitioning.

Though research has concentrated much on the relationship between globalization and environmental degradation, the need for greater attention has been felt more on the aspects of energy consumption and demand, coupled with clean cooking technologies. High costs and lack of financing also remain some of the major barriers to the transition of renewable energy technologies into wide use, underpinning the need for supportive institutions to ensure effective implementation and collaboration. Transitioning into renewable energies offers some considerable environmental, economic, and social benefits: decreased emissions, improved public health, and jobs created—especially for youth and women. While the renewables' price is nearing parity, even undercutting in some areas with solar and wind in the lead, there are elements of deploying clean energy that will require financial intervention. Shifting subsidies away from fossil fuels and increasing the level of transparency in funds would definitely help accelerate the pace of change toward renewable energy, long-term sustainable growth, and just opportunities for vulnerable groups.

This highly non-unproblematic transition in global energy does call for the advance of sustainability in technologies, economies, and policy. While innovations in microgrids and renewable energy systems show promise, there are some obstacles that need to be overcome, including infrastructure limitations, financing constraints, and social equity of the process—one that affects developing nations quite often. Clean energy, its storage integrated with the transition towards decentralized energy systems, comes with opportunities and challenges that can only be judiciously managed if the benefits from them have to be equitably distributed among all communities. From the leading countries of Denmark, Uruguay, and Namibia, down to the laggards like China, India, and the United States, the energy transition is gathering great momentum. These three latter countries face specific problems that put the world in jeopardy of achieving sustainability in all aspects. A delicate balance between environmental sustainability, energy security, and economic growth should be aimed at, with equity and reduction in vulnerability to climate change impacts. While the transition to renewable energy is becoming more viable economically, it can ensure job creation and economic development—provided that regulatory barriers are dealt with and long-term stability is guaranteed.

28 Future Scope

The future scope of energy transition is bright, comprising the continuous development of renewable technologies, better energy storage facilities, and modernization of the energy grid. In all these areas, research and development are expected to yield significant improvement in efficiency and cost-effectiveness. Indeed, innovative energy storage systems and services include high-capacity batteries, along with pumped hydro storage—each one is very important in ensuring the intermittency challenge relative to using renewable energy sources is sorted out. In addition, Smart Grid integration will provide the key to optimize energy distribution and reduce energy losses while seamlessly transitioning into decentralized energy solutions. The policymaking and regulatory framework is ever-evolving, but the development of incentives, carbon pricing, and subsidies has to be further extended in support of clean technologies. International collaboration and knowledge exchange will help the entire world accelerate toward renewable energy implementation and mitigate climate change faster. By the same token, the emphasis is given to public engagement and education as part of the sustainable culture, making the communities create initiatives to support energy transition. Implemented effectively in tandem, these will lead to a sustainable, robust, and economically viable energy future.

References

Alhaddi, H. (2015). Triple bottom line and sustainability: A literature review. *Business and Management Studies, 1*(2), 6–10. https://doi.org/10.11114/bms.v1i2.752

Alnaser, W. E., & Alnaser, N. W. (2011). The status of renewable energy in the GCC countries. *Renewable and Sustainable Energy Reviews, 15*(6), 3074–3098. https://doi.org/10.1016/j.rser.2011.03.021

Awasthi, A., Charan Pattnayak, K., Dhiman, G., & Raj Tiwari, P. (Eds.). (2024). *Artificial intelligence for air quality monitoring and prediction* (1st ed.). CRC Press

Azadani, L. N. (2023). Vertical axis wind turbines in cluster configurations. *Ocean Engineering, 272*, Article 113855. https://doi.org/10.1016/j.oceaneng.2023.113855

Benoit, P. (2019). *Energy and development in a changing world: A framework for the 21st century*. SIPA Center on Global Energy Policy, Columbia University. https://www.energypolicy.columbia.edu/research/report/energy-and-development-changing-world-framework-21st-century

Caiafa, C., Hattori, T., Nam, H., & de Coninck, H. (2023). International technology innovation to accelerate energy transitions: The case of the international energy agency technology collaboration programmes. *Environmental Innovation and Societal Transitions, 48*, 100766. https://doi.org/10.1016/j.eist.2023.100766

Chen, L., Gao, L., Xing, S., Chen, Z., & Wang, W. (2024). Zero-carbon microgrid: Real-world cases, trends, challenges, and future research prospects. *Renewable and Sustainable Energy Reviews, 203*, 114720. https://doi.org/10.1016/j.rser.2024.114720

Chipangamate, N. S., & Nwaila, G. T. (2024). Assessment of challenges and strategies for driving energy transitions in emerging markets: A socio-technological systems perspective. *Energy Geoscience, 5*(2), 100257. https://doi.org/10.1016/j.engeos.2023.100257

Correia, M. S. (2019). Sustainability: An overview of the triple bottom line and sustainability implementation. *International Journal of Strategic Engineering (IJoSE), 2*(1), 29–38.

Debnath, R., Mittal, V., & Jindal, A. (2022). A review of challenges from increasing renewable generation in the Indian power sector: Way forward for electricity (Amendment) Bill 2020. *Energy & Environment, 33*(1), 3–40. https://doi.org/10.1177/0958305X20965526

Di Santo, N., Russo, I., & Sisto, R. (2022). Climate change and natural resource scarcity: A literature review on dry farming. *Land, 11*(12), 2102. https://doi.org/10.3390/land11122102

Harichandan, S., Kar, S. K., Bansal, R., Mishra, S. K., Balathanigaimani, M. S., & Dash, M. (2022). Energy transition research: A bibliometric mapping of current findings and direction for future research. *Cleaner Production Letters, 3*, 100026. https://doi.org/10.1016/j.clpl.2022.100026

Khojasteh, D., Shamsipour, A., Huang, L., Tavakoli, S., Haghani, M., Flocard, F., Farzadkhoo, M., Iglesias, G., Hemer, M.,Lewis, M., Neill, S., Bernitsas, M. M., & Glamore, W. (2023). A large-scale review of wave and tidal energy research over the last 20 years. Ocean Engineering, 282, 114995. https://doi.org/10.1016/j.oceaneng.2023.114995

Kumar, M. (2020). Social, economic, and environmental impacts of renewable energy resources. In *Wind solar hybrid renewable energy system* (Vol. 1).

Laccourreye, O., & Maisonneuve, H. (2019). French scientific medical journals confronted by developments in medical writing and the transformation of the medical press. *European Annals of Otorhinolaryngology, Head and Neck Diseases, 136*(6), 475–480. https://doi.org/10.1016/j.anorl.2019.07.005

Markard, J. (2018). The next phase of the energy transition and its implications for research and policy. *Nature Energy, 3*(8), 628–633. https://doi.org/10.1038/s41560-018-0171-7

McNamara, R. S. (1972). *Address to the United Nations Conference on the Human Environment, Stockholm,* 8 Jun 1972.

Mensah, J. (2019). Sustainable development: Meaning, history, principles, pillars, and implications for human action: Literature review. *Cogent Social Sciences, 5*(1), 1653531. https://doi.org/10.1080/23311886.2019.1653531

Ritchie, H., Roser, M., & Rosado, P. (2024, January). *Renewable energy*. Our World in Data. https://ourworldindata.org/renewable-energy

Slaria, S. (2024). *Exploring climate change impacts and natural resource scarcity*. Doctoral dissertation, University of Kansas.

Sokołowski, M. M., & Heffron, R. J. (2022). Defining and conceptualising energy policy failure: The when, where, why, and how. *Energy Policy, 161*, 112745. https://doi.org/10.1016/j.enpol.2021.112745

Tandon, A., Awasthi, A., Pattnayak, K. C., Tandon, A., Choudhury, T., & Kotecha, K. (2025). Machine learning-driven solar irradiance prediction: Advancing renewable energy in Rajasthan. *Discover Applied Sciences, 7*(2), 107.

Tseng, M. L., Chang, C. H., Lin, C. W. R., Wu, K. J., Chen, Q., Xia, L., & Xue, B. (2020). Future trends and guidance for the triple bottom line and sustainability: A data-driven bibliometric analysis. *Environmental Science and Pollution Research, 27*, 33543–33567. https://doi.org/10.1007/s11356-020-09497-1

Yu, M. (2021, March). Assessment on the environmental impact of conventional energy forms. In *IOP Conference Series: Earth and Environmental Science* (Vol. 680, No. 1, p. 012002). IOP Publishing. https://doi.org/10.1088/1755-1315/680/1/012002

Zohuri, B. (2023). Navigating the global energy landscape balancing growth, demand, and sustainability. *Journal of Materials Science and Applied Engineering, 2*(7).

Sustainable Material Selection of Metallic Bipolar Plates for Polymer Electrolyte Fuel Cells: A Pythagorean RE-CoCoSo Approach

Debashis Ghosh and B. S. Mahapatra

Abstract

Clean energy solutions are essential for addressing the most prevalent global challenges, including energy security, environmental degradation, and climate change. Utilizing metallic bipolar plates (MBPs) in polymer electrolyte fuel cells (PEFCs) proves necessary for sustainable green energy generation. However, the selection process becomes a complex and uncertain decision-making process due to the linguistic criteria used to describe the quality of MBPs. This chapter introduces an uncertain multi-criteria decision-making (MCDM) method to address this issue in a Pythagorean fuzzy context. The robust fuzzy extension Pythagorean fuzzy sets quantify the conflicting nature of the factors in MBPs, allowing for the incorporation of varying degrees of human ambiguity. The quality of the MBPs is analyzed using the Pythagorean extension of the combined compromise solution (PF-CoCoSo) approach. This study employed the rank-sum approach to ascertain the importance of the standards criteria to mitigate the decision expert's subjectivity. The proposed technique indicates that molybdenum is the optimal material for MBPs, whereas aluminium, nickel, and copper are the least suitable options.

Keywords

Sustainable material selection • MCDM • CoCoSo • Pythagorean fuzzy set

D. Ghosh (✉)
Department of Mathematics, School of Applied Science, UPES, Dehradun, India
e-mail: debashis.ghosh@ddn.upes.ac.in

B. S. Mahapatra
School of Applied Science and Humanities, Haldia Institute of Technology, Haldia, India

T. Choudhury et al. (eds.), *Open Access as a Knowledge Platform for Implementing the United Nations Sustainable Development Goals*, World Sustainability Series,
https://doi.org/10.1007/978-3-031-99605-4_9

Abbreviations

MBP	Metallic bipolar plate
PEFC	Polymer electrolyte fuel cell
MCDM	Multi-criteria decision-making
CoCoSo	Combined compromise solution
PF-CoCoSo	Pythagorean fuzzy combined compromise solution
IFS	Intuitionistic fuzzy set
PFS	Pythagorean fuzzy set
BWM	Best-worst method
TOPSIS	Technique for order preference by similarity to the ideal solution
MOORA	Multi-objective optimization on the basis of ratio analysis
CRITIC	CRiteria importance through inter-criteria correlation
LBWA	Level-based weight assessment
FUCOM	Full consistency method
DM	Decision maker

1 Introduction

The PEFCs represent a viable technology for sustainable energy solutions, especially as hydrogen production and storage techniques become increasingly economical and environmentally sustainable (Ye et al., 2022). These cells consist of three components: the first part contains hydrogen, the middle part contains electrolytes, and the final part contains oxygen. Hydrogen splits into protons and electrons through an electrolyte, while the proton exchange membrane facilitates the transfer of protons to oxygen, forming water. The flow of electrons against the moment generates electricity in this process. The advantages of utilizing PEFCs for electricity generation include low operating temperatures, high efficiency, environmental sustainability, and lightweight characteristics (Moreno et al., 2015). However, challenges persist regarding durability, elevated material costs, and the limited infrastructure for hydrogen storage. This study wishes to mitigate the challenges of utilizing PEFCs by analyzing their characteristics in a systematic way of decision-making.

The selection of materials for MBPs primarily involves considerations of corrosion resistance, electrical conductivity, mechanical strength, cost, availability, and environmental impact (Song et al., 2020). The materials used to construct PEFCs are stainless steel, titanium, nickel-based alloys, coating materials of different combinations, and many more. These materials have standard features that can not be characterized quantitatively from their qualitative description. The sustainability of these materials is influenced by strength, stiffness, and ductility, which directly affect their long-term performance, durability, and recyclability in fuel cell applications (De et al., 2024). High-strength materials such as stainless steel and titanium maintain structural integrity under compression in fuel cell stacks, enhancing durability. The material's durability guarantees consistent electrical conductivity and uniform dispersion of reactants. Conversely, low-cycle fatigue may lead to coating delamination, exposing the base metal to corrosive environments. Elevating the temperature of the annealing treatment enhances the fatigue lifespan of the material and induces alterations in its microstructure. Utilising cost-effective materials for fuel cell production significantly impacts longevity and sustainability. By using recovered substances, developers may reduce their reliance on costly, traditional materials that often harm ecosystems. Innovations such as waste reuse reduce manufacturing costs and advance a circular economy, which is essential for renewable energy sources. Incorporating biomaterials in fuel cell designs facilitates the development of rural areas by enabling on-site biofuel production with minimal capital investment, thereby providing a source of renewable energy.

Classical MCDM techniques cannot handle qualitative data in their present form; hence, modifying these MCDMs is necessary to handle qualitative information regarding the criteria descriptions (Mahapatra et al., 2024). Fuzzy set (Zadeh, 1965) has proven to help handle uncertain information. However, fuzzy sets cannot handle the degree of rejection of an uncertain measure. Intuitionistic fuzzy sets (IFSs) (Atanassov, 1986) are formulated to represent acceptance and rejection degrees; hence, they are considered superior options to fuzzy sets. However, IFSs are limited in representing uncertain information, as the sum of both degrees is bound to unity. In comparison with IFSs, Pythagorean fuzzy sets (PFSs) (Yager, 2013) have the liberty of representing uncertain information since they cover a broader area by both degrees. Because of this, PFSs are better at representing uncertain information with simple operations than the new fuzzy set extensions. A significant amount of research has been conducted on decision-making issues to address uncertainty through PFS, including applications in forest management model selection (Mondal et al., 2024), disease identification (Baranidharan et al., 2024), and pollution assessment (Xian et al., 2025).

The preference allocation to the attributes of an MCDM technique involves subjective and objective considerations of their quality. The subjective preference allocations are only possible when the decision expert clearly knows the credibility of the attributes (Mahapatra et al., 2025). Due to the limitation of funds for the remuneration of experts and the delay of judgement about the attributes, researchers prefer objective preference allocations to the attributes (Chang, 2024). The determination of objective weights minimises the subjectivity of the decision expert, leading to a data-centric evaluation that may conceal the actual importance of the attributes. This study uses a rank-exponent approach to blend subjective and objective methods in the weight assignment procedures. An expert in the subject evaluates attributes according to their perceived significance, subsequently employing an algebraic calculation to allocate weights by the rank-exponent method. The weight assignment procedure employing the rank-exponent approach is clear and straightforward.

The CoCoSo (Yazdani et al., 2019) method is an effective MCDM technique due to its flexibility, computational simplicity, and ability to manage trade-offs and compromise solutions efficiently. The CoCoSo techniques are inadequate for managing qualitative descriptions of attributes and allocating preferences to those attributes (Ghosh et al., 2024). Therefore, it is essential to modify CoCoSo techniques in two aspects: first, to integrate with a preference allocation procedure for weight determination and second, to extend the approach within the Pythagorean fuzzy framework to address uncertainties related to criteria descriptions. This study introduces a combined Pythagorean RE-CoCoSo model for decision-making, designed to independently address uncertain MCDM problems while offering a hybrid framework for evaluating decision quality.

This study organises the subsequent sections as follows: The literature review pertinent to the methodology and the application of MCDM in material selection is presented in Sect. 2. MBPs for polymer electric fuel cells, along with their relevant properties, are detailed in Sect. 3. The Pythagorean CoCoSo technique is detailed in Sect. 4. The following section details the selection of MBP using the proposed methodology. Section 6 provides a summary of the findings of this study.

2 Literature Review

This section assesses the advancement of the CoCoSo method through imprecise settings and the utilization of MCDM in the selection of diverse materials.

2.1 Development of Fuzzy CoCoSo Model

The classical CoCoSo approach cannot compute qualitative criteria information in an MCDM problem. Hence, researchers develop the CoCoSo approach in fuzzy settings to incorporate the descriptions of the qualitative criteria. Ecer and Pamucar (2020) proposed an integrated fuzzy Best Worst Method (BWM)-CoCoSo method using the Bonferroni mean for supplier selection. This approach modified the Bonferroni mean through normalization and geometric weight accumulation. Ulutaş et al. (2020) extended the CoCoSo method in a fuzzy environment for location selection of logistic centres, where they used the SStep-Wise Weight Assessment Ratio Analysis to determine the criteria preference. The CoCoSo method is enhanced in q-rung ortho-pair fuzzy environments for evaluating financial risks (Peng & Huang, 2020) and in Pythagorean settings for assessing the 5G industries (Peng et al., 2020). Both CoCoSo enhancements compute the criteria weights using the CRiteria Importance Through Inter-criteria Correlation (CRITIC) approach. Peng et al. (2021) proposed a modified CoCoSo approach using an interval-valued fuzzy soft set for healthcare management, where the CRITIC approach computes the criteria weights. Deveci et al. (2021) assessed traffic management systems through the fuzzy CoCoSo approach using the Heronian mean operator. The Pythagorean extension of the CoCoSo approach is combined with the Pythagorean analytic hierarchy process for a successful circular supply chain management system (Lahane & Kant, 2021). Torkayesh et al. (2021) proposed a hybrid decision-making model by combining the CoCoSo approach with the BWM and level-based weight assessment (LBWA) for healthcare facility assessment. Demir et al. (2022) combined fuzzy Full Consistency Method (FUCOM) and fuzzy CoCoSo approaches to solve transportation problems in urban areas. Pamucar and Görçün (2022) integrated LBWA and fuzzy extension of the CoCoSo approaches using the Bonferroni mean to identify the efficient container seaport. Qiyas et al. (2022) suggested a picture fuzzy extension of the CoCoSo approach for identifying effective vaccines to manage COVID-19's symptoms. Lai et al. (2022) suggested a hesitant Fermatean fuzzy CoCoSo approach to develop group decision-making and applied it to blockchain platform selection. Ogundoyin and Kamil (2023) combined the fuzzification of BWM, LBWA, and CoCoSo methods to determine the preference of criteria of fog gateways. Fernandez-Portillo et al. (2023) identified suitable strategies for organic farming using a combination of fuzzy BWM and CoCoSo approaches. Rasoanaivo et al. (2024) identified the limitations of the classical CoCoSo method, proposed an improved version of the CoCoSo method, and established its applicability. Chatterjee and Chakraborty (2024) extended the CoCoSo approach in five different fuzzy contexts, including Pythagorean, to optimize the drilling performance of aluminium metal. Dhruva et al. (2024) integrated logarithmic percentage change-driven objective weighting and CoCoSo methods in fermatean fuzzy for cloud vendor selection. Choosing a site for electric vehicle charging stations is investigated through the extended CRITIC-CoCoSo approach in the spherical fuzzy environment (Xian et al., 2025). The combination of failure mode and effects analysis with the fermatean fuzzy CoCoSo approach is applied to determine the risk analysis of liquid natural gas leakage from storage tanks (Yan et al., 2024).

The literature review reveals several research gaps regarding the extension of the CoCoSo approach in various fuzzy environments, its integration with weight determination procedures, and its application across different decision-making fields.

1. The extension of the CoCoSo approach to various fuzzy sets is underway. Nevertheless, the current fuzzy CoCoSo approaches cannot aggregate fuzziness from various sources, maintain a simple calculation, and represent uncertainty throughout the computation process.
2. The CoCoSo approach is required to incorporate a weight determination procedure for the criteria, as it cannot determine the weight of the criteria. Therefore, it is imperative to incorporate a weight determination technique into the CoCoSo approach to make the combined approach self-sufficient. Numerous researchers attempted to integrate the CoCoSo approach with the weight determination procedure; however, they could not demonstrate the combined approach's dependability, durability, and applicability.
3. The fuzzy CoCoSo approach, which is currently in use, practically penetrated every aspect of ambiguous decision-making. However, to our knowledge, the

fuzzy modification of the CoCoSo technique was not implemented in the material selection of MBPs.

2.2 Application of MCDM in Material Selection

The use of MCDM in material selection entails systematically evaluating and ranking materials according to various conflicting criteria, including cost, strength, weight, sustainability, and longevity. MCDM methods are essential in engineering and design, as selecting the optimal material significantly impacts a product's or structure's functionality, productivity, and economic sustainability. Shanian and Savadago (2006) employed the TOPSIS approach for the material selection problem of PEFCs. Santiago et al. (2019) employed four MCDM approaches to select the thermoplastic polymer used as MBPs for methanol fuel cells. Talebi-Ghadikolaee et al. (2024) employed MCDM to ascertain the appropriate process element values for rubber pad making. Raja et al. (2024) used WSM methods for MBP selection for PEFC in electric vehicles. Khatir et al. (2024) used MOORA and TOPSIS to ascertain the optimal mixture of maximum filling depth and minimal thickness dispersion.

Researchers have not extensively investigated MCDM approaches, which provide substantial potential for analyzing the quality of MBPs used in PEFCs via systematically evaluating their contradictory qualitative qualities. Fuzzy MCDM approaches may effectively represent the MBP selection process as an uncertain decision-making procedure by appropriately quantifying the credibility of the MBP's attributes. Therefore, it might be advantageous for the industry to use sophisticated MCDM methodologies for MBP selection.

3 Description of Criteria and Alternatives

The choice of sustainable materials for MBPs in PEFCs is essential for enhancing performance and durability while minimising environmental impact. MBPs play a crucial role in fuel cells, acting as pathways for gas flow and as current collectors. The subsequent sections describe the criteria and possible materials for MBPs.

3.1 Criteria of Materials for MBPs

The choice of material for PEFC is influenced by various considerations, including electrical conductivity, thermal conductivity, operational operator or durability, strength and stiffness, corrosion resistance, efficiency and longevity, permeability, surface characteristics, cost and sustainability. The following are important considerations for choosing a sustainable PEFC material:

Thermal Conductivity: Thermal conductivity plays a significant role in ensuring the efficient operation of the fuel cell by managing heat distribution and dissipation. High thermal conductivity materials ensure adequate heat transfer to integrated cooling channels or external systems. Ensures the fuel cell stack does not get too hot, which can diminish its performance and lifespan.

Electrical Conductivity: High electrical conductivity is critical for efficient power generation. While metals generally exhibit high conductivity, surface treatments, such as coatings of gold, platinum, or conductive carbides, can further optimize this property.

Operational operator: Bipolar plates are essential elements in fuel cells, enabling gas transfer, distributing electrical current, and offering structural support. Lightweight materials, such as composites or thin metal plates, are essential for portable and automobile applications. This criterion identifies how various materials and techniques affect the environmental analysis of the production processes' carbon footprint and the bipolar plates' recyclability at the end of their life.

Strength and stiffness: Strength and stiffness are essential mechanical attributes for evaluating materials for MBP applications like fuel cells. High-strength materials reduce the likelihood of cracking or failure over the plate's operational lifetime. Pressure-induced deformation can modify the geometry of flow channels, impacting gas distribution and cooling efficiency. In portable and lightweight fuel cells, elevated strength and stiffness facilitate the use of thinner plates, hence diminishing weight and enhancing energy efficiency.

Corrosion Resistance: The MBPs must endure harsh conditions, including humid and acidic environments. High corrosion resistance ensures a longer lifespan and reduces maintenance. Metals like stainless steel and titanium have good corrosion resistance, but surface treatments (e.g., coatings) are often applied to enhance their performance further.

Efficiency and longevity: The efficacy and durability of bipolar plates are essential in choosing suitable materials and designs for applications like fuel cells or electrolyzers. The resistance to oxidation and other chemical processes prevents the plates from degrading or releasing impurities that could damage the fuel cell. Effective thermal management ensures a consistent temperature across the cell. That leads to the electrochemical reaction rate and mitigates the formation of hot spots.

Surface characteristics: Less expensive metals like stainless steel are often coated with conductive and corrosion-resistant materials like graphite, gold, or chromium-based to balance performance and cost. These coatings improve durability and conductivity without compromising the material's affordability.

Permeability: The material must withstand compression in the fuel cell stack, along with thermal and mechanical stresses. Stainless steel and titanium offer excellent strength and formability, making them ideal for MBPs.

Cost: Sustainable materials should be economically viable and widely available. While metals like titanium perform well, they are expensive. Cheaper alternatives like stainless steel are often preferred to enhance their performance, especially when treated with protective coatings.

Sustainability: The material's sustainability depends on its environmental footprint, including energy consumption during production, recyclability, and potential for reuse. Stainless steel is highly recyclable, making it a more sustainable option. On the other hand, materials like titanium, though effective, require energy-intensive processing, which may raise environmental concerns.

3.2 Materials for MBPs

1. **Graphite**
   Advantages: Graphite is relatively lightweight compared to metals, which is advantageous for portable and automotive fuel cell applications.
   Challenges: Fabricating graphite plates involves energy-intensive procedures and difficulties acquiring essential components, affecting their sustainability.
2. **Titanium**
   Advantages: Excellent corrosion resistance and lightweight.
   Challenges: High cost and energy-intensive manufacturing process, though it can be mitigated with protective coatings.
3. **Aluminium-Nickel-Copper Alloys**
   Advantages: Strong corrosion resistance and high conductivity.
   Challenges: It is expensive, and like titanium, surface treatment may be necessary.
4. **Stainless Steel (SS316L or SS304)**
   Advantages: It is readily available, cost-effective, has good corrosion resistance when coated, and is recyclable.
   Challenges: Requires surface treatments to improve conductivity and further enhance corrosion resistance.
5. **Nickel**
   Advantages: Nickel offers robust mechanical integrity and anti-wear properties, prolonging the surfaces' operational endurance.
   Challenges: Nickel is denser than other materials, such as carbon materials, which may affect scalability and performance for specific purposes.
6. **Boron-Containing Carbon Film**
   Advantages: Gold, platinum, or carbides (e.g., TiC, CrC) to improve conductivity and durability.
   Challenges: Coatings can be resource-intensive but prolong the life of cheaper and more sustainable base materials.
7. **Molybdenum**

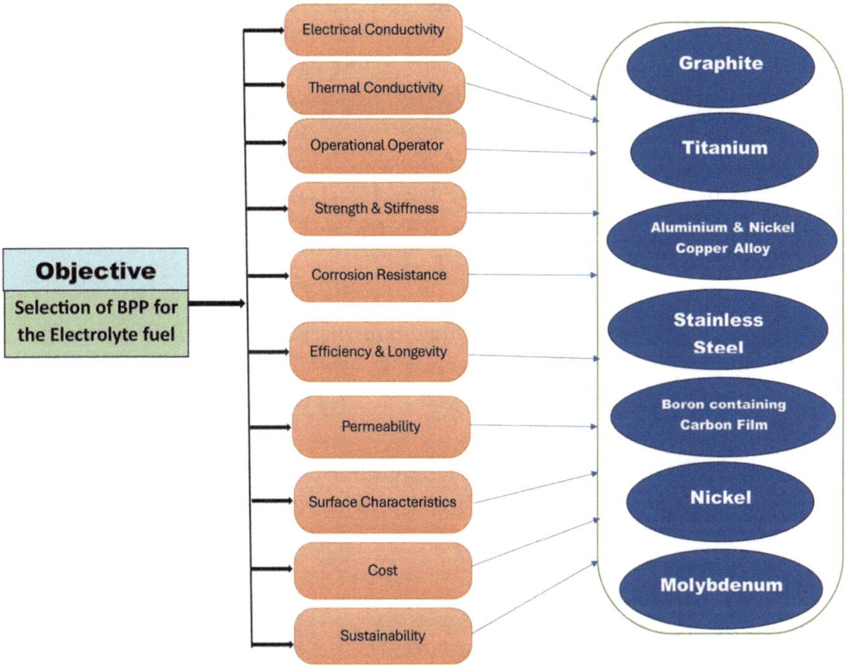

Fig. 1 Character modelling of a MBP decision-making challenges: Alternatives and criteria

Advantages: Molybdenum provides substantial benefits for exceptionally well and resilient bipolar plates, particularly in mechanisms performing under harsh temperatures.

Challenges: In fuel cell settings, molybdenum hydrides, formed when molybdenum reacts with hydrogen at high enough temperatures, might affect mechanical qualities and performance.

Considering the advantages and disadvantages of metallic and alloy architectures, it is essential to identify the most sustainable alternative. Figure 1 illustrates the important factors for selecting the optimal option.

4 Proposed Pythagorean CoCoSo Approach

The primary objective of this section is to offer a comprehensive decision-making methodology that employs the rank-exponent and CoCoSo techniques in a Pythagorean context to address ambiguous MCDM issues.

4.1 Preliminaries of Pythagorean Fuzzy Sets

This section introduces the discussion on the basic concept of PFS and the necessary arithmetic operations on PFS for this article.

Definition 1 (Devici et al., 2021) *Intuitionistic Fuzzy Set: The IFS defined on a set* $S \neq \phi$ *is an expression given by* $\tilde{I} = \{s, \iota_{\tilde{I}}(s), \kappa_{\tilde{I}}(s) : s \in \mathcal{S}\}$, *where* $\iota_{\tilde{I}}(s) : S \to [0, 1]$, $\kappa_{\tilde{I}}(s) : S \to [0, 1]$, *where* $0 \leq (\iota_{\tilde{P}}) + (\kappa_{\tilde{P}}) \leq 1 \forall s \in \mathcal{S}$. *The grade of indeterminacy of* $\tilde{I}$ *is* $\chi_{\tilde{I}} = 1 - (\iota_{\tilde{I}}) - (\kappa_{\tilde{I}})$.

Definition 2 (Dhruva et al., 2024) *Pythagorean fuzzy set (PFS): A PFS is defined as* $\tilde{P} = \{u, \iota_{\tilde{P}}(u), \kappa_{\tilde{P}}(u) : u \in \mathcal{U}\}$, *on a set* $X \neq \emptyset$, *where* $0 \leq (\iota_{\tilde{P}})^2 + (\kappa_{\tilde{P}})^2 \leq 1 \forall u \in \mathcal{U}$. *The grade of indeterminacy of* $\tilde{P}$ *is* $\chi_{\tilde{P}} = \sqrt{1 - (\iota_{\tilde{P}})^2 - (\kappa_{\tilde{P}})^2}$.

Definition 3 (Zhou et al., 2020) *Basic operations on PFS: Let the three PFS be* $\tilde{P} = (\iota_{\tilde{P}}, \kappa_{\tilde{P}})$, $\tilde{P}_1 = (\iota_{\tilde{P}_1}, \kappa_{\tilde{P}_1})$, *and* $\tilde{P}_2 = (\iota_{\tilde{P}_2}, \kappa_{\tilde{P}_2})$ *and the operations on them are as follows*:

- $\tilde{P}^c = (\kappa_{\tilde{P}}, \iota_{\tilde{P}})$
- $\tilde{P}_1 \oplus \tilde{P}_2 = \left(\sqrt{\iota_{\tilde{P}_1}^2 + \iota_{\tilde{P}_2}^2 - \iota_{\tilde{P}_1}^2 \iota_{\tilde{P}_2}^2}, \kappa_{\tilde{P}_1}\kappa_{\tilde{P}_2})\right)$
- $\tilde{P}_1 \otimes \tilde{P}_2 = \left(\iota_{\tilde{P}_1}\iota_{\tilde{P}_2}, \sqrt{\kappa_{\tilde{P}_1}^2 + \kappa_{\tilde{P}_2}^2 - \kappa_{\tilde{P}_1}^2 \kappa_{\tilde{P}_2}^2}\right)$
- $\tilde{P}^\theta = \left(\iota_{\tilde{P}}^\theta, \sqrt{1 - \left(1 - \kappa_{\tilde{P}}^2\right)^\theta}\right)$

4.2 Description of the Proposed Approach

Each phase of the proposed technique is elucidated as follows:

Step-1: Formulation of MCDM problem: Let the MCDM problem have $'p'$ options $\mathrm{O} = \{\mathrm{O}_1, \mathrm{O}_2, \ldots, \mathrm{O}_p\}$ and $'q'$ criteria $\mathrm{T} = \{\mathrm{T}_1, \mathrm{T}_2, \ldots, \mathrm{T}_q\}$, where beneficiary and non-beneficiary criteria are denoted by T_b and T_{nb} respectively, such that $\mathrm{T}_b \cup \mathrm{T}_{nb} = \mathrm{T}$ and $\mathrm{T}_b \cap \mathrm{T}_{nb} = \emptyset$.

Step-2: Pythagorean fuzzy decision matrix: Gather DM's opinions to form a Pythagorean fuzzy decision matrix $\beth = (\mathrm{a}_{ij})_{p \times q}$, such that $\mathrm{a}_{ij} = (\varphi_{ij}, \varpi_{ij}, \varrho_{ij})$, $i = 1(1)p; j = 1(1)q$. The evaluation of the ith possibility against the jth objective is denoted by a_{ij}, determined by the Pythagorean fuzzy linguistic measure shown in Table 1.

Preference of the alternative: Determine the primary preference based on the alternate choice $\mathrm{P}_{\mathrm{O}_i}$. Since the DM views the options as having an equal chance

Table 1 Criteria assessment scale

Linguistic terms	PFN (φ, ϖ)
Worst case acceptable (WCA)	(0.181, 0.913)
Very poorly acceptable (VPA)	(0.326, 0.798)
Poorly acceptable (PoA)	(0.423, 0.671)
Acceptable (Acc)	(0.538, 0.513)
Fairly acceptable (FaA)	(0.696, 0.423)
Moderately acceptable (MoA)	(0.823, 0.327)
Exceptionally acceptable (ExA)	(0.972, 0.163)

of occurring:

$$\mathrm{P}_{\mathrm{O}_i} = \frac{1}{p} \text{such that} \sum_{i=1}^{p} \mathrm{P}_{\mathrm{O}_\mathrm{i}} = 1.$$

Rank-exponent approach to determine the criteria weight: Follow these procedures for establishing the criteria weight using the rank-exponent method:

Utilize the decision matrix $\beth$ from Table 1 to organize the criteria based on the preferences indicated by the DM. The Eq. (1) delineates the formula for the weights of the criterion.

$$\omega_j = \frac{(n - r_j + 1)^p}{\sum_j (n - r_j + 1)^p} \tag{1}$$

n denotes the number of criteria, r_j indicates the rank of the jth criteria according to the DM's preference, and p is the process parameter describing the weight. The value of the parameter p lies between 0 and 1.

Normalization of Pythagorean decision matrix: Normalize the Pythagorean decision matrix (]) using Eq. (2).

$$r_{ij} = \begin{cases} \frac{x_{ij} - \min\limits_j x_{ij}}{\max\limits_j x_{ij} - \min\limits_j x_{ij}} \; for\, j \in \mathrm{T}_b \\ \frac{\max\limits_j x_{ij} - x_{ij}}{\max\limits_j x_{ij} - \min\limits_j x_{ij}} \; for\, j \in \mathrm{T}_{nb} \end{cases} \tag{2}$$

Real evaluation of the alternatives: Determine the accurate evaluation of the Pythagorean weighted sum measure (WSM) (L_i) and the weighted product measure (WPM) (M_i) corresponding to performance indexes using Eq. (3).

$$\mathrm{L}_i = \sum_{j=1}^{q} \omega_j \otimes r_{ij} \;\; and \;\; \mathrm{M}_i = \sum_{j=1}^{q} r_{ij}^{\omega_j} \tag{3}$$

Pythagorean appraisal score: Calculate the Pythagorean appraisal scores for the alternatives using Eq. (4).

$$\begin{cases} \kappa_{i\alpha} = \frac{L_i + M_i}{\sum_{i=1}^{m} (L_i + M_i)} \\ \kappa_{i\beta} = \frac{L_i}{\min_i (L_i)} + \frac{M_i}{\min_i (M_i)} \\ \kappa_{i\gamma} = \frac{\lambda L_i + (1-\lambda) M_i}{\lambda \max_i (L_i) + (1-\lambda) \max_i (M_i)}, \quad \lambda \in (0, 1) \end{cases} \tag{4}$$

Overall appraisal score: Compute overall appraisal score κ_i of each alternatives using Eq. (5).

$$\kappa_i = \left(\kappa_{i\alpha}\kappa_{i\beta}\kappa_{i\gamma}\right)^{1/3} + \frac{1}{3}\left(\kappa_{i\alpha} + \kappa_{i\beta} + \kappa_{i\gamma}\right), i = 1(1)p \tag{5}$$

Ranking order: Determine the ranking of each choice in decreasing order depending on the κ_i values. Figure 2 describes the MBP selection problem by a sequential structure.

5 Numerical Simulation of MBP Selection

To illustrate the proposed decision-making model, materials such as graphite (O_1), titanium (O_2), aluminium, nickel, copper (O_3), stainless steel (O_4), nickel (O_5), boron-containing carbon film (O_6), and molybdenum (O_7) are selected as options for the MBP.

The MCDM framework of the MBP selection has the set of alternatives and criteria as follows: $O = \{O_1, O_2, O_3, O_4, O_5, O_6, O_7\}$ and $T = \{T_1, T_2, T_3, T_4, T_5, T_6, T_7, T_8, T_9, T_{10}\}$ such that $T_b = \{T_1, T_2, T_3, T_4, T_5, T_6, T_7, T_8\}$ and $T_{nb} = \{T_9, T_{10}\}$. Table 2 provides the construction of the Pythagorean decision matrix $(\beth)_{7\times 10}$ in accordance with the Pythagorean fuzzy number (PFN) rating from Table 1.

Since this MCDM issue of for MBP have seven alternatives, $P_{O_i} = \frac{1}{7}$ such that $\sum_{i=1}^{7} P_{O_i} = 1$.

Proposed Pythagorean RE method for criteria weight computation:

The Pythagorean decision matrix of Table 2 is generated using the criteria rating from Table 1. The rank of the criteria for determining weights is provided in Table 3.

The criteria weights are evaluated by the Eq. (1) and are given in Table 4.

The normalized Pythagorean decision matrix of the decision matrix (Table 2) is computed using Eq. (2) and is displayed in Table 5.

The Pythagorean WSM and WPM are computed using Eq. (3). Table 6 provides Pythagorean WSM and Pythagorean WPM.

The comparative importance of the options is assessed by computing the appraisal score using Eq. (4), as shown in Table 7. The overall appraisal scores for the options are calculated using Eq. (5) and presented in Table 7.

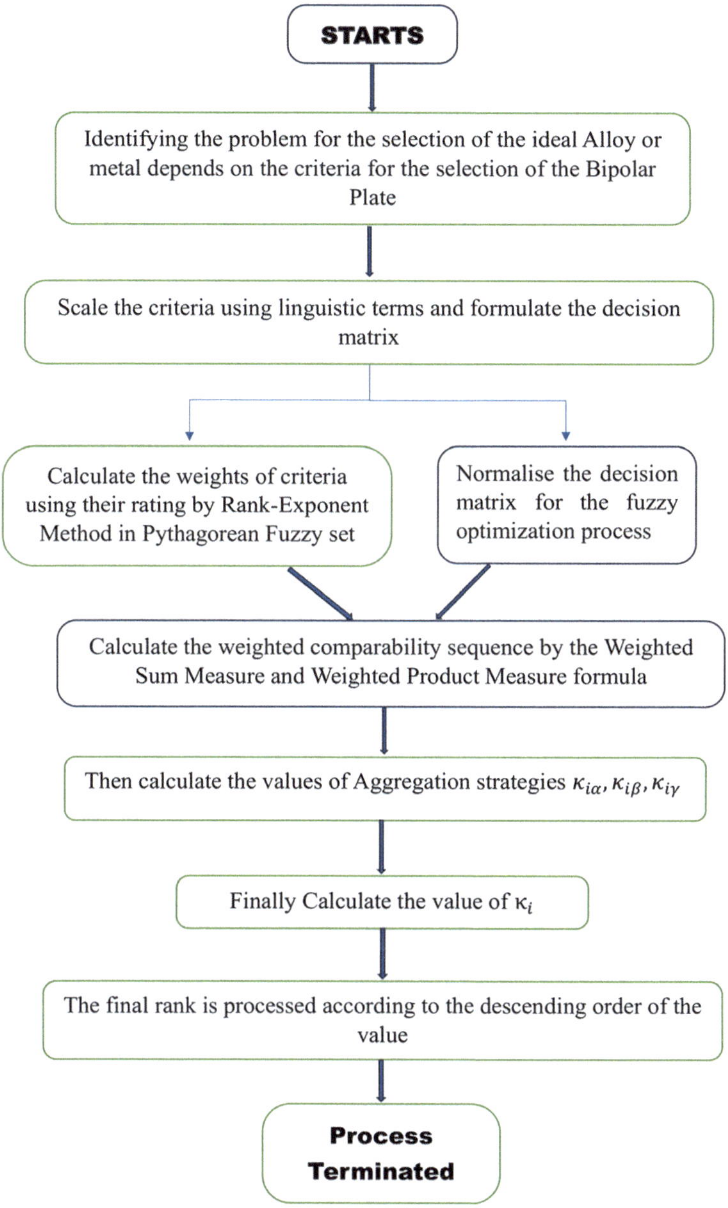

Fig. 2 MBP selection through the suggested PF-CoCoSo hierarchical framework

Table 2 Linguistic Pythagorean decision matrix

Alt./Cr	T_1	T_2	T_3	T_4	T_5	T_6	T_7	T_8	T_9	T_{10}
O_1	Acc	VPA	Acc	VPA	FaA	ExA	VPA	MoA	Acc	PoA
O_2	FaA	Acc	FaA	Acc	ExA	MoA	Acc	Acc	VPA	PoA
O_3	MoA	Acc	MoA	Acc	PoA	Acc	FaA	PoA	FaA	Acc
O_4	MoA	FaA	MoA	FaA	Acc	FaA	Acc	MoA	Acc	Acc
O_5	Acc	MoA	Acc	MoA	MoA	FaA	PoA	PoA	MoA	FaA
O_6	MoA	ExA	MoA	ExA	Acc	PoA	VPA	FaA	WCA	VPA
O_7	ExA	MoA	ExA	MoA	VPA	Acc	MoA	MoA	VPA	WCA

Table 3 Criteria rank by DM's preference

Criteria	T_1	T_2	T_3	T_4	T_5	T_6	T_7	T_8	T_9	T_{10}
ω_i	8	6	10	7	4	9	3	5	2	1

Table 4 Criteria weights determined through rank-sum approach

Criteria	T_1	T_2	T_3	T_4	T_5	T_6	T_7	T_8	T_9	T_{10}
ω_i	0.087	0.101	0.0623	0.094	0.112	0.077	0.116	0.107	0.120	0.124

For standardization, the coefficient of the compromise decision system is indicated by $\lambda \in (0, 1)$ and is set to 0.7 to assign preference of weights for the sum measures. The decreasing order of the overall appraisal scores determines the ranking order of the MBP locations, which is $\mathcal{O}_7 \succ \mathcal{O}_2 \succ \mathcal{O}_6 \succ \mathcal{O}_4 \succ \mathcal{O}_1 \succ \mathcal{O}_3 \succ \mathcal{O}_5$. Figure 3 shows the final ranking of the MBP against their score value obtained from the PF-CoCoSo method.

Hence the ranking order of the alternatives is $\mathcal{O}_7 \succ \mathcal{O}_2 \succ \mathcal{O}_6 \succ \mathcal{O}_4 \succ \mathcal{O}_1 \succ \mathcal{O}_3 \succ \mathcal{O}_5$. The parameter q is not significantly affected by the ranking of the MBP selection, even though the overall performance scores are sensitive to variations in p.

Table 5 Normalized decision matrix according to proposed PF-CoCoSo method

Metals	T_1	T_2	T_3	T_4	T_5
O_1	(0.150, 0.800, 0.880)	(1.000, 0.000, 0.000)	(0.429, 1.000, 0.437)	(0.000, 1.000, 0.676)	(0.573, 0.409, 0.823)
O_2	(0.573, 0.409, 0.823)	(0.150, 0.643, 1.000)	(0.745, 0.743, 0.248)	(0.328, 0.551, 1.000)	(1.000, 0.000, 0.000)
O_3	(1.000, 0.000, 0.000)	(0.769, 0.258, 0.553)	(1.000, 0.469, 0.000)	(0.328, 0.551, 1.000)	(0.150, 0.800, 0.880)
O_4	(0.328, 0.551, 1.000)	(0.000, 1.000, 0.632)	(1.000, 0.469, 0.000)	(0.573, 0.409, 0.823)	(0.328, 0.551, 1.000)
O_5	(0.769, 0.258, 0.591)	(0.573, 0.409, 0.769)	(0.429, 1.000, 0.437)	(0.769, 0.258, 0.591)	(0.769, 0.258, 0.591)
O_6	(0.000, 1.000, 0.676)	(0.328, 0.551, 0.935)	(1.000, 0.469, 0.000)	(1.000, 0.000, 0.000)	(0.328, 0.551, 1.000)
O_7	(1.000, 0.000, 0.000)	(0.769, 0.258, 0.553)	(0.000, 0.000, 1.000)	(0.769, 0.258, 0.591)	(0.000, 1.000, 0.676)
Metals	T_6	T_7	T_8	T_9	T_{10}
O_1	(1.000, 0.000, 0.000)	(0.000, 1.000, 0.207)	(1.000, 0.000, 0.000)	(0.444, 0.683, 0.000)	(0.530, 0.494, 0.198)
O_2	(0.729, 0.323, 0.591)	(0.427, 0.395, 1.000)	(0.288, 0.541, 1.000)	(0.774, 0.196, 0.534)	(0.530, 0.494, 0.198)
O_3	(0.209, 0.689, 1.000)	(0.744, 0.204, 0.566)	(0.000, 1.000, 0.707)	(0.198, 0.836, 0.292)	(0.307, 0.816, 0.000)
O_4	(0.497, 0.512, 0.823)	(0.427, 0.395, 1.000)	(1.000, 0.000, 0.000)	(0.444, 0.683, 0.000)	(0.307, 0.816, 0.000)
O_5	(0.497, 0.512, 0.823)	(0.195, 0.730, 0.707)	(0.000, 1.000, 0.707)	(0.000, 1.000, 0.674)	(0.000, 1.000, 0.292)
O_6	(0.000, 1.000, 0.880)	(0.000, 1.000, 0.207)	(0.683, 0.279, 0.566)	(1.000, 0.000, 1.000)	(0.718, 0.235, 0.534)
O_7	(0.209, 0.689, 1.000)	(1.000, 0.000, 0.000)	(1.000, 0.000, 0.000)	(0.774, 0.196, 0.534)	(1.000, 0.000, 1.000)

Table 6 Pythagorean weighted sum and weighted product measure

Alternatives	O_1	O_2	O_3	O_4
WSM(i)	(0.627, 0.658, 0.480)	(0.604 ,0.460, 0.577)	(0.553, 0.660, 0.435)	(0.545, 0.606, 0.465)
WPM(i)	(2.388, 2.367, 0.551)	(2.711, 2.904, -0.041)	(2.564, 2683, 0.188)	(2.573, 2.708, 0.144)
Alternatives	O_5	O_6	O_7	
WSM(i)	(0.491, 0.729, 0.355)	(0.642, 0.611, 0.520)	(0.790, 0.408, 0.729)	
WPM(i)	(2.899, 2.160, 2.370)	(2.378, 2.394, 0.463)	(2.134, 2.798, -1.139)	

Table 7 Overall appraisal scores of the alternatives

Alt	$\kappa_{i\alpha}$	$\kappa_{i\beta}$	$\kappa_{i\gamma}(\lambda = 0.7)$	κ_i	Rank
Graphite	0.571	0.870	0.411	1.206	5
Titanium	0.570	1.661	0.320	1.523	2
Aluminium and Nickel Copper	0.466	1.061	0.295	1.134	6
Stainless steel	0.488	1.183	0.302	1.216	4
Nickel	0.744	0.008	0.785	0.680	7
Boron containing carbon film	0.596	1.058	0.411	1.326	3
Molybdenum	0.542	3.055	0.138	1.857	1

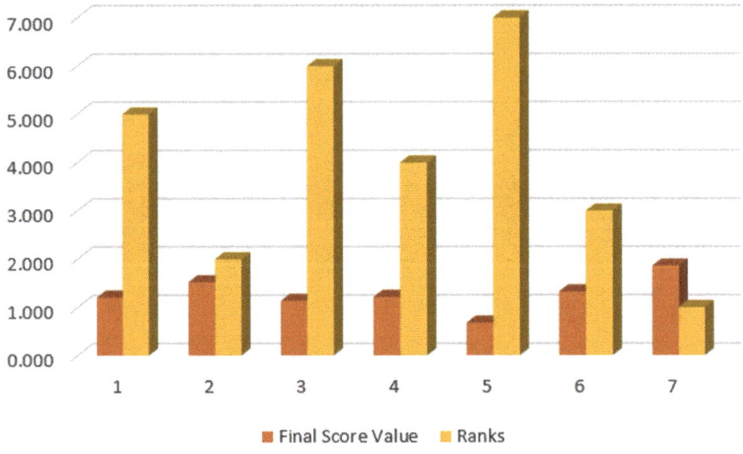

Fig. 3 Description of the rank of MBP selection

6 Conclusion

This study presents a comprehensive MCDM technique that effectively addresses uncertain MCDM issues. This study presents the following key theoretical contributions: (i) Determination of criterion weights using the rank exponent method. This contribution significantly improves the ranking of alternatives relative to current crisp and Pythagorean decision-making models. (ii) Application of the Pythagorean fuzzy weighted sum measure and the Pythagorean fuzzy weighted product measure. The measures integrate Pythagorean information during computation, leading to a detailed analysis of the criteria's credibility relative to each alternative. (iii) Extension of the integrated compromised solution within a Pythagorean framework to evaluate the alternatives.

This study examines the numerical evaluation of MBP option selection. The evaluation process comprises ten criteria and seven alternatives. The proposed PF-CoCoSo method indicates that alternative $\mathcal{O}_7$ is the most preferred option, while other $\mathcal{O}_5$ is the least preferred. The identified criteria, including electrical conductivity, thermal conductivity, operational operator, strength and stiffness, corrosion resistance, efficiency and longevity, permeability, surface characteristics, cost, and sustainability, are deemed more significant according to their order.

Future research may focus on the following aspects: (i) The data-driven weight assignment of the materials of fuel cells can be improved by applying the FUCOM or by employing simultaneous evaluation of criteria and alternatives. (ii) The alternatives of MBPs may consider conductive polymers, which offer a lightweight and environmentally friendly alternative to traditional coatings. (iii) Recent fuzzy extensions, such as hesitant bi-fuzzy, type-II fuzzy, and spherical fuzzy, are applicable for criteria evaluation. The suggested procedure can be applied to real-life MCDM challenges across various domains, including supplier selection, location selection, artificial intelligence, treatment procedures, and medical waste removal alternatives.

References

Atanassov, K. T. (1986). Intuitionistc fuzzy sets. *Fuzzy Sets and Systems, 20*(1), 87–96.

Baranidharan, B., Liu, J., Mahapatra, G. S., Mahapatra, B. S., & Srilalithambigai, R. (2024). Group decision on rationalizing disease analysis using novel distance measure on Pythagorean fuzziness. *Complex & Intelligent Systems, 10*(3), 4373–4395.

Chang, K. H. (2024). Combining subjective and objective weights considerations to solve the emergency location selection problems under spherical fuzzy environments. *Applied Soft Computing, 153*, 111272.

Chatterjee, S., & Chakraborty, S. (2024). A comparative study on combined compromise solution (CoCoSo)-based optimization of drilling of aluminium metal matrix composites in fuzzy environments. *International Journal on Interactive Design and Manufacturing (IJIDeM), 18*(7), 4737–4763.

De, A., Ramasubramian, B., Ramakrishna, S., & Chellappan, V. (2024). Advances in additive manufacturing techniques for electrochemical energy storage. *Advanced Materials Technologies, 9*(4), 2301439.

Demir, G., Damjanović, M., Matović, B., & Vujadinović, R. (2022). Toward sustainable urban mobility by using fuzzy-FUCOM and fuzzy-CoCoSo methods: The case of the SUMP podgorica. *Sustainability, 14*(9), 4972.

Deveci, M., Pamucar, D., & Gokasar, I. (2021). Fuzzy Power Heronian function based CoCoSo method for the advantage prioritization of autonomous vehicles in real-time traffic management. *Sustainable Cities and Society, 69*, 102846.

Dhruva, S., Krishankumar, R., Zavadskas, E. K., Ravichandran, K. S., & Gandomi, A. H. (2024). Selection of suitable cloud vendors for health centre: A personalized decision framework with fermatean fuzzy set, LOPCOW, and CoCoSo. *Informatica, 35*(1), 65–98.

Ecer, F., & Pamucar, D. (2020). Sustainable supplier selection: A novel integrated fuzzy best worst method (F-BWM) and fuzzy CoCoSo with Bonferroni (CoCoSo'B) multi-criteria model. *Journal of Cleaner Production, 266*, 121981.

Fernández-Portillo, L. A., Yazdani, M., Estepa-Mohedano, L., & Sisto, R. (2023). Prioritisation of strategies for the adoption of organic agriculture using BWM and fuzzy CoCoSo. *Soft Computing*, 1–32.

Ghosh, D., Muhuri, B. R., Mahapatra, G. S., & Mahapatra, B. S. (2024). Integrated MEREC-CoCoSo approach using score function ofNeutrosophic for location selection of special economic zone. *Neutrosophic Sets and Systems, 73*(1), 55.

Khatir, F. A., Barzegari, M. M., & Ghadikolaee, H. T. (2024). Optimal geometry selection for the microchannel in bipolar plates of PEM fuel cells from a multi-criteria decision analysis point of view. *International Journal on Interactive Design and Manufacturing (IJIDeM)*, 1–12.

Lahane, S., & Kant, R. (2021). A hybrid Pythagorean fuzzy AHP–CoCoSo framework to rank the performance outcomes of circular supply chain due to adoption of its enablers. *Waste Management, 130*, 48–60.

Lai, H., Liao, H., Long, Y., & Zavadskas, E. K. (2022). A hesitant Fermatean fuzzy CoCoSo method for group decision-making and an application to blockchain platform evaluation. *International Journal of Fuzzy Systems, 24*(6), 2643–2661.

Mahapatra, B. S., Bera, M. B., Mondal, M. K., Smarandache, F., & Mahapatra, G. S. (2024). Health insurance provider selection through novel correlation measure of neutrosophic sets using TOPSIS. *Contemporary Mathematics*, 4497–4522.

Mahapatra, B. S., Ghosh, D., Pamucar, D., & Mahapatra, G. S. (2025). Dynamic group decision-making for enterprise resource planning selection using two-tuples Pythagorean fuzzy MOORA approach. *Expert Systems with Applications, 263*, 125675.

Mondal, M. K., Mahapatra, B. S., Bera, M. B., & Mahapatra, G. S. (2024). Sustainable forest resources management model through Pythagorean fuzzy MEREC–MARCOS approach. *Environment, Development and Sustainability*, 1–32.

Moreno, N. G., Molina, M. C., Gervasio, D., & Robles, J. F. P. (2015). Approaches to polymer electrolyte membrane fuel cells (PEMFCs) and their cost. *Renewable and Sustainable Energy Reviews, 52*, 897–906.

Ogundoyin, S. O., & Kamil, I. A. (2023). An integrated Fuzzy-BWM, Fuzzy-LBWA and V-Fuzzy-CoCoSo-LD model for gateway selection in fog-bolstered Internet of Things. *Applied Soft Computing, 143*, 110393.

Pamucar, D., & Görçün, Ö. F. (2022). Evaluation of the European container ports using a new hybrid fuzzy LBWA-CoCoSo'B techniques. *Expert Systems with Applications, 203*, 117463.

Peng, X., & Huang, H. (2020). Fuzzy decision making method based on CoCoSo with critic for financial risk evaluation. *Technological and Economic Development of Economy, 26*(4), 695–724.

Peng, X., Zhang, X., & Luo, Z. (2020). Pythagorean fuzzy MCDM method based on CoCoSo and CRITIC with score function for 5G industry evaluation. *Artificial Intelligence Review, 53*(5), 3813–3847.

Peng, X., Krishankumar, R., & Ravichandran, K. S. (2021). A novel interval-valued fuzzy soft decision-making method based on CoCoSo and CRITIC for intelligent healthcare management evaluation. *Soft Computing, 25*, 4213–4241.

Qiyas, M., Naeem, M., Khan, S., Abdullah, S., Botmart, T., & Shah, T. (2022). Decision support system based on CoCoSo method with the picture fuzzy information. *Journal of Mathematics, 2022*(1), 1476233.

Raja, C., Ramachandran, M., Prasanth, V., & Nanjundan, P. (2024). Metallic bipolar plates for polymer electrolyte fuel cell (PEFC) used in electric vehicles using the WSM method. *Journal on Applied and Chemical Physics, 3*, 1.

Rasoanaivo, R. G., Yazdani, M., Zaraté, P., & Fateh, A. (2024). Combined Compromise for Ideal Solution (CoCoFISo): A multi-criteria decision-making based on the CoCoSo method algorithm. *Expert Systems with Applications, 251*, 124079.

Santiago, Ó., Raso, M. A., Navarro, E., & Leo, T. J. (2019). Selection of thermoplastic polymers for use as bipolar plates in direct methanol fuel cell applications. *Materials & Design, 183*, 108148.

Shanian, A., & Savadogo, O. (2006). TOPSIS multiple-criteria decision support analysis for material selection of metallic bipolar plates for polymer electrolyte fuel cell. *Journal of Power Sources, 159*(2), 1095–1104.

Song, Y., Zhang, C., Ling, C. Y., Han, M., Yong, R. Y., Sun, D., & Chen, J. (2020). Review on current research of materials, fabrication and application for bipolar plate in proton exchange membrane fuel cell. *International Journal of Hydrogen Energy, 45*(54), 29832–29847.

Talebi-Ghadikolaee, H., Modanloo, V., Elyasi, M., & Khatir, F. A. (2024). Multiple criteria decision support analysis for manufacturing process parameters selection of metallic bipolar plates for polymer electrolyte membrane fuel cells. *Proceedings of the Institution of Mechanical Engineers, Part l: Journal of Materials: Design and Applications, 238*(5), 929–942.

Torkayesh, A. E., Pamucar, D., Ecer, F., & Chatterjee, P. (2021). An integrated BWM-LBWA-CoCoSo framework for evaluation of healthcare sectors in Eastern Europe. *Socio-Economic Planning Sciences, 78*, 101052.

Ulutaş, A., Karakuş, C. B., & Topal, A. (2020). Location selection for logistics center with fuzzy SWARA and CoCoSo methods. *Journal of Intelligent & Fuzzy Systems, 38*(4), 4693–4709.

Xian, S., Xu, N., Lan, X., & Tu, F. (2025). Z-Pythagorean adjusted fuzzy cloud-asymmetric regret decision-making model for environmental pollution assessment. *Expert Systems with Applications*, 127309.

Yager, R. R. (2013, June). Pythagorean fuzzy subsets. In *2013 Joint IFSA world congress and NAFIPS annual meeting (IFSA/NAFIPS)* (pp. 57–61). IEEE.

Yan, R., Han, Y., Zhang, H., & Wei, C. (2024). Location selection of electric vehicle charging stations through employing the spherical fuzzy CoCoSo and CRITIC technique. *Informatica, 35*(1), 203–225.

Yazdani, M., Zarate, P., Kazimieras Zavadskas, E., & Turskis, Z. (2019). A combined compromise solution (CoCoSo) method for multi-criteria decision-making problems. *Management Decision, 57*(9), 2501–2519.

Ye, M., Sharp, P., Brandon, N., & Kucernak, A. (2022). System-level comparison of ammonia, compressed and liquid hydrogen as fuels for polymer electrolyte fuel cell powered shipping. *International Journal of Hydrogen Energy, 47*(13), 8565–8584.

Zadeh, L. A. (1965). Fuzzy Sets. *Information and Control, 8*(3), 338–353.

Zhou, J., Li, K. W., Baležentis, T., & Streimikiene, D. (2020). Pythagorean fuzzy combinative distance-based assessment with pure linguistic information and its application to financial strategies of multi-national companies. *Economic research-Ekonomska istraživanja, 33*(1), 974–998.

Green Technology Intervention: Its Adaptation and Sustainability in the Automobile Sector Under Sustainable Development Goals

Vidushi Sangwan, Satyam Prakash Srivastava, Rupa Khanna Malhotra, and Tanupriya Choudhary

Abstract

Green technology covers all sets of technology that are based upon all kinds of production activity which is environment-driven and friendly. It is a consistently evolving form of technology that is not under a stringent taxation system as compared to humans who are consuming resources more than in expected amount. Green technology tries to conserve the basic nature of the environment and mitigate the influences of uneven climate changes, further mitigating dependence upon renewable resources. Green technology has precedence over all modes of technology that are detrimental to the environment and dependent upon fossil fuel usage. Consumption of fuel more than anticipated can result to decrease the natural resources, keeping less inclination towards conservation. Industries like recycling industries, solar energy, transportation, and logistics are pooling exorbitantly in green innovations. It is obvious that such investments in such have become a requisite for the existence of mankind and for securing the environment which has become pruned to risky conditions. Though the country is bringing up a paradigm shift by adopting the concept of using electric vehicles, solar cells, and windmills but still the concern for technology transfer is inevitable as a lot of countries are contributing to scaling up the percentage of carbon emissions and greenhouse gases just because of improper implementation of educational policy.

V. Sangwan · S. P. Srivastava (✉) · R. K. Malhotra
Graphic Era Deemed to be University, Dehradun, Uttarakhand, India
e-mail: satyamprakash07@gmail.com

T. Choudhary
School of Computer Science, UPES, Dehradun, Uttarakhand, India

T. Choudhury et al. (eds.), *Open Access as a Knowledge Platform for Implementing the United Nations Sustainable Development Goals*, World Sustainability Series,
https://doi.org/10.1007/978-3-031-99605-4_10

Keywords

Adaptation • Green technology • Intervention • Sustainability

1 Introduction

Green technology intervention and its adaptation need to be frequently done in most of the sectors that are dependent upon types of equipment that consume a lot of fossil fuel which emits lots of carbon particles and is hazardous to human life. In due course of time, the intervention of green technology should be adopted in all kinds of methods or processes, supply chain, logistics etc. As we know there will be a dearth of Non-renewable resources in the entire world that would be depleting with due course of time, human life wound perishes due to acute of shortage these resources. As per the report of Global Footprint Network in 2018, mankind is consuming natural resource at the rate of 1.7% faster than the Earth can recoup the deficiency. The entire market of green technology is already touching new heights and the progression is observed in most of the sectors. The policies came into inception since eighteenth and nineteenth century when Industrial revolution took place. Producers tried to mitigate the poor elements existing in the surrounding in the nineteenth century by doing changes in waste by-products. Green innovation as a separate division was not planned until 1990s. Investments at the global level in green technology and renewable sources exceeded beyond $200 billion in the year 2017, as per the data released by United Nations which got published in 2018. $3 trillion has been invested in the sources such as solar energy and wind energy since 2004 followed by China that did the investment of $127 billion reported by United Nations.

1.1 Benefits of Green Technology

Green technology is envisaged as major driver for climate controls and green energy transition, but technology won't be adequate to handle issues of carbon emissions which are predominant in society at large. The governance needs to adopt mechanism which can furnish solutions that are based upon the nature and could be acting as integral part of reducing the decarbonisation plans.

i. **Electric Vehicle**: Around one-third of the carbon emissions are discharged by transportation activities as per the Environmental Protection Agency. Particulate matter of small size emitted which have serious effect on the health of the individuals (Roy et al., 2024). Most of automobile companies from foreign natives to regional level have switched from fuel-dependent engines to hybrid and electric transmissions though the basic requisite infrastructure such as charging stations, replacement of batteries or rechargeable batteries need to be set up that can ease most of the automobile companies to be decisive with strong intent to participative in the welfare of the society. The companies have limited choices to escalate the prices of the vehicle due to shortage of the lithium component required for backing up the battery. The basic infra and footprints should be viable enough that cater the major chunk of the population that is looking forward to giving their contribution to the national and social cause.

ii. **Sustainable farming**: Most of the agricultural produce are still dependent upon conventional methods and cropping the field in every season is moreover a cumbersome task for the farmers as they have to dependent upon redundant techniques. It needs to updated with techniques where biological or animal waste and should be less pruned to usage of hazardous chemical. Moreover, farming should be based more on organic nature which enrich and retain the productivity of the soil for long.

iii. **Solar panels**: within the span of last 5-year companies most of companies are venturing in solar panels to set up at places where there can be stored and further disseminated in the electric grids. The government has taken several steps to lend all kinds of support systems that can be beneficial for companies to take full leverage of the subsidies to set the infrastructure and set up at unproductive places apart from the residential areas. The bureaucracy through multiple schemes like Production Linked Incentive Scheme, viability gap funding, Grid-connected Rooftop Solar Scheme, PM Surya Ghar: Muft Bijli Yojana, IREDA Financing Schemes, International Solar Alliance (ISA) provides abundant opportunities and options to initiate ventures in such areas with the motive to scale up reservoir of energy saving and limiting the consumption of fuel that releases exhaustive carbon elements. As per the report, around 70% of the energy comes from solar and wind in the current scenario. Small solar panels may supply electricity as compared to large panels on the terrace can significantly cut a family's energy costs. Passive panels are more detrimental as compared to solar concentrators which may spoil the life of birds and natural habitants residing at that place. Pooling investments in solar panels is a cost concern for companies to venture. Places where the sun radiations are not adequate then wind turbines can be beneficial.

iv. **Upcycling**: This is also known as upcycling waste management wherein most of the waste is transformed into usable stuff or newly created that has not been so famous in the market. In most of the companies, certain ways have been found to upcycle the existing waste that are occurring in from fuel, fertilizer or attires and cycles. To protect the scarce resources, it is expected that one should make eco-friendly bags and avoid the usage of plastic, glass or metal that takes many years to dissolve in soil or difficult to be recycled. Further auto parts should be reusable to recover expensive material from e-waste or automobile parts.

v. **Capturing Carbon Elements**: It involves engaging technologies that can capture carbon elements at the point of combustion or directly from the atmosphere. Such technologies are usually engaged by the fossil fuel industries. The Mammoth plant in Iceland, the world's largest carbon capture facility can absorb up to 36,000 tons of carbon dioxide annually. Industries are looking forward to avoiding supply chain interruptions to mitigate cost and follow legitimate regulatory system. Further industries are shifting towards new technologies and implementing strategies that will stay ecofriendly and fulfil the aspirations of a technology-driven economy.

vi. **Efficient Logistics**: Vehicles engaged in logistics services are now inclined to make a switch over to electric vehicles to ensure all kinds of transportation activities in an efficient way. As per the report of Proceedings of the National Academy of Sciences reported around 37% of the products are accountable for contributing carbon emissions. Electric vehicles can deliver better mileage within the proximity of the city area. Cities, towns, and villages can engage electric vehicle for routine nature of deliveries.

vii. **Prediction of demand**: Now the company can approximately project what can be demand projections through the assistance of artificial intelligence. In the post COVID-19 due to market unstable conditions the demand used to remain unfulfilled and organisations used to bear hefty losses as the supplies were stucked at most places and it was threatening for the survivors as well. Knowing what users are looking forward so that material can be delivered as and when it was required.

viii. **Optimizing the navigation**: Usually the companies are curious to deliver the goods and services as and when the supply is probable to get short. When the company is dragged behind the planned schedule, then the drivers try to navigate the shortest route to deliver the product at the defined places. This saves lot of fuel from wastage and ultimately the energy is saved. Through the assistance of green technology navigation tools can assist in deciding the optimum route and logistics activities. The companies can choose the most ideal route for delivering things at the customer's doorstep by taking orders in bulk amount increasing the amount of sales and revenue for the company and mitigating the amount of fuel uses in the heavy vehicle.

Figure 1 shows sustainable route optimization through green technology. The figure representation shows how the optimization eases just-in-time management, routing for full delivery area and no sub districts in between, powered by an algorithm, saving up to 20% as compared to standard route planning, lower carbon footprints by minimum kilometers driven, and individual solutions for same-day delivery.

ix. **Cost-effective**: Electric Vehicles secure the environment and follow regulatory norms, and they also benefit most companies working at the bottom level. They are envisaged to be cost advantage as they some parts for moving material from one place to another and comparatively cheaper to labour cost as well.

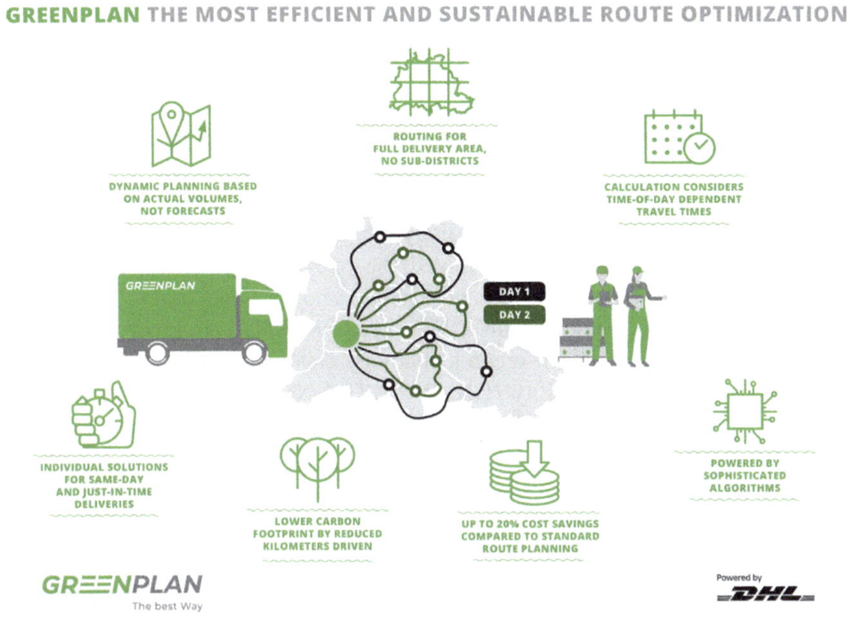

Fig. 1 Sustainable route optimization. *Source* https://bolt.earth/blog/indian-ev-charging-infrastructure-by-2030

According to CNN report, Ford has estimated that EV will be taking 30% less hours of labour than conventional gasoline-powered car.

x. **Regulatory framework**: Besides keeping a sustainable business purpose and being cost-effective for the automobile sector, governments across the world are framing rules and regulations to follow for designing vehicles that must be eco-friendly and fuel efficient. For example, the European Green Deal aims to transforming EU into a competitive economy by 2050, which must be backed by an ample number of resources with the least greenhouse emissions.

xi. **Other Sources of Fuel**: With the evolution of green technology and stringent rules and regulations for securing the environment, truck manufacturers are looking forward to design and develop electric vehicles that must be capable of being used inside the factory area for loading and unloading materials or which are operable through renewable energy. This will help in saving substantial amounts of fuel when these trucks are idle consuming 1500 gallons of fuel in 1800 h every year, according to US Department of Energy. Switching from alternative fuel will overcome the issue of overall cost and wipe additional burden of fuel cost.

2 Literature Review

In recent times, scholars, policy developers and practitioners have warmed up to the idea of integrating green technologies into sustainable development and its benefits. This review departs from others in that a cross-section of significant scholarship on the costs and benefits of the transfer of green technology for sustainable development is provided. Several works done in several regions of the world have indicated how green technology are very important to attain the SDGs. The latest report of the IPCC underlined the relevance of renewable technologies in relation to problems of climatic changes and the reduction of greenhouse emissions. Moreover, studies by the UNEP i.e. United Nations Environment Programme show that adoption of green technology enhances both economic growth and environmental sustainability (UNEP, 2017). Several methods of technology transfer have been studied to enhance the adoption and spread of green technologies. It has been found that Public–Private Partnerships (PPPs) are an appropriate vehicle to implement as a strategy in increasing the uptake of clean energy technology to address climate change across communities, industries, and nations (CEM, n.d.). In the same framework, the US government-initiated technology transfer programs. Transfer of international programs and licensing of technology and its agreement are very important for innovation of environmentally friendly techniques and technology (UNICEF, n.d.; WIPO, 2020). Not ignoring the advantages that might come with this, many disadvantages stay on this side of bridge which include not only the transfer but implementation of green technology (IRENA, 2019; Kedia and Sood). The difference in geographical features and issues of society and culture can be a subject of hindrance, especially area of small infrastructure and little resources like North east India Green technology is important and not avoidable. It elaborates the process of goods production with a small degree of negative environmental impact. In India and other developing nations, green technology has emerged as a key idea and is seen as a necessary strategy to support sustainable development. Which results in, producers being more environmentally conscious and create eco-friendly goods, which could lead to a rise in their market share and a boost in the company's reputation for being environmentally conscious. To communicate the green worth of their goods and services, marketers today.

Moreover, they do not hesitate to use the term 'green' to refer to ecofriendly, recyclable and other natural products. In the last few decades, the shift in consumers purchasing power and demand towards green products has been tremendous. Also, it has made the consumers eager to know more about the novelty of green products. This is an opportunity for the businesses to take advantage of and publicly 'demonstrate' the superiority and environmental safety of their goods over similar goods available on the market. Current marketers have acknowledged the importance of environmental management as a marketing strategy to minimize ecological concerns and widen the market scope. Our society has become more complex aggravating competition because of business irregularities as well as environmental conservation. And for consumers as well as companies, it is important to sustain the environment. Therefore, change takes place in the purchasing pattern of the person. Consumers tend to replace traditional products by more ecologically safe ones. Such social "problems" that have appeared because of this change are now dealt with a corporate organization. Today's businesses consider that simply introducing a green technology strategy is not sufficient to succeed in fierce competition. On the other hand, sustainability requires trackable management over a long period.

One of the most widely used models for analyzing the acceptance and use of technology is the Technology Acceptance Model, which has been adapted over time to suit different situations, that includes of green technologies. In essence, TAM holds that a person's intention to adapt to the technology is dependent upon perceived usefulness and perceived ease of use of that technology, and hence, how such a technology is used in practice.

For instance, it is reasonable that adoption of green technologies will be slow as risk and costs are perceived to be high, yet these innovations enhance the environment. By observing these factors TAM becomes more enabling, especially in determining the ways through which firms harness green innovations such as green AI energy management innovations or green supply chain management.

Elevating green technologies' extended adoption and integration requires an understanding of how they might improve operational effectiveness and sustainability. The two main approaches to accomplishing green innovation are minimising waste and the use of environmentally friendly and hazardous resources. In addition, it has been shown that deliberately investing in modern facilities and green technologies increases organisational efficiency. By fostering the growth of environmental abilities, these active measures also facilitate organisations by eliminating environmental risks. Additionally, there is currently no globally recognised definition of the broad, multidisciplinary term "green technology," which is constantly evolving.

The creation and implementation of green technologies that lessen environmental harm throughout the production, usage, and consumption of products is the primary objective of green innovation. These developments can monitor, govern, and prevent ecological harm before it occurs. Conservation of energy, decreased emissions, proper management of adverse environmental effects, and eventually, sustainability improvement may all be assisted by innovation in green technologies. As it is challenging to accomplish with the present nongreen technology, we can only provide realistic affirmation for the commencement of sustainable development and certainly achieve green expansion by boosting green innovation improvements. Hottenrott et al. define green technology deployment of innovations intended for reducing carbon emissions. This may involve both decreasing the use of fossil fuels as well as establishing additional GHG mitigation measures.

This additionally encompasses green inventions that emphasise on energy and the preservation of resources. Green hydrogen is a carbon-neutral fuel that can substantially decrease emissions of greenhouse gases, is crucial to the fight against global warming. An essential part of green technology initiatives, the production of electricity from renewable resources, including solar and wind may speed up the transition to renewable sources of energy systems. Morocco may enhance its current renewable energy efforts and accomplish its environmental goals by incorporating green hydrogen. Due to the OECD (Organisation for Economic Co-operation and Development), "green/sustainable growth" has been defined by the World Bank as the objective of economic expansion and advancement while ensuring that it does not negatively impact the ecological community or the long-term sustainability of renewable resources. To reduce the effect on the ecosystem involves expanding efficiency, introducing new technologies and bringing legislations into effect that enhance conservation of resource. This can be achieved through promoting investment and innovation, which will maintain perpetual growth and extend fresh possibilities for business. Considering advanced green technology lowers greenhouse gas emissions, optimises resource utilisation, and stimulates eco-innovation, it is crucial for fostering sustainable investment. Since green technology encourages cleaner manufacturing methods and achieves national and international sustainability objectives, its deployment is vital to preventing ecological damage. Green technology offers an approach to equitable development and growth as it preserves socioeconomic security as countries strive to achieve the Sustainable Development Goals (SDGs), particularly those pertaining to resource conservation and climate action. Renewable sources of energy sources which include solar, wind, hydropower, and geothermal power might supply a significant amount of the world's energy needs without damaging the environment. Implementation of green technology, would help in conservation of energy in renewable form and would encourage sustainable business practices to attain better resilience in environment and bring more economic sustainability. By minimising the impact on the environment and the depletion of supplies, these technologies contribute to developing a future that is more sustainable. Since green energy innovation can ease to shift from petrol and coal usage towards renewable energy, it is important for sustainable growth. Nevertheless, there are lot of hindrances to all kind of innovation in terms finances, regulations or public coordination. Beside envisaging all these hindrances green energy has significant opportunities for future growth in the economy and employment potential. Green consumption of energy may additionally guarantee the planet's continued existence and reduce the impacts of global warming. Several studies have been done to define "green growth" because, due to its growing acceptance and popularity, the notion continues to struggle with the absence of an obvious conceptual framework. Particularly, the OECD characterized green potential in 2009 as an approach that fosters future opportunities for growth potential and addresses challenges such as biodiversity loss, global warming, and environmental impact while assuring sustainable resource consumption. This description prompted the establishment of a measurement approach that included numerous green growth indicators. This study is particularly significant as it shows how economic strength and sustainability may coexist together. By demonstrating how green technology investment techniques can result in a "win-win" situation that enhances both corporate performance and surroundings, this study reinforces the idea that long-term investment in green technologies may be an essential element of international competition and economic growth. Adding to this, Shen et al. looked thoroughly into how environmental pollution in the

G-7 economies was affecting green investment, growth financially and economically. He stumbles on that green investments largely decrease pollution through enhancing manufacturing processes, encouraging advancements in technology, and ameliorating energy conservation. This discovery is like that of Antonietti and Marzucchi, which describes those green expenditures facilitates results of export and export Increases company productivity. It is seen that there is a correlation between environmental pollution and both positive and negative shocks to green investment in the G-7 context, underlining the importance that the dynamism in investments affects environmental quality. Green finance specially outlines the financial mechanisms needed to support environmentally conscious operations. When it comes to sustainable investment looks at a variety of ESG concerns. Green funding has notably impacted the sector of renewable energy by fostering to the transition to clean energy sources and advancing energy efficiency, as described by Nepal et al. A change like this also changes and increases the renewable energy system's subsistence, which makes it an important part of sustainable yield. According to Perry, to complete its climate change (NDC) goals, Morocco needs implementation policies more moderate strategies to decrease the quantity of carbon in the energy sector and continue to work towards a successful transition to a sustainable economy. To grant a consistent supply of energy and sustainable growth, it also needs to boost its investments in these energies. Employing of policies including direct regulation, subsidies, and environmental levies to moderate the gloom-ridden effects of global warming. Implementing direct regulation, exemption, and environmental taxes to reduce the impacts caused by global warming. There exist contradictory studies that argue the case for environmental taxes allows enhancement to the environment while improving how resources are utilized which fosters green growth. There are managers who focus on sustainability and tend to do better as they are perceived as poised to gain from putting money into green technology that helps in improving energy efficiency, minimizing pollution, and even recycling waste materials (Asadi et al., 2020; Chen et al., 2006). The dire need for these technologies is further emphasized with the current environmental situation that requires more commitment from the corporate world to help meet the sustainable development goals. Entering contracts to invest into green technologies emerges as a response in order to meet such targets while also fostering sustainable development and promoting economic revivals. These technology approaches are essential for achieving many environmental goals and improving human quality of life. AI powered Indoor Environmental Quality (IEQ) systems makes a big impact towards causing behavioral changes that positively affect productivity and health.

Sustainable ordinances are crucial in giving motivation to the companies to pool in such technology, which is a very crucial and a pinnacle strategy for resolving a variety of environmental hazards. These limits in regulation often serve as stimulants, encouraging businesses to cling to the broader sustainable growth objectives set by policymakers. By providing rules, offering funding, and enacting comprehensive legislation, governments create favourable conditions for these investments. As a powerful instrument for promoting investments in green technologies, it is essential for encouraging the adoption of green advances. By raising the price of non-sustainable resources, this fiscal strategy incentivizes businesses to adopt to the practices that are environmentally friendly. Henceforth acts as a guiding light for them to make or prefer more environmentally friendly choices.

Following recent studies, green technologies reduce anthropogenic emissions to enhance environmental sustainability, which makes them an essential component of sustainable development practices that are typically associated with decreasing environmental degradation (Pei et al., 2019). Based on an evaluation of advances in technology industrial economies by Bashir et al. (2024), breakthroughs in technology are crucial for mitigating ecological pressure during instances characterised by rapid industrial and economic transformation. Beise and Rennings (2003) evaluated the significant technologies which are environment conducive and would assist in minimising environment depletion and determined that they support environmental cleanup initiatives on the same premise. Mäkitie et al. (2023) retained this connection, claiming that green environmental technologies contribute to minimising energy consumption and enhance industrial production processes, which will, in return, decrease environmental externalities gradually. The findings of Huang et al. (2022), who reviewed crucial industrial emissions in China's industrial industry, advances in green technology infrastructure have been beneficial to the solution of environmental problems. To demonstrate that creative and sustainable innovations or technologies minimise environmental and ecological challenges, Rout et al. (2022) examined how innovations in technology could potentially be used to address the challenges of urban environmental sustainability in BRICS economies. In their investigation of the significance of green technologies in environmental performance, Haldar and Sethi (2022) highlighted that for emerging developing nations to make progress towards sustainable development goals, technological incorporation must be accompanied by ecological policy formulation.

By minimising emissions, reducing environmental degradation, and boosting atmospheric capacity, green technologies may help enhance air quality. Experts like Carrión-Flores and Innes claimed that GET may reduce air pollution while also improving environmental conditions. In accordance with Zhou et al., countries that utilise greater amount amounts of eco-friendly technology and sustainable power are also experiencing lower levels of sulphur dioxide (SO_2) and nitrogen oxide (NO_x). While Gao et al. (2022) emphasised GET's function in mitigating Airborne particulate pollution, Shi et al. (2021) similarly established a definitive link between GET adoption and diminished carbon emissions. Nevertheless, an array of studies indicates that while technological advances enhance productivity, they may also adversely boost the consumption of energy and environmental impact. Schmidt and Sewerin (2019) demonstrated that environmental pollution may rise over time regardless of advancements in technologies and that technological advancements rarely translate towards cleaner production. Adoption of GET certainly impacts energy consumption as well as the environment, whether positively or negatively.

In academic research and legislative debates, the significance of energy from renewable sources to mitigate the adverse effects of environmental degradation has become the forefront. Recent studies have demonstrated the relevance of eco-friendly energy alternatives to mitigate the harmful consequences of environmental damage (Chen, 2023). The complex procedure of converting to renewable energy relies on striking a balance between traditional and green energy sources to eliminate fossil fuels gradually. Evidence that is presently available implies that since renewable energy potential expands, global economies are going to be able to sustain advanced economic reform, optimise energy efficiency, and limit ecological damage (Chen et al. (2021) reinforced these assertions by noting that non-EU economies are utilising leveraging renewable resources to encourage sustainable future, breakthroughs due to a stronger perceptual and legal framework. The growth hypothesis was investigated by Bhattacharya et al. and Shafiei and Salim (2014) to explain whether policymakers might favour fossil fuels to facilitate swift industrialisation, which is hurting RE in short-term energy policies. On the other hand, in the words of Koçak and Şarkgüneşi (2017), RE is essential for achieving a carbon-neutral industrial reform, and assigning energy transition rather than economic-driven output permits RE sources to gradually eradicate traditional energy market dependency and overcome variations in prices of fossil fuel (Antimiani et al., 2023; Bashir et al., 2024).

However, striking an approach to a more environmentally friendly future is not without its difficulties. The complex nature of this evolution is illustrated by Ahmad et al., who emphasise that even though the eventual goal is apparent, the path is challenging due to economic variables, infrastructure challenges, and every factor from technological developments to policy drawbacks. Every component requires thorough preparation and execution. Similarly, Koçak and Şarkgüneşi (2017) observed that although the economic impacts of REN adoption are complicated, the justification for such an evolution goes beyond environmental issues. On the contrary, in situations that promote its basic adoption, REN has demonstrated itself to be more affordable. However, Valente et al. (2019) claimed that supplying renewable energy is comparable to financing an obstacle against the financial implications of climate change. Global trends and systemic stresses also additionally impact the discourse in the wider context. Energy demands will probably increase as the international population grows. The energy sector has gained greater scrutiny as an outcome of this boost in energy appetite (Rehman et al., 2022). However, Menyah and Wolde-Rufael (2010) highlight that the increasing overreliance on carbon-based energy sources promotes harmful carbon compounds driving environmental deterioration.

2.1 The Current Environmental Crisis in the World

Now, the environment is at dire straits and is facing numerous complex issues. Pollution, loss of resources, decline of biodiversity, and climate change are all interrelated problems, and all these call for changes in structure and policies.

Because of climate change, the world's temperature is ever-increasing, giving way to extreme weather, droughts, flooding, and rising sea levels, which have serious repercussions for ecosystems and the people and economies around the globe.

It is, however, pertinent to note that evaporation of glaciers and ice caps, thinning of the ozone layer, expansion of deserts etc. are but the tip of the iceberg. Warming of the oceans and the atmosphere have affected the weather across the globe also making phenomena like monsoons, winter temperatures and hurricane intensity increasingly difficult to predict. While now 8.5% of the oceans are covered by coral reefs, estimates show that if climate instability persists in the future, that percentage will decrease to almost 0%.

Speaking of which, an estimated 6 billion individuals have unduly consumed natural resources in excess of the Earth's regenerative capacity, which is what resulted to the depletion of resources which in this case happens to be aquatic resources including animals, nuts, algae, oil, etc. When that occurs, entire species can become extinct as a result of ecosystem disruption.

However, Aqua Farming is a sustainable growth model for the future as it regards farming aquatic creatures such as fish, crustaceans, mollusks, and seaweed as an agricultural activity as it can help avoid global warming, enhance air quality, and expand coastal lands among a myriad of other benefits.

3 Link Between Sustainability, Adaptation, and Green Technologies

The escalating concerns regarding climate change, environmental degradation, and resource depletion necessitate innovative solutions that harmonize environmental conservation with economic prosperity. In all these the solutions rest with the intersection of sustainability, adaptation, and green technologies. A comprehensive understanding of the interrelationships among these three concepts is essential for fostering a sustainable future.

3.1 The Interconnection of Sustainability, Adaptation, and Green Technologies

There is a positive and mutually reinforcing dynamic of inter-dependencies between sustainability, adaptation and green technologies. Green technologies are critical for achieving sustainability in different sectors and for adapting to changes in the environment.

Climate change is a reality and green technologies will aid as mode of adaptation. With the dynamism in climate conditions, adaptation will be a critical option. There is more to green technologies than reducing greenhouse emissions; they are also good to have in case the impacts of climate change occur. The following shows the use of green technologies in adaptation to climate change impact:

i. **Resilient Infrastructure**: Green technologies are vital in the development of infrastructure to withstand the effects of climate change. For example, permeable pavements and green roofs can mitigate urban flooding in areas that are prone to it by absorbing excess moisture. Communities threatened by sea level rise can use sea walls along with natural buffers such as restored mangroves and built structures that are elevated to prevent flooding and storm surges.

ii. **Water Management**: Water is becoming increasingly scarce as climate change-driven drought is on the rise, green technologies have the potential to mitigate this by increasing water supply in stressed areas. Other solutions include rainwater harvesting systems and desalination, which aim to produce drinkable water, To see material impacts or benefits from these solutions. Also, the adoption of improved irrigation methods such as drip irrigation and smart irrigation systems enhances the efficiency of water use. Developments in Irrigation Technology. Topics that deal with water management in agriculture are diverse, ranging from irrigation techniques to more complex water supply systems. This first one deals with simple methods such as irrigation scheduling, irrigation compact, drip irrigation and many more.

4 The Role of Green Technologies in Tackling Environmental Challenges

Green technologies do play a decisive role in response to some of the most crucial environmental challenges that the world is currently facing. These innovations bid solutions that are effective which can remediate environmental damage at the same time promoting sustainability, while tackling concerns such as climate change, pollution, depletion of resources and loss of biodiversity.

(a) **Diminishing Greenhouse Gas Emissions and Mitigating Climate Change**: One of the most complex and significant challenges of environment that is confronting humanity is the climate change. The prime driver of global warming, which is responsible for surging temperatures, heightened sea levels, extreme weather events frequently, and interruptions to the biodiversity and the ecosystems, industrial burning of fossil fuels for energy purposes and transportation. In the effort to lower the emissions of the greenhouse gases green technologies are very crucial. The key contributions include:

   i. **Renewable Resource Technologies**: The usage of energies such as hydro power, solar, wind and geothermal consequently reduces dependencies on the fossil fuels. These types of technologies do not radiate any greenhouse gases or carbon dioxide (CO_2) during the course of electricity generation.

   ii. **Electric Vehicles (EVs)**: Internal combustion engines powered vehicles (ICEVs) produces a considerable volume of CO_2, making the transport sector a significant contributor of emissions globally. In contrast, the EVs offers more cleaner alternative, which do not emit pollutants of the tailpipe and function on electricity which are stored in the batteries. The environmental perks of Electric Vehicles are anticipated to expand as the electrical framework increasingly integrates energy sources which are renewable. As per the study by International Energy Agency (IEA), the transportation related emissions can nearly be reduced by 50% by 2050 with the adoption of EVs.

(b) **Storing Energy Technologies**: Solution to energy storage, like batteries, are crucial and essential for maintaining reliable supply of energy, particularly sources of renewable energies like wind and solar have become much more prevalent.

   i. **Water Pollution Reduction**: Green technologies play a very vital role in the implementation of improving management of water practices and water pollution mitigating.

   ii. **Technologies for the Treatment of Wastewater**: Advanced methods such as the membrane filtration, the biological filtration, and the green infrastructure are crucial for wastewater safe rendering and for environmental discharge by eliminating substances which are harmful effectively. Natural process, such as constructed wetlands or wetland for floating treatment, are there to enhance the quality of water through the treatment of wastewater.

   iii. **Techniques of Zero-Waste and Circular Economy**: The adoption of the principles of the circular economy, which includes the recycling of wastewater and reducing of chemical runoffs, can substantially diminish the water pollution in industrial as well as the agricultural sectors both. The Closed-loop systems of water facilitate the treatment, the reuse, and the reintegration of the water with minimal contamination in the ecosystem.

(c) **Conservation of Natural Resources**:

   i. **Depleting natural resources**: It includes fertile soil, minerals, fossil fuels, and water, poses a compelling risk to the sustainability of human development. Green technologies are crucial for the preservation of these resources, which are finite, as they reinforce productivity and enable the reuse and recycling of materials.

   ii. **Energy Efficiency Technologies**: Green technologies commit to improvement in energy efficiency across different sectors, which includees construction, agriculture, manufacturing, and transportationn. Many iinnovations such as smart thermostats, LED lighting, high-performing insulation, and appliances which are energy-efficient reduce the consumption of energy. Therefore, this will lead to decreased dependencies on energy which are based on fossil fuel, thereby diminishing resource depletion and emissions.

   iii. **Technologies for Preserving Water**: Worldwide many of the regions are experiencing a downturn in freshwater resources which is due to increased pollution, overly extracting, and climatic changes. The Water-conserving technologies which are green, such as harvesting rain water systems, agricultural drip irrigation, and smart meters of water for the detection of leak, are very crucial for maximizing the efficiency of water.

5 Significance of the Study

Due to the existing environmental barriers and the need to achieve sustainable development goals, it is critical to appreciate the opportunities and challenges that exist with the adoption and diffusion of green technologies. As such, it shows the potential of green technologies help in solving current problems experienced in the entire world such as dynamism in climatic conditions, deficiency of resource and acute issues of pollution. This to help further clarify what their role is, by focusing on mitigation of GHG emissions, conservation of natural resources and cleaner production processes, To what extent do green technologies improve cleaner production processes and how does this affect the overall goal of sustainable development. Herein lies the significance of the study as it contributes to the existing scholarship on the interrelationship between green technologies, adaptation and sustainable development. This includes the understanding of cross border (i.e. between developed and developing countries) green technologies diffusion.

6 The Potential Influence of Green Technologies on Sustainable Development

By focusing on how these green technologies are expected to impact the end goal of sustainable development by focusing on environmental, social, and economic metrics.

i. **Environmental Influence**: Green technologies have the potential of significantly reducing the overall environmental burden through the reduction in greenhouse gas emissions, conservation of natural resources, and lowering pollution. As such, solar and wind farms have shown their potential to reduce the reliance on fossil fuels and also reduce carbon emissions. Innovations like energy-efficient devices and the use of sustainable building materials are equally important in reducing energy consumption and minimizing the generation of waste (UNEP, 2017).
ii. **Social Influence**: Introduction of green technologies can also have enjoyed social impact by improving the provision of essential services such as clean water, health, and education. For example, integrated water purification units powered by solar energies enhance rural communities' hydrations whilst averting possibilities of water borne diseases and improving general health. Further, projects that tend to promote "sustainable practices in agriculture" also contribute to food availability and income adequacy especially in developing countries.
iii. **Economic Influence**: Green technologies offer economic opportunities in employment, creativity and reducing costs. Outlays on renewable energy and energy efficiency efforts lead to employment in industries like manufacturing, construction, and research (IRENA, 2019). In addition, the application of green

technologies can provide relief on energy expenses and boost service delivery in both businesses and government for years to come.

Green technologies have the potential to employable vast amounts of resources in effort to achieving sustainable development goals, improving the state of the environment, social relations and economic growth.

However, it can be noted that the recognition of its promotion is likely to influence the civil society innovations and other works being carried out for improvement and support of government is greatly needed and is welcome.

7 Green Technology Model

See Fig. 2.

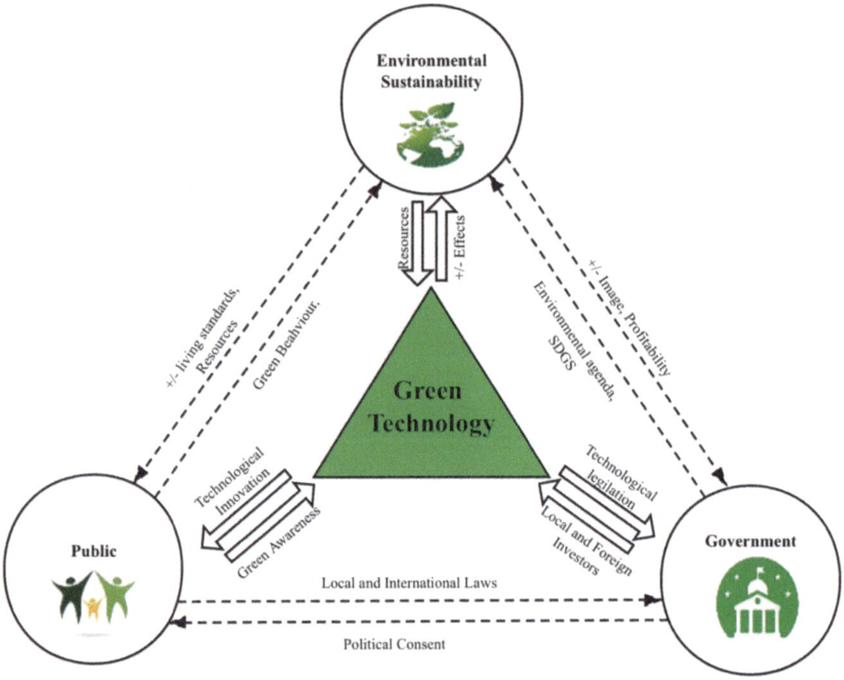

Fig. 2 Framework of integrated green technology flows

8 Role of Education in the Implementation of Green Technology

As crucial as education is for implementing green technology, it also fosters awareness about the environment, necessary skill development, also promotion and innovation technology, In the long as expected by electric vehicles (EVs) a significant reduce fossil fuel consumption by transitioning to cleaner, renewable energy sources.

i. **Fostering Change and Guiding Towards Sustainability**: Education plays a vital role in promoting environmental awareness and sustainable efforts focused on individual and community levels.
ii. **Creating Skills and Talents**: With proper education, students can be trained to design, develop, and implement green technologies like renewable systems, sustainable transport, and waste management.
iii. **Encouraging Social Change and Entrepreneurship**: As economies grow and more jobs are made available in the green sector, education can help create a culture where individuals invent and develop new technologies.
iv. **Creating Social Awareness and Responsible Consumption**: Education can enable individuals to comprehend their environmental footprint and encourage them to adopt sound ecological decision-making patterns.
v. **Training and Fostering Future Leaders**: By educating future leaders on green technology and environmental issues, decision-makers are empowered during climate change.

9 Effect of Electric Vehicle on Consumption of Fossil Fuel

Electric vehicles use less oil, coal, and natural gas because they are powered by electricity not powered fuelled engines.

i. **Lower Greenhouse Gas Emissions**: EVs have zero tailpipe emissions, implying they do not emit harmful emissions such as nitrogen oxides and particulate matter, which translates into cleaner air and lower greenhouse gas emissions.
ii. **Transition to Renewable Energy**: As the use of EVs increases, there will be increased demand for electricity, which will be fulfilled by renewable sources such as solar and wind power, further decreasing the use of fossil fuels.
iii. **Long-Term Environmental Benefits**: The shift towards EVs can result in an environmentally cleaner and more sustainable transportation system with profound long-term environmental advantages, such as less air pollution, reduced carbon footprints, and a healthier climate.
iv. **Economic Opportunities**: The EV market is anticipated to generate new economic prospects and employment opportunities in the manufacturing, infrastructure, and other sectors.

10 Green Technology Intervention Its Adaptation and Sustainability in the Automobile Sector

The sole focus is on electric vehicles and giving alternative fuels, and sustainable manufacturing, is adapting and driving sustainability in the automotive sector, reducing environmental impact and promoting eco-friendly mobility.

i. **Sustainable Manufacturing**: As the automotive industry evolves, a symphony of innovation is underway. The quest for sustainability is driving a harmonious convergence of technology, design, and environmental stewardship.
ii. **Light as a Feather**: Materials like aluminium and carbon fibre are being used to craft vehicles that are not only agile but also fuel-efficient. The result is a weight reduction, which translates to a lighter carbon footprint.
iii. **Recycling Renaissance**: The industry is embracing the art of recycling, where plastics, fabrics, and metals are given a new life. This approach reduces waste and conserves resources, creating a more circular economy.
iv. **Energy Efficiency**: Production methods are being optimized to minimize energy consumption, while resource usage is being fine-tuned to reduce waste. The outcome is a more streamlined and environmentally friendly manufacturing process.
v. **Waste Reduction**: The focus is on minimizing production waste and developing end-of-life vehicle recycling programs. This forward-thinking approach ensures that the industry is not only reducing its ecological impact but also creating new opportunities for growth.
vi. **Smart Mobility**: The integration of technology is revolutionizing the way we move. Connected cars, autonomous driving, and data analytics are working in tandem to create a more efficient, safe, and sustainable transportation ecosystem.
vii. **Industry-Wide Initiatives**: Government regulations, industry collaboration, and consumer demand are driving the adoption of sustainable practices. The result is a collective effort to reduce emissions, promote green technologies, and create a better future for generations to come.

11 Challenges and Opportunities

While there are challenges to overcome, such as cost and infrastructure, the opportunities are vast. The transition to green technologies is creating new business opportunities in areas like battery technology, charging infrastructure, and sustainable manufacturing.

12 Role of Government Policies in the Promotion and Adoption of Green Technology in Industries

For the past few years, the impact of sustainable energy has found significant attention as the world now is in the boiling pot of the challenges of climate change and a crucial need for reducing emissions of greenhouse gases. The Government plays an important role in advancing sustainable energy by controlling policies that sensitise the adoption and development of renewable and sustainable energy sources. The key aspect of this article is to explore the role that government policies play in promoting sustainable energy and underline some positive initiatives that have given it an exemplary start.

i. **Establishing Renewable Energy Targets**: One of the most important methods through which governments encourage sustainable energy is by establishing renewable energy targets. These goals are then used as measures for the percentage of energy derived from renewable resources within a short span period. What these targets do is it establish government insight and a visible picture and set goal towards shifting to renewable energy.
ii. **Financial Incentives and Subsidies**: Governments can actuate the use of sustainable energy through financial subsidies and incentives. These are done to make renewable energy technologies more affordable and competitive with conventional fossil fuel-based resources. Subsidies may be in the form of tax relief, feed-in tariffs, or grants that can cover the initial cost of adopting renewable energy systems.
iii. **Regulatory Frameworks and Power Purchase Agreements**: Government's successful policies need to be supported by improved regulatory environments to promote a level field for competition and stimulate investment in renewable energy projects. Governments can set up power purchase agreements (PPAs) that commit to the purchase of renewable energy at a predetermined price for a fixed period. Such certainty allows long-term planning by investors and minimizes risks, drawing in private sector investment.
iv. **Investments in Research and Development**: Governments can encourage innovation and technological progress in sustainable energy through strategic investments in research and development (R&D). Governments can help fund R&D programs, thereby promoting the invention of new technologies, the development of existing ones, and cost reduction for renewable energy production.
v. **Capacity Building and Awareness**: Encouraging sustainable energy also entails educating and sensitizing citizens, industry players, and policymakers. Governments can hold training sessions, workshops, and public campaigns to raise awareness about the advantages of sustainable energy, energy efficiency, and conservation. By empowering individuals and organizations, governments promote a culture of sustainability and inspire action towards cleaner energy options.

vi. **International Cooperation and Agreements**: They can cooperate on the global level by engaging in international efforts and agreements that address climate change and promote sustainable energy. Some examples are the Paris Climate Agreement, in which the global warming limit is pursued, and the United Nations Sustainable Development Goals, that underscore the significance of sustainable energy for social, economic, and environmental well-being.

Government policies have a crucial role in advancing sustainable energy at the global and national levels. Through target setting, financial incentives, developing enabling regulatory environment, research investments, capacity development, and active engagement in global cooperation, governments can speed up the shift to a cleaner, more sustainable energy future. It is only through these efforts combined that we can combat climate change, curtail reliance on fossil fuels, and create a healthier and sustainable world for the next generation.

13 Regulations and Incentives Can Drive Businesses to Implement Eco-Friendly Practices

In the business mindset of things, a very subtle shift is being worked on. Regulations and incentives are weaving a pattern of sustainability, where eco-friendly practices are no longer a 'nicety but a necessity'. These policies are innovations that are taking us to a new era of the environment for decades to come.

14 The Rhythm of Regulation

Governments are setting the standards with environmental handling that leave an echo through the corridors of industry. Emission of gases, efficiency of energy, management of waste, and resource usage are all being fine-tuned to create a symphony of sustainability. Stringent regulations and penalties for non-compliance are motivating businesses to adopt eco-friendly practices, lest they face the discordant notes of costly fines and reputational damage.

i. **The Incentive Melody**: Tax credits, subsidies, and grants are being offered to businesses that invest in renewable energy, energy-efficient technologies, and sustainable infrastructure. Green procurement is creating a market demand for sustainable solutions, encouraging businesses to harmonize their practices with the environment. Carbon pricing mechanisms are internalizing the environmental costs of carbon emissions, incentivizing businesses to reduce their carbon footprint.

ii. **Benefits of Harmony**: Sustainable practices are leading to reduced energy consumption, lower waste disposal costs, and more efficient resource use. This, in turn, is resulting in significant cost savings. Businesses that prioritize sustainability are improving their brand image, attracting environmentally conscious customers and investors, and gaining a competitive advantage. The market is shifting towards sustainable products and services, creating new opportunities for businesses that can meet these demands.
iii. **The Talent Attraction**: Employees are increasingly drawn to companies that prioritize sustainability, making it easier to attract and retain top talent. This is creating a virtuous cycle of innovation and environmental stewardship, where businesses are not only reducing their ecological footprint but also fostering a culture of sustainability. Sustainability is no longer just a moral imperative, but a sound business strategy. By embracing eco-friendly practices, companies can reap a range of benefits that improve their bottom line, enhance their reputation, and drive growth.
iv. **Cost Reduction**: Sustainable practices can lead to significant cost savings. By reducing energy consumption, lowering waste disposal costs, and using resources more efficiently, businesses can cut their operational expenses and improve their profitability. For example, installing solar panels or investing in energy-efficient lighting can reduce energy costs, while implementing recycling programs can minimize waste disposal costs.
v. **Enhanced Reputation**: Businesses that prioritize sustainability can improve their brand image and attract environmentally conscious customers and investors. In today's socially responsible market, consumers are increasingly looking for companies that share their values and are committed to making a positive impact. By embracing sustainability, businesses can build trust, loyalty, and a positive reputation that drives long-term growth.
vi. **Competitive Advantage**: Companies that adopt eco-friendly practices can gain a competitive advantage by offering innovative and sustainable solutions. Whether it's developing sustainable products, reducing carbon emissions, or implementing environmentally friendly supply chains, businesses that prioritize sustainability can differentiate themselves from their competitors and establish a leadership position in their industry.
vii. **Access to New Markets**: Consumers are increasingly demanding sustainable products and services, creating new market opportunities for businesses that can meet these demands. From sustainable fashion to eco-friendly home goods, the market for sustainable products is growing rapidly. Businesses that can adapt to these changing consumer preferences can tap into new markets and drive growth.
viii. **Attracting and Retaining Talent**: Employees are increasingly drawn to companies that prioritize sustainability, making it easier to attract and retain top talent. By embracing sustainability, businesses can create a positive work environment that attracts employees who share their values and are motivated to make a positive impact. This can lead to improved employee engagement, retention, and productivity, which can drive business success.

15 Artificial Intelligence and Data Analytics Enhance the Efficiency of Green Technology Applications

i. **Optimization of energy and installing smart grids**: AI generated demand projections can help to mitigate the demand and supply conditions in real time and mitigating the requirements of energy consumption. Through smart grids load is balanced through automated AI and reduced outages. Further projections in maintenance through machine learning minimize maintenance cost.
ii. **Smart approach for energy management**: AI-enabled system in infrastructure optimizes heating, cooling, and lighting. Internet of things with AI enables appliances to to work more efficiently. Data analytics assist business to minimize carbon elements by controlling energy consumption.
iii. **Predicting renewable energy**: Artificial Intelligence forecast better weather status which can be helping the reliability of solar power (Tandon et al., 2025). More accurate predictions overcome the dependence upon fossil fuels
iv. **Precision farming**: With the help of AI it has become easy to do precision farming more efficiently by using water, fertilizer and pesticides. Drone camera keeps updating data through real time system for sustainable agriculture practices.
v. **Monitoring climate and environmental conditions**: Through machine learning the accuracy of climate models can be ensured. AI trackers data for deforestation apart from checking ocean health and analysing air quality. It also supports in waste recycling through image recognition and role of robotics beside optimizing collection routes.
vi. **Green logistics**: AI minimize the traffic congestion and smooth transportation system, miniming the carbon discharge beside scaling up the life span of battery. Forecasting analytics assist in route planning to overcome the issue of fuel requirements and delivery time consumption.
vii. **Carbon emission**: AI tracks all carbon footprints and supports compliances under regulatory framework. It also assist in optimizing energy consumption with planned transportation activities. Apart from these roles it assists planners to decide smart cities project with minimum environmental influences.

16 Consumer Preferences Impact the Demand for Electric Vehicles and Other Green Technologies

i. **Changing consumers preferences**: In order to protect climate from drastic changes consumers are now willing to shift from fuel consumption vehicle towards electric vehicle. According to reports, Gen Z prefer to purchase things that have sustainability.

ii. **Perception**: People have started to perceive that electric vehicles have low fuel consumption and maintenance costs, followed by subsidies, relaxation, and free charging booths, which can lead to adoption. Rather, consumers prefer to use electric vehicle as they are luxurious, tech-savvy and fast as well. Strong features like automated driving, AI-driven front ends turned consumers to make such wise decisions. In Norway, electric vehicles contributing around 80% of the sales due to environmental concern. In India, high prices and limitation of infrastructure hindered mass adoption beside the interest is rising.
iii. **Environmental concern**: Consumers are willing to make use of solar panels and smart devices and energy-efficient appliances. Apart from this demand for smart lights and LEED-certified infrastructure is designed due to market driven conditions. Consumers liking to use eco-friendly products, or biodegradable with ethical concern as well. This is the reason companies like Unilever and Nestle are restructuring thieir supply chain management and minimising plastic usage.

17 Potential Barriers to the Widespread Adoption of Green Technologies in Developing Countries

i. **High entry cost**: Initiating and promoting green technologies through solar panels, electric vehicles involve major capital investments that are difficult for small and medium enterprises and individuals. Inadequate financing with reasonable interest rates, green bonds lead to restrain people towards adoption. These projects are envisaged riskier just because of lengthy recovery period.
ii. **Inadequate charging infrastructure**: In the rural areas there is shortage of power supply or internet connectivity restricting the usage of green technologies and smart devices. Pathetic roads and redundant transportation technologies again restrain green technologies models. Acute shortage of charging stations is major deterrent to adopt these services.
iii. **Challenges of regulatory framework**: Lack of government support in terms of unstable policies for carbon footprints and renewable energy reduces investors' confidence for investments. At times even though green regulations exist but due to corruptions effective implementation is not ensured or overlooked in areas of adhering the environmental standards.
iv. **Insufficient expertise**: Most of the users are not familiar about the advantages of green technologies and therefore they are not confident to have firm reliability. There is also acute shortage of professionals who can look after the repairs and maintenance of green technologies. Most of the institutions still unable to create the rigor to design course curriculum on green tech or extend research and development activities.
v. **Dependence upon global companies**: Countries that are into developing have to depend upon global firms and market to leverage such technologies through imports costing the pockets of many locals and limiting the adaptation towards

it. Licensing fees and stringent patent rights restrain such form of green energies.

vi. **Sociological hindrances**: conventional practices and routines can discourage adoption in the rural and semi-rural areas. Farmers may not be inclined to adopt solar system for irrigation or AI-based precision agriculture system without checking the outcome or benefits.

18 Rising EVs May Influence the Design of Cities, Transport Systems, and Public Spaces to Accommodate Sustainable Practices

Increase in demand of electric vehicle is bringing paradigm shift in automobile industry and updating the design and development of cities, convenience services and public spaces (Mhatre & Shukla, 2024). As more electric vehicles are on roads now, the regulatory framework are resigning into smart cities with eco-friendly transportation system supporting sustainable least emission lifestyle.

There are lot of urban development activities done through installing charging stations in public places, complex, official places. Beside it certain lamp post or parking meters are also installed at public places. In huge rush areas of malls fast charging stations are involved for better facilities. Cities which are converting into smart cities and engaging charging stations which are backed by solar powered and battery storage to furnish the demand for energies. Through biodirectional charging facilities Electric Vehicles disseminate power back to the grid, making vehicles as mobile energy units. Lot of cities are also restructuring transportation services by transitioning vehicles like buses, taxis which are polluting and moving to one that are eco-friendly. Electric buses are emitting zero pollution and improving the quality of atmosphere. These electric vehicles are also promoting the concept of shared ride through e-car sharing, e-scooter, or e-biking. Convenient integration into applications give access to people to use multimodal trips that could be carrying low emissions. To plan for smart city projects the government is creating low-emissions area wherein only such electric cars, bicycles are permitted to park for their charging. This will mitigate the problem of congestion and develop urban areas. Few cities are creating exclusive lane for dedicating EVs movement and promoting adoption. Sustainable solutions in this respect need to develop for recycling battery, disposing EV with due to course of time. Used battery need to be disposed of for recycling purposes. The concept of green construction should be followed to make of recyclable materials in building constructions.

19 Logistics and Supply Chain Management Practices Be Optimized for Sustainability

The companies are making strong shift in terms of adopting green transportation and furnishing logistics services through route optimization, minimizing emissions. Using AI driven tools can be optimizing the pathways and facilitating the most fuel-efficient routes to cover the planned destination in idle time. Dynamic routing helps to adjust road congestion and weather conditions, lending more efficiency in terms of mileage. Transition fleet towards hybrid or biofuel vehicles encourages people to explore other options from road to rail or sea which could be more energy efficient for distant places to cover. This can assist in delivering small shipments into full loads to minimize trips and carbon footprints. Companies switching towards collaborative logistics across different companies in the same areas. The companies also looking forward for eco-friendly packaging rather than making use of plastics to maintain sustainable, re-useable packaging. They also check the packaging volume to minimise wastage, convenience cost, and space required. In the B2B supply chain, recyclable containers are used. To develop sustainable storage houses, it needs to be backed by the installation of solar panels, smart HVAC systems. Through robotics and artificial intelligence, the consumption of energy can be mitigated at a reasonable labour cost. To work out logistics and supply chain management, companies are looking after product lifecycle management by designing products with durability, recyclability. Further tracking for the carbon footprints and inefficient process in supply can be thoroughly checked and comparing performances by monitoring through standards set out. Green policies are designed for suppliers that fulfil ethical environment benchmarks and build sustainable business practices through supply chain.

20 Conclusion

Green technology adoption in the automobile sector is not only a passing craze. Instead, it is a vital requirement to address the major environmental problems of today. Green technologies offer an achievable path to a better sustainable and environmentally conscious and sustainable future as we battle global challenges, which include contamination, depletion of resources, and climate change. Given its significant effect on global carbon emissions and its ability to encourage transformation through the widespread adoption of electric automobiles, green renewable energy, and improved production procedures, the automotive sector in particular, is at the forefront of this transformation. Green technology offers an opportunity to revolutionise the economy in a way that not only serves client's requirements but also preserves the environment by prioritising environmental responsibility and minimising dependencies on fossil fuels.

The automotive industry is undoubtedly shifting towards cleaner and greener solutions, as demonstrated by increasing investments in energy from renewable resources like solar and wind electricity, along with advancements in battery technologies and energy-efficient systems. However, this shift stretches beyond developments in technologies. It implies an essential shift in the concept of sustainability, with domains including recycling, logistics, and transportation at the forefront of this change. These industries may contribute to minimising the general environmental impact of the automotive sector and, thus benefiting, the global economy by reconsidering patterns of energy production and consumption.

There are persisting challenges to face in these developments. Due to inadequate or ineffective policy implementation along with frequently delayed or unbalanced transfer of technology across economies, the amount of greenhouse gas emissions continues to rise in numerous locations. Due to inadequacies in educational facilities, infrastructure, and financing, many developing nations find it challenging to effectively implement green technologies. Despite the adoption of renewable energy sources and electric cars is establishes in an appropriate direction, these technologies cannot achieve their true potential until there is an integrated worldwide initiative to encourage innovation, share expertise and knowledge, and contribute to the development of infrastructure across national boundaries. This reinforces the significance of global coordination and initiatives that give sustainability and safeguarding the environment on highest priority.

In addition, it is crucial to acknowledge that no single solution or one-size-fits-all method for dealing with green technology. The pursuit of sustainability offers unique opportunities, possibilities and difficulties for numerous countries, regions, and fields. Consequently, even while international efforts are crucial, they must be tailored to local contexts while considering social, cultural, and economic aspects into account. To ensure that environmentally friendly innovations are implemented in a manner that is both efficient and equitable, governments, organisations, and consumers need to work hand in hand.

In a nutshell, the implementation of green technology throughout the auto sector is an essential component of the wider effort towards sustainability. It is an essential step towards alleviating the environmental issues that we are currently experiencing while creating a more sustainable, resilient future. The heading, nevertheless, will require ongoing creativity, planned financial expenses, and above all, an international commitment to changing regulations, exchanging information, and assuring that green solutions are accessible to everyone. The way that green technology has been integrated into the international economy and whether we can tackle the environmental challenges of the twenty-first century will be contingent on the collaborative effort of governments, companies, and individuals as we proceed.

References

Aboelmaged, M., & Hashem, G. (2019). Absorptive capacity and green innovation adoption in SMEs: The mediating effects of sustainable organisational capabilities. *Journal of Cleaner Production, 220*, 853–863. https://doi.org/10.1016/j.jclepro.2019.02.150

Ahmed, M., & Shimada, K. (2019). The effect of renewable energy consumption on sustainable economic development: Evidence from emerging and developing economies. *Energies (Basel), 12*, 2954. https://doi.org/10.3390/en12152954

Antimiani, A., Costantini, V., & Paglialunga, E. (2023). Fossil fuels subsidy removal and the EU carbon neutrality policy. *Energy Economics, 119*, 106524. https://doi.org/10.1016/j.eneco.2023.106524

Antonietti, R., & Marzucchi, A. (2014). Green tangible investment strategies and export performance: A firm-level investigation. *Ecological Economics, 108*, 150–161.

Asadi, S., Pourhashemi, S. O., Nilashi, M., Abdullah, R., Samad, S., Yadegaridehkordi, E., & Razali, N. S. (2020). Investigating influence of green innovation on sustainability performance: A case on Malaysian hotel industry. *Journal of Cleaner Production, 258*, 120860. https://doi.org/10.1016/j.jclepro.2020.120860

Bangwal, C., Kumar, R., Suyal, J., & Ghouri, A. M. (2023). Does AI-technology-based indoor environmental quality impact occupants' psychological, physiological health, and productivity? *Annals of Operations Research*, 1–19.

Bashir, M. F., Shahbaz, M., Ma, B., & Alam, K. (2024). Evaluating the roles of energy innovation, fossil fuel costs and environmental compliance towards energy transition. *Journal of Environmental Management, 351*, 119709. https://doi.org/10.1016/j.jenvman.2023.119709

Beise, M., & Rennings, K. (2003). *Lead markets of environmental innovations: A framework for innovation and environmental economics* (ZEW Discussion Papers, No. 03-01). https://ftp.zew.de/pub/zew-docs/dp/dp0301.pdf

Ceci, F., & Razzaq, A. (2023). Inclusivity of information and communication technology in ecological governance for sustainable resources management in G10 countries. *Resource Policy, 81*, 103378.

Chen, C., Pinar, M., & Stengos, T. (2021). Determinants of renewable energy consumption: Importance of democratic institutions. *Renewable Energy, 179*, 75–83.

Chen, G. Q., Jiang, M. M., Chen, B., Yang, Z. F., & Lin, C. (2006). Emergy analysis of Chinese agriculture. *Agriculture, Ecosystems & Environment, 115*(1–4), 161–173. https://doi.org/10.1016/j.agee.2006.01.005

Chen, H. (2023). Coupling high natural resources and carbon emission efficiency on economic growth in China. *Resources Policy, 85*, 103709. https://doi.org/10.1016/j.resourpol.2023.103709

Cheng, B., Ren, X., Dong, K., Dong, X., & Wang, Z. (2021). How does technological innovation mitigate CO_2 emissions in OECD countries? *Journal of Environmental Management, 280*, 111818.

Chiou, T. Y., Chan, H. K., Lettice, F., & Chung, S. H. (2011). The influence of greening the suppliers and green innovation on environmental performance and competitive advantage in Taiwan. *Transportation Research Part E, 47*(6), 822–836. https://doi.org/10.1016/j.tre.2011.05.016

Diamond, J. (2005). *Collapse: How societies choose to fail or survive.*

Greenly Earth. (2022). *Everything you need to know about green technology in 2022.* https://greenly.earth/en-us/blog/ecology-news/everything-you-need-to-know-about-green-technology-in-2022

Eltayeb, K. T., & Zailani, S. (2011). Green supply chain initiatives among certified companies in Malaysia. *Resources, Conservation & Recycling, 55*(5), 495–506. https://doi.org/10.1016/j.resconrec.2010.09.003

Gao, P., Wang, Y., Zou, Y., Su, X., Che, X., & Yang, X. (2022). Green technology innovation and carbon emissions nexus in China. *Frontiers in Psychology, 13*, 951172. https://doi.org/10.3389/fpsyg.2022.951172

Guo, M., Nowakowska-Grunt, J., Gorbanyov, V., & Egorova, M. (2020). Green technology and sustainable development: Assessment and frameworks. *Sustainability, 12*(16), 6571. https://doi.org/10.3390/su12166571

Gupta, S., Kumar, R., & Kumar, A. (2024). Green hydrogen in India: Prioritization of its potential. *International Journal of Hydrogen Energy, 50*, 226–238.

Haldar, A., & Sethi, N. (2022). Environmental effects of information and communication technology—Exploring the roles of renewable energy, innovation, trade and financial development. *Renewable and Sustainable Energy Reviews, 153*, 111754. https://doi.org/10.1016/j.rser.2021.111754

Hickel, J., & Kallis, G. (2019). Is green growth possible? *New Political Economy, 25*(4), 469–486. https://doi.org/10.1080/13563467.2019.1598964

Hottenrott, H., Rexhauser, S., & Veugelers, R. (2016). Organisational change and the productivity effects of green technology adoption. *Resource and Energy Economics, 43*, 172–194.

Huang, S., Lin, H., Zhou, Y., Ji, H., & Zhu, N. (2022). The influence of the policy of replacing environmental protection fees with taxes on enterprise green innovation. *Sustainability, 14*, 6850.

Ikram, M., et al. (2021). Assessing green technology indicators for cleaner production and sustainable investments in a developing country context. *Journal of Cleaner Production, 322*, 129090. https://doi.org/10.1016/j.jclepro.2021.129090

International Energy Agency (IEA). (2021). *World energy outlook 2021*. https://www.iea.org/reports/world-energy-outlook-2021

International Renewable Energy Agency (IRENA). (2019). *Renewable energy market analysis: Southeast Asia*. https://www.irena.org/publications/2019/May/Renewable-Energy-Market-Analysis

Jiang, Z., Lyu, P., Ye, L., & Zhou, Y. W. (2020). Green innovation transformation, economic sustainability and energy consumption. *Journal of Cleaner Production, 273*, 123044.

Karmaker, C. L., Al Aziz, R., Palit, T., & Bari, A. M. (2023). Analyzing supply chain risk factors in SMEs under fuzzy environment. *Sustainable Technology and Entrepreneurship, 2*(1), 100032. https://doi.org/10.1016/j.stae.2022.100032

Kashav, V., Garg, C. P., & Kumar, R. (2023). Ranking strategies to overcome maritime supply chain barriers. *Annals of Operations Research*, 1–46.

Katebi, A., Homami, P., & Najmeddin, M. (2022). Acceptance model of precast components based on TAM and TOE. *Journal of Building Engineering, 45*, 103518.

Koçak, E., & Şarkgüneşi, A. (2017). Renewable energy and economic growth nexus in Black Sea and Balkan countries. *Energy Policy, 100*, 51–57. https://doi.org/10.1016/j.enpol.2016.10.007

Mahmood, A., Eqan, M., Pervez, S., Alghamdi, H. A., Tabinda, A. B., Yasar, A., & Pugazhendhi, A. (2020). COVID-19 and hand sanitizers: Health and environmental hazards. *Science of the Total Environment, 742*, 140561.

Mäkitie, T., Hanson, J., Damman, S., & Wardeberg, M. (2023). Digital innovation's contribution to sustainability transitions. *Technology in Society, 73*, 102255. https://doi.org/10.1016/j.techsoc.2023.102255

Menyah, K., & Wolde-Rufael, Y. (2010). Energy consumption, pollutant emissions and economic growth in South Africa. *Energy Economics, 32*, 1374–1382. https://doi.org/10.1016/j.eneco.2010.08.002

Mhatre, A. S., & Shukla, P. (2024). A review of energy harvesting technologies for sustainable EVs. *Environmental Science and Pollution Research*, 1–24.

Min, S., So, K. K. F., & Jeong, M. (2021). *Consumer adoption of Uber app: Innovation theory and TAM* (pp. 2–15). Routledge.

Nepal, R., Zhao, X., Liu, Y., & Dong, K. (2024). Can green finance strengthen energy resilience? *Technological Forecasting and Social Change, 202*, 123302.

Pei, Y., Zhu, Y., Liu, S., Wang, X., & Cao, J. (2019). Environmental regulation and carbon emission: Mediation by technical efficiency. *Journal of Cleaner Production, 236*, 117599. https://doi.org/10.1016/j.jclepro.2019.07.074

Perry, W. (2020). Social sustainability and the argan boom in Morocco. *World Development Perspectives, 20*, 100238.

Rajani, R. L., Heggde, G. S., & Kumar, R. (2022). Service redesign strategies for performance impact. *Vision*, 09722629221107238.

Rehman, M. A., Fareed, Z., & Shahzad, F. (2022). Financial inclusion and clean environment: Role of governance. *Environmental Science and Pollution Research, 29*, 27651–27663.

Rout, S. K., Gupta, M., & Sahoo, M. (2022). Technological innovation, energy use, and ecological footprint in BRICS. *Environmental Science and Pollution Research*, 1–18.

Roy, A., Mandal, M., Das, S., Kumar, M., Popek, R., Awasthi, A., Giri, B. S., Mondal, K. C., & Sarkar, A. (2024). Non-exhaust particulate pollution in Asian countries: A comprehensive review of sources, composition, and health effects. *Environmental Engineering Research, 29*(3), 230384.

Sarkodie, S. A., Owusu, P. A., & Taden, J. (2024). Green growth assessment across 203 economies. *Sustainable Horizons, 10*, 100083.

Saunila, M., Ukko, J., & Rantala, T. (2018). Sustainability as a driver of green innovation. *Journal of Cleaner Production, 179*, 631–641.

Schmidt, T. S., & Sewerin, S. (2019). Measuring temporal dynamics of renewable policy mixes. *Research Policy, 48*, 103557. https://doi.org/10.1016/j.respol.2018.03.012

Shafiei, S., & Salim, R. A. (2014). Energy use and CO_2 emissions in OECD countries. *Energy Policy, 66*, 547–556. https://doi.org/10.1016/j.enpol.2013.10.064

Sharif, A., Kartal, M. T., Bekun, F. V., Pata, U. K., Foon, C. L., & Depren, S. K. (2023). Green tech, taxes, and energy in Nordic countries. *Gondwana Research, 117*, 194–206.

Shen, Y., Ur Rahman, S., Hafiza, N. S., Meo, M. S., & Ali, M. S. E. (2024). Green investment and environmental pollution: Asymmetric ARDL. *PLoS ONE, 19*(4), e0292260.

Shi, T., Si, S., Chan, J., & Zhou, L. (2021). Tech innovation's carbon reduction impact in transport. *Atmosphere (Basel), 12*, 1169. https://doi.org/10.3390/atmos12091169

Singh, S., Sahni, M. M., & Kovid, R. K. (2020). What drives FinTech adoption? *Management Decision, 58*(8), 1675–1697.

Song, M., Fisher, R., & Kwoh, Y. (2019). Green innovation challenges and large data. *Technological Forecasting and Social Change, 144*, 361–368.

Tandon, A., Awasthi, A., Pattnayak, K. C., Tandon, A., Choudhury, T., & Kotecha, K. (2025). Machine learning–driven solar irradiance prediction. *Discover Applied Sciences, 7*(2), 107.

The Washington Post. (2024). *Climeworks' Mammoth carbon capture plant removes CO_2 from air.* https://www.washingtonpost.com/climate-solutions/2024/05/09/climeworks-mammoth-carboncapture

United Nations Environment Programme (UNEP). (2017). *Resource efficiency: Potential and economic implications.* https://www.unep.org/resources/resource-efficiencypotential-and-economic-implications

Valente, D., Miglietta, P. P., Porrini, D., Pasimeni, M. R., Zurlini, G., & Petrosillo, I. (2019). Integrating ecological aspects into financial insurance. *Ecological Modelling, 392*, 117–127. https://doi.org/10.1016/j.ecolmodel.2018.11.009

World Intellectual Property Organization (WIPO). (2020). *Patents and clean energy technologies.* https://www.wipo.int/patents/en/topics/clean_energy_technologies.html

Yuen, K. F., Cai, L., Qi, G., & Wang, X. (2021). Factors influencing autonomous vehicle adoption. *Technology Analysis & Strategic Management, 33*(5), 505–519.

Empowering Economies: The Intersection of Innovation, Industry, and Infrastructure for SDG 9

Tanya Rawal, Shipra Agarwal, Malika Pahwa, Tanupriya Choudhury, and Pooja Kanojia

Abstract

Sustainable Development Goal 9 (SDG 9) aims to build resilient infrastructure, promote inclusive and sustainable industrialization, and foster innovation as an important driver of economic growth and development. The analysis underscores the important role of SDG 9 in driving economic empowerment as well as economic and sustainable development more generally. It looks into the extent to which industry, innovation, and infrastructure are the pillars of job creation and connectivity. The report also reviewed targets falling under SDG 9, global-level synchronization, and interlinkages with other SDGs, besides discussing issues such as challenges regarding funding gaps and technology access, the need for sustainable industrialization, digital transformation, and resilient infrastructure. The regional case studies provide programs and best practices that facilitate the management and adoption of efficient policies and governance models. Finally, concrete recommendations are provided on how stakeholders can fast-track progress toward SDG 9, including in areas of collaboration on green innovation and global sustainable practices that enable economies through green innovation.

Keywords

SDG 9 • Economic growth • Sustainability • Infrastructure

T. Rawal (✉) · S. Agarwal · M. Pahwa · P. Kanojia
Department of Commerce, Graphic Era Deemed to be University, Dehradun, India
e-mail: tanyarawal3@gmail.com

T. Choudhury
School of Computer Science, UPES, Dehradun, Uttarakhand, India

T. Choudhury et al. (eds.), *Open Access as a Knowledge Platform for Implementing the United Nations Sustainable Development Goals*, World Sustainability Series,
https://doi.org/10.1007/978-3-031-99605-4_11

1 Introduction

Sustainable development has gained momentum as a defining philosophy with which policymakers are addressing various economic inequalities, degrading environmental disasters, and turbulent social unrest. Of the seventeen SDGs set out by the United Nations, the SDG 9 speaks equally to industry, innovation, and infrastructure. This goal will show how good infrastructure, inclusive and sustainable industrialization, coupled with innovation, go hand-in-hand with the economy in all aspects of sustainable development and environmental sustenance.

The next section is Empowering Economies: Innovation, Industry and Infrastructure for SDG 9, wherein further inter-linkage of the SDGs towards Just and Sustainable economies will be done. It mentions the fact that countries like developing ones might employ industrialization, deployment of technology, and infrastructure scaling their economies to be on par in line with Global sustainability goals. In this context, the section discusses a prospective path toward meeting these challenges through an interdisciplinary look into a series of key challenges, opportunities, and selected real examples and thus offers a vision of how economies might be changed to build toward the better futures set out for 2030 as envisioned by SDG 9.

1.1 Understanding Sustainable Development Goal 9

SDG 9: Build resilient infrastructure, promote inclusive and sustainable industrialization, and encourage innovation. Our commitment to such initiatives is underpinned by our vision for strengthening economies, protection of the environment, and advancement of equity in society. Besides, SDG 9 is connected with the attainment of other goals like the eradication of poverty, which is SDG 1; quality education, which is SDG 4; and taking action against climate change, which is SDG 13, and thus becomes very important in developing a sustainable and better future for all.

1.2 Key Targets of SDG 9

1.2.1 Establishing Strong, Solid Infrastructure

Infrastructure—roads, bridges, and power systems—enable people to connect, trade, and access crucial services (healthcare and education, for example) (SDG, 2019). SDG 9 aims to develop durable, low-impact physical infrastructures that are accessible to all so that those further away or disadvantaged are not left behind.

1.2.2 Supporting Fair and Sustainable Industries

Industries create jobs, boost incomes, and grow the economy. SDG 9 aims to increase the role of industries in improving people's lives, especially in developing and poorer countries, by encouraging more job opportunities and fair economic growth (do Rosário Cabrita et al., 2023).

1.2.3 Encouraging New Ideas and Technology

Innovation is key to solving big problems like climate change and improving daily life. SDG 9 promotes investing in science, research, and entrepreneurship to bring fresh ideas and new technology, especially in countries that need them most (Sinha et al., 2020).

1.2.4 Closing the Technology Gap

Not everyone has equal access to technology and digital tools. SDG 9 aims to make sure that everybody, no matter where we live, can access modern tools, such as the internet and financial services, which enhances equality of opportunity and strengthens the economy for us all.

1.3 The Importance of Industry, Innovation, and Infrastructure in Economic Growth

1.3.1 Industry: Building a Strong Economy

Industries are the engines of economic development. They provide jobs for people, grow incomes, and enable nations to compete in the world marketplace. But in the world, we live in today, it's not about making more; it's about making better and cleaner. Sustainable industrialization puts weight on efficiently using resources with minimal wastage and harming the environment (Fei et al., 2021). The renewable energy industry like solar, wind turbines, along with sustainable practices like the circular economy and recycling, contribute to a solid and green future. The economy is not only moving forward but also taking care that the planet remains healthy for future generations.

1.3.2 Innovation: Solving Problems and Creating Opportunities

Innovation is the discovery of new ideas, technologies, and solutions making life better. Besides being a key driver of economic growth, innovation can allow enterprises to become more creative, efficient, and competitive (Mavuri et al., 2019). Innovation can also battle some of the huge global challenges in areas such as climate change, poverty, and using resources more efficiently. Changes in technology, such as state-of-the-art renewable energy and AI, are altering the way industries work, making their operations smoother and more efficient. Investing in research, fostering entrepreneurship, and igniting creativity will drive forward progress, creating new opportunities for all.

1.3.3 Infrastructure: Connecting People and Opportunities

Infrastructure is indeed the backbone of every society, since it involves everything that makes one go: roads, bridges, power grids, and communication-facilitating infrastructure. Without infrastructure, businesses can barely operate, while people struggle to get to health and educational services, and even access clean water. Good infrastructure links people up with markets and jobs, fueling community growth in return. And resilient infrastructure helps to minimize damage from disasters or climate change; long-term security and well-being will thrive on such investment.

2 Objectives and Scope

2.1 Objectives of the Chapter

This chapter seeks to explore the critical interlinkages of industry, innovation, and infrastructure in achieving SDG 9 for sustainable development and economic growth. The report, conducted on the basis of previous research, explores barriers such as limited access to technology, financial hurdles, and policy gaps but also highlights opportunities such as international cooperations and public-private partnerships. The chapter will be able to showcase how industrialization and innovation are being done successfully through the presentation of real success stories. It also looks at how policies and technological innovations accelerate progress toward SDG 9. It further highlights the different challenges developing and least-developed countries face, from poor infrastructure to lack of resources, and provides best practices on tackling these discrepancies to help ensure that microfinance is an inclusive catalyst for sustainable growth everywhere.

3 Understanding SDG 9: Boosting Economies Through Innovation, Industry, and Infrastructure

SDG 9: Build resilient infrastructure, promote inclusive and sustainable industrialization, and encourage innovation. These three aspects are interdependent and have giant impacts on reduction in poverty and economic development, eventually improving the general well-being of people, which matters so much where there is retarded development and a scarcity of prospects.

This goal acknowledges that infrastructure that is good, environmentally friendly industrial practices, and technological progress are most effective when taken in unity (do Rosário Cabrita et al., 2023). Combined, they can help countries grow in ways that work for all people and for the planet.

Specific countries, especially developing countries, face huge challenges to growth due to a lack of infrastructure, outdated technologies, and weak industries, which underscores the importance of SDG 9. It presents a roadmap for addressing these challenges through collaboration among industry, government, and innovators.

4 Key Targets and Indicators of SDG 9

SDG 9 aims to build resilient infrastructure, promote inclusive and sustainable industrialization, and foster innovation as enablers for our economies to grow in an equitable, sustainable, and environmentally sound manner. Here, are the main targets and their objectives, as well as the indicators of progress.

4.1 Building Strong, Reliable, and Sustainable Infrastructure

Infrastructure is defined as the basic frameworks built to ensure the functionality of our society, such as roads, electricity, water supply, and telecommunication. These are critical for connecting people, spurring economic development, and enhancing access to critical services such as schools and hospitals. SDG 9 focuses on building eco-friendly, resilient infrastructure to prepare for challenges such as natural disasters or economic collapse.

Example: Imagine farmers struggling to get their produce to market because they can't transport it easily to the nearest town. If a good road is constructed, their lives would be transformed: it will be much easier to sell goods, get medical care, or send their children to school.

Indicator: The share of the rural population living within 2 km of a road that is passable in all seasons.

This measures how many people living in remote areas can travel easily for access to services and opportunities that can enhance their quality of life.

4.2 Supporting Fair and Sustainable Industrial Growth

Industries, like factories and manufacturing plants, are essential for creating jobs and producing goods. However, for industrial growth to be beneficial for everyone, it needs to be fair and environmentally sustainable. SDG 9 focuses on developing industries that include all communities, reduce pollution, and use resources wisely.

Example: In some countries, industrial growth only benefits cities while rural areas remain underdeveloped. SDG 9 encourages industries to grow in ways that include everyone and minimize harm to the environment, like using clean energy sources.

Indicator: The share of a country's economy that comes from manufacturing (called Manufacturing Value Added or MVA) and how much this contributes to each person's income.

This helps track whether industrial growth is boosting the economy and improving people's living standards.

4.3 Helping Small Businesses Access Money and Technology

Small businesses, like local shops, farms, and small factories, are vital for communities because they provide jobs and services. But many small businesses have trouble obtaining loans or modernizing their tools to scale. SDG 9 emphasizes ensuring access to credit, finance, and new technologies for small businesses to prosper.

Example: A small bakery, for instance, might need a loan to get better equipment or expand its shop. Without credit, the bakery might not scale up, limiting its ability to hire additional workers or serve more customers.

Indicator: Share of small businesses that can access loans or credit.

This monitors the financial support of local small businesses that are the seeds of growth, helping them expand and provide further local economic support.

4.4 Expanding Access to Digital Technology

In the modern technological era, access to technology like the internet and mobile phones has become one of the most important tools for education, business, and communication. Example: A student in a remote village without internet access might struggle to learn or access online resources due to the lack of technology. Expanding mobile networks and internet access can enable individuals in such regions to remain connected and access new opportunities.

Indicator: The percentage of people covered by a mobile network, including advanced technologies like 4G and 5G.

This measures how well digital connectivity is improving, bringing more people into today's economy.

4.5 Making Industries More Environmentally Friendly

Industries can cause damage to the environment by emitting pollution, wasting resources, or producing excessive CO_2. SDG 9 encourages the adoption of greener technology and practices to minimize the impact of industries on the environment while continuing to produce goods and promote economic growth.

Example: A factory that uses solar power rather than burning coal is still able to produce the necessary outputs while generating less CO_2. This assists in fighting climate change and keeps the industry sustainable.

Indicator: CO_2 emissions per unit of economic output.

This reflects how effectively sectors are minimizing their carbon footprint while still expanding the economy.

4.6 Why These Targets Are Important

These targets combine to address critical issues such as weak infrastructure, limited support for small businesses, and environmental degradation. SDG 9 aims to build resilient infrastructure, promote inclusive and sustainable industrialization, and foster innovation to create a stronger, fairer, and greener economy for all, no matter where they live. It's about creating a future in which growth and progress aren't achieved at the expense of people or the planet.

5 Links to Other Sustainable Development Goals

5.1 How SDG 9 Connects to Other Global Goals

Many of the other Goals intersect with SDG 9 as shown in Table 1 and Fig. 1, which targets the building of resilient infrastructure, promotes inclusive and sustainable industrialization, and fosters innovation (Pradhan et al., 2017). While advances on SDG 9 do, of course, have co-benefits across other SDGs, progress on this goal also has the potential to risk doing harm elsewhere—and leaving people behind—by creating a world that is simply less equitable, less sustainable, and less prosperous. Here's how.

5.1.1 SDG 1: No Poverty

Better infrastructure and industries create more jobs and economic opportunities. This empowers people to make money, find stable employment, and pull themselves out of poverty.

5.1.2 SDG 2: Zero Hunger

Well-maintained roads, storage, and transport systems assist farmers in growing, processing, and distributing food. This ensures that people have enough to eat and that food security increases.

5.1.3 SDG 4: Quality Education

Schools and online learning are supported by reliable infrastructure in terms of electricity and the internet. This has the effect of making learning more accessible and increasing the quality for the learner where access might be restricted.

5.1.4 SDG 5: Gender Equality

Creating opportunities in an open field of innovation in industries creates space for women to engage and thrive. This reduces the gender gap and further empowers women with more authority in the workplace and leadership.

Table 1 SDG 9 interlinkages with other SDGs and potential co-benefits

SDG	Interlinkage with SDG 9	Potential co-benefits
SDG 1: no poverty	Improved infrastructure and industrialization reduce poverty	Economic diversification and increased job opportunities alleviate poverty
SDG 2: zero hunger	Enhanced agricultural supply chains improve food distribution	Reduced post-harvest losses and better food security
SDG 3: good health and well-being	Better infrastructure improves access to healthcare and clean resources	Reduced health risks and better emergency medical responses
SDG 4: quality education	Digital infrastructure enables access to education	Lifelong learning opportunities, especially in remote areas
SDG 5: gender equality	Inclusive policies empower women in industry and innovation	Improved financial access for women and gender-inclusive employment
SDG 6: clean water and sanitation	Infrastructure supports sustainable water management	Enhanced water quality and availability reduce disease
SDG 7: affordable and clean energy	Supports renewable energy infrastructure and access	Cleaner energy reduces environmental impact and boosts energy efficiency
SDG 8: decent work and economic growth	Promotes sustainable industries and technological innovation	Increased employment and economic expansion through industrial growth
SDG 10: reduced inequalities	Expands infrastructure in underserved areas	Greater economic inclusion for marginalized communities
SDG 11: sustainable cities and communities	Supports smart city development and urban resilience	Improved urban living conditions and environmental sustainability
SDG 12: responsible consumption and production	Promotes sustainable manufacturing and resource efficiency	Reduced industrial waste and promotion of circular economies
SDG 13: climate action	Encourages low-carbon technologies and sustainable practices	Reduced greenhouse gas emissions and climate resilience
SDG 14: life below water	Minimizes industrial pollution affecting marine environments	Protection of marine biodiversity and sustainable ocean use
SDG 15: life on land	Reduces land degradation through sustainable industry practices	Conservation of terrestrial ecosystems and biodiversity

(continued)

Table 1 (continued)

SDG	Interlinkage with SDG 9	Potential co-benefits
SDG 16: peace, justice, and strong institutions	Supports transparent and accountable industrial policies	Stronger governance and regulatory frameworks for fair trade practices
SDG 17: partnerships for the goals	Fosters global collaboration for technology and knowledge sharing	Accelerated progress through international cooperation and innovation

Fig. 1 Sustainable development goals

5.1.5 SDG 7: Affordable and Clean Energy

The use of clean energies—solar or wind—is imperative for sustainable industry. It reduces pollution and contributes to SDG 9 on work, which is to create sustainable industries.

5.1.6 SDG 8: Decent Work and Economic Growth

Growing industries are indicative of an increase in job opportunities and overall economic growth. These factors will, in turn, improve the quality of life for individuals while providing excellent job opportunities to a large percentage of the workforce—a key goal within SDG 8.

5.1.7 SDG 13: Climate Action

Industries can reduce harmful emissions through green technologies and sustainability practices. This helps combat climate change and is in line with international efforts to save the planet.

6 The Current Global Progress on SDG 9

6.1 Infrastructure Development

6.1.1 Developed Regions

In First World countries, such as North America and Europe, infrastructure is usually well developed. These include good roads, power, water, and communication systems. All these are fundamental systems since they facilitate the smooth operations of economies. They also spur technical evolution and innovation. For instance, when the roads are well-maintained, it is easier to deliver goods, and when the power grids are reliable, businesses do not have to worry about disruptions.

6.1.2 Developing Regions

Most developing or low-income countries still have underdeveloped infrastructure, especially in the rural parts. It means that most people lack dependable roads, electricity, and clean water (Sharma et al., 2024). It is this lack of basic services that makes lives harder and limits access to work, education, and health care. In fact, such basic facilities render it quite difficult for the masses to take off. For instance, farmers in the most isolated parts of the world cannot take their crops to market without decent roads, while children cannot study at night without electricity.

6.2 Inclusive and Sustainable Industrialization

6.2.1 Success Stories

Countries like China and India, along with other parts of Asia, have seen huge industrial growth. This has resulted in big gains in their economies. More people have jobs now, and standards of living have risen. Factories have employed previously unemployed workers, while industries have provided much-needed products to enhance people's quality of life—such as clothing, electronics, and household items. These countries are successfully building industries that generate wealth and lift people out of poverty.

6.2.2 Challenges

Conversely, many of the world's most impoverished nations, known as Least Developed Countries (LDCs), continue to face significant industrialization challenges. Their industries are small or obsolete, and productivity is low. This means these countries don't have the same opportunities for economic growth. They must aim for inclusive growth, ensuring that everyone—including marginalized groups like women and young people—benefits from industrialization. When industrial growth only benefits a small portion of the population, it increases inequality and limits overall progress.

6.3 Fostering Innovation

6.3.1 Advances

Investments in R&D have increased all over the world. That means countries have invested more to seek new ideas and technologies. Major countries in the world, such as Japan, South Korea, and the US, have made considerable progress in strategic areas like artificial intelligence, biotechnology, and renewable energies. These, in turn, are creating a new industry of goods and services and can be solutions for some of humankind's most daunting problems. Patent filings, which give legal protection to new ideas, are flooding in—the surest sign yet that innovation is soaring.

6.3.2 Gaps

There is a huge gap between high-income countries, such as the U.S. or Germany, and low-income countries, many of which are found in Africa or Southeast Asia, in terms of access to technology and innovation. Wealthy countries are often more accessible in terms of the use of technology, and so more funds become available for research studies. In turn, poorer countries face poor opportunities.

Spending more on education, facilitating technological exchange between nations, and establishing innovation-friendly environments—all these will make up for a great deficiency if pursued in poor countries. We cannot have the wealthy countries keep on churning out new ideas and technologies.

6.4 Environmental Impact

6.4.1 Progress

Industries are changing day by day to implement greener technologies that can help reduce their emissions. This is quite a positive step in the direction of taking care of climate change. Many companies are now starting to utilize solar energy and avail other renewable resources, hence producing less pollution. These changes do show that it is possible to build industries and take care of the needs of the planet hand in hand.

6.4.2 Challenges

On the other hand, it is sad that industries all around the world keep being major contributors to greenhouse gas emissions, which in turn are contributors to climate change. Most of the sectors of industries across the world burn fossil fuel or otherwise pollute both air and water, even with cleaner technologies present. In the real sense of working towards combating climate change, more needs to be done in the promotion of green technologies that are low-impact on the environment. We also need to head in the direction of a circular economy where resources would be used much more efficiently, with a minimum amount of waste.

6.5 ICT Accessibility

6.5.1 Improvements

There has been great progress in the availability of Information and Communication Technology (ICT), like the internet and mobile networks. Hundreds of millions of people have gained access to mobile phones and the internet over the last few years. This has brought many new opportunities, such as online education, e-commerce, and remote working. There is more connectivity than ever before, which improves economic participation and access to information. For example, farmers can receive real-time weather updates, and small businesses can sell their products online.

7 SDG 9 and India

In July 2024, NITI Aayog unveiled the 4th edition of the SDG India Index for 2023–24, a vital instrument for tracking India's progress toward the United Nations' Sustainable Development Goals (SDGs). First introduced in 2018, the index assesses the performance of Indian States and Union Territories (UTs) on 16 SDGs.

With 113 indicators, the index assigns scores ranging from 0 to 100, where 100 signifies full attainment of the targets. Beyond merely evaluating progress, the index promotes collaboration among States and UTs, encouraging the exchange of best practices to foster improvement. This initiative ensures that sustainable development remains aligned with India's national priorities, advancing unified efforts to achieve these global objectives. Source: SDG India Index 2023–24.

1. Under the Pradhan Mantri Gram Sadak Yojana, nearly 99.7% of the targeted villages are now linked by all-weather roads—a significant leap from just 47.38% in 2017–18.
2. Around 93.3% of households now have at least one mobile phone.
3. About 95% of villages have access to 3G or 4G mobile internet services.
4. In 2023–24, 45 departments worked on 151 schemes and programs with a total budget of ₹9,571.35 crore under SDG 9.
5. For 2024–25, 42 departments are set to implement 183 schemes and programs with a planned budget of ₹12,011.98 crore under the same goal.
6. The budget for 2024–25 shows a 13% increase compared to the revised estimates (RE) of 2023–24.

8 Learning from Maharashtra: Steps Toward Sustainable Industrial Growth

Land acquisition is a crucial part of building infrastructure, but the Land Acquisition Bill has made the process more complex. Maharashtra, a key industrial hub, has taken steps to address these challenges and promote sustainable industrialization.

8.1 Simplified Land Acquisition

Previously, state laws made it tough for private parties to buy rural land. With the introduction of the Land Acquisition, Rehabilitation, and Resettlement (LAAR) Act, these rules have been relaxed. Now, agricultural land can be purchased without special permissions, and district-level authorities handle zone conversions to industrial use.

8.2 Improved Industrial Infrastructure

In Aurangabad, part of the Delhi-Mumbai Industrial Corridor, the government is implementing better infrastructure. This includes stormwater drainage, sewage treatment plants, and waste collection systems—initiatives not previously available in the MIDC industrial estates.

8.3 Water Recycling Initiatives

As a water-scarce state, Maharashtra is adopting innovative solutions for industrial water needs. For example, Aurangabad follows Singapore and Israel's models of recycling sewage water. The city plans to treat and recycle 20 million liters of sewage daily, using it for industrial purposes and potentially even drinking. The state is also drafting a policy to ensure all municipal corporations recycle sewage for various uses.

9 Meghalaya's Development Efforts

Meghalaya has set up the **Meghalaya Basin Development Authority** to address infrastructure and development challenges and to promote livelihood initiatives tailored to the state's needs.

9.1 Key Initiatives

9.1.1 MGNREGA's Impact

The Mahatma Gandhi National Rural Employment Guarantee Act (MGNREGA) is a vital program in Meghalaya, reaching 90% of households. It has significantly improved connectivity across the state.

Over the past four years, a special program has focused on **last-mile connectivity**. Roads built or upgraded under MGNREGA are now being black-topped using state funds. These roads connect villages to markets, enabling farmers to sell their produce, such as cashew nuts, more easily.

A sub-program called **Bottom 20** focuses on identifying the 20 poorest households in each village. These families are supported in building assets like small farms or gardens, creating sustainable livelihoods and helping them market their agricultural produce.

9.1.2 Promoting Fisheries and Honey

The **Aqua Mission** aims to produce fish locally instead of relying on imports. To achieve this, 20,000 fish ponds have been created across the state.

Through the **Apiculture Mission**, the state is tapping into its potential for organic honey production, free from antibiotics, and positioning it as a premium product.

9.1.3 Boosting Exports

Meghalaya is leveraging its location along the Indo-Bangladesh border to export agricultural produce through eight land customs stations. This opens up new markets for local farmers and boosts cross-border trade.

10 Learning from Punjab: Sustainable Industrial Practices

Punjab is focused on providing affordable technologies to protect the environment and improve energy efficiency in its small-scale industries.

10.1 Key Initiatives

10.1.1 Cleaner Cupola Furnaces

Punjab has around 1000 small-scale cupola furnaces, which consume ₹200 crore worth of fuel annually and emit significant pollutants, including 800 tonnes of particulate matter and 3 lakh tonnes of carbon dioxide.

A **technical advisory cell** developed air pollution control devices to tackle this issue. So far, 700 of these devices have been installed in Punjab, and the technology has been shared with other states like Bihar, Jharkhand, and Haryana.

This intervention has doubled the efficiency of cupola furnaces. Previously, 6 parts of metal required 1 part of coke (fuel). Now, 12 parts of metal can be processed with the same amount of coke, cutting costs and emissions.

10.1.2 Improving Brick Kilns

Punjab's 3000 brick kilns use 2 million tonnes of coal and 1.15 million tonnes of biomass annually, releasing 0.12 million tonnes of particulate matter into the air.

Innovative operational changes have been introduced to reduce energy consumption and pollution.

These improvements align with the **National Emission Standards** for the brick sector, designed in collaboration with the Central Pollution Control Board.

10.1.3 Sustainable Steel Rerolling Mills

Punjab is home to 300 of the country's 1200 steel rerolling mills, which play a key role in the steel supply chain.

To date, 60 mills have adopted technologies to reduce pollution and improve energy efficiency. The program aims to expand this effort across more mills.

11 A Local Success Story: The Kandi Belt

The Shivalik foothills, or Kandi belt, are less productive due to challenging terrain and poor soil. A government-supported facility was set up to process local bio-resources, starting with a ₹37 lakh grant and the involvement of a local cooperative society.

1. This initiative has grown significantly, with the cooperative now generating ₹10 crore in turnover.
2. It has created employment equivalent to 1 lakh man-days, boosting livelihoods in the local community.

12 Methodology

This book chapter on industry, innovation, and infrastructure in achieving SDG 9 involves a critical review of existing literature to establish the theoretical framework. It incorporates detailed case studies to highlight successful strategies, along with both quantitative and qualitative analyses to identify trends and themes. Policy analysis examines current frameworks, and an interdisciplinary approach integrates diverse insights from various fields. Specific recommendations are then provided to help stakeholders accelerate progress toward achieving SDG 9.

13 Industrialization as a Driver of Economic Growth

Industrialization has always been a key factor in driving a country's economic growth. When countries move from farming-based economies to industrial ones, they unlock a lot of potential for growth and improve how much they can produce. Key industries like manufacturing, mining, and construction are the main sources of wealth and economic activity.

As nations industrialize, they don't depend only on agriculture or raw materials to earn revenue. Instead, they begin expanding into new fields, such as tech, services, and infrastructure. This shift creates many different job opportunities, from low-tech manual labor to high-skill tech jobs. Industries like cars, textiles, electronics, and energy don't just employ a huge number of people; they also increase wages, raising overall living standards.

Industrialization also benefits other sectors of the economy, like transport, logistics, and finance. These industries have made trading and markets accessible with much more ease. The presence of a strong industrial base lets countries compete not only in home ground markets but also in most international markets to secure better export earnings and to strike a balance in international trade. With more industries, their operational efficiency grows, production becomes cheaper, and consequently, goods become more competitive globally.

Large-scale industrialization also stimulates technological innovation. New technology, procedures, and ideas increase the efficiency of production and enhance the process of production altogether. In such a manner, with this level of technological development, countries are capable of producing value-added products and services, thereby competitively advancing themselves in the global economy.

14 Promoting Inclusive and Sustainable Industrial Practices

While industrialization is a powerhouse for economic growth, it has to be inclusive and green for the benefit of all. This implies that all will have an equal chance to participate in and benefit from industrial development, including those normally marginalized, such as women, those living in rural areas, and poor communities. Removing barriers will be key in allowing more participants. Equipping the underrepresented groups with wherewithal to enable their participation—education, training, and financial support—are some of the ways.

Inequality will lessen as more and more people start sharing the benefits of industrialization, and jobs and lives improve. This thus leads to fair distribution of wealth and helps in narrowing the income gaps that at times widen up during periods of rapid industrial growth.

At the same time, we have to make sure that industrialization does not harm the environment. If not taken with care, industrialization can lead to pollution, deforestation, and destruction of ecosystems that dent both the economy and the quality of life of its people. The way to avoid this is the use of sustainable practices. These include the utilization of renewable energy sources such as solar and wind, the use of energy-efficient technologies, and finding ways of manufacturing with minimal emissions.

In a circular economy, some examples of practices involved include recycling, reduction of waste, protection of natural resources, and added value to the economy (Shukla et al., 2024). Also, this ascertains with the United Nations Sustainable Development Goal 9: Build resilient infrastructure, promote inclusive and sustainable industrialization, and foster innovation. Inclusive and sustainable industrialization will go a long way to create economic growth that benefits everyone while helping the planet and future generations (Gera et al., 2023).

15 Challenges to Industrial Growth in Developing Economies

Industrialization can bring great rewards in the form of more jobs and more robust economies, but getting it right isn't always a piece of cake for developing nations. They are faced with myriad challenges that slow progress and make it hard for people to realize the full benefit.

The big issue: Poor infrastructure: Infrastructure such as roads, electricity, and access to clean water are needed for factories and businesses to operate (Liu, 2020). Without infrastructure, it is more expensive to make goods, harder to reach customers, and less likely that foreign investors will step in. This problem weighs heavily on rural areas, since a lack of infrastructure stifles growth and makes the gap between cities and villages even greater (Opoku & Yan, 2019).

A major issue is money—or the lack of it. SMEs are highly important for job creation and industrial development in many parts of the world. However, their access to bank credit or finance is hard to come by for these entities to expand their business activities. It also dampens the government's capacity, as most governments cannot afford mega projects for better infrastructure development. Without good infrastructure, countries cannot compete well on a global scale.

The second issue is a lack of skilled labor. In developing countries, most training and education programs do not address the needs of modern industries. With a lack of skilled labor, firms are unable to apply advanced technologies or increase productivity (Khan et al., 2021). As such, this keeps them confined to low-value-added activities, which contribute very little to economic growth.

The existence of political barriers, corruption, and weak policies ails businesses, while inconsistent rules plant uncertainty that keeps investors away. However, at the same time, rapid industrialization causes environmental damage such as pollution and deforestation, hence making it even more challenging to balance economic development with the protection of natural resources.

Clear and just policies on the part of governments would help attract investors besides encouraging practices that are sustainable and do not harm the environment. Co-operation with other countries and sharing of technology will help the developing nations to overcome their constraints in building industries for service of the citizens and people at large.

16 Incentive of Sustainable Development Through Innovation

Innovation is a car that drives us along big global problems towards a better future. It fuels growth in our economies, boosts our environment, and makes societies more equitable. Innovation is critical if we are to reach the United Nations' Sustainable Development Goals (SDGs), especially Goal 9 (Hales & Birdthistle, 2022). Making this a reality means building solid infrastructure, supporting industries and growth that are fair and environmentally responsible, and developing creativity and new technologies. When new ideas are applied to industries and infrastructure, they grow in such a way that the present is met without compromising the environment and resources for future generations.

It is all about innovation for sustainable growth. This paper identifies how this is doable by setting an enabling environment for innovation to thrive, with digital tools and technology at the forefront in investment for research and development that enables economic growth (Denoncourt, 2020). These also nurture the creation of inclusive and resilient industries and infrastructures in line with overarching global sustainability objectives like SDG 9.

17 Innovation Ecosystems

What Drives Creativity and Problem Solving.

Innovation ecosystems are like groups where various players from all backgrounds—businesses, universities, governments, and entrepreneurs—work in synergy to address problems and develop new ideas. These ecosystems convert creative ideas into tangible products or solutions that can enhance people's lives and protect the planet.

These ecosystems are principally built for collaborating and sharing know-how, tools, and resources. The best solutions would come from working together with many people having various skills and expertise (Küfeoğlu, 2022). In the case of renewable energy or sustainable farming, the ecosystems create venues for people to coordinate, experiment with new ideas, and make them feasible. Hence, solutions merge money, technology, and social effort towards scale and touching many more lives for much more good of society and the planet.

The other interesting thing about ecosystems of innovation: they also squarely put the accent on this ability of equity—developing nations in terms of capacity for access. This would actually ensure progress to be truly, genuinely shared, not just confined to little "privileged" ghettos of excellence in the world.

Innovation ecosystems hence connect people; they create an economic incentive until solutions to at least any given challenge are attained for practical attainment of sustainable development typified by environmentally friendly technologies in industries that have to be encouraged because of the ecological environment and not enabled to destroy. Similarly, sound and inclusive improvement of infrastructure is about supporting international aspiring levels expressed through SDG 9.

18 Main Barriers to Achieving SDG 9 Targets in Developing Countries

Achieving SDG 9 in growing nations faces several limitations, consisting of insufficient infrastructure, lack of funding, and technological gaps, which restrict industrial growth and innovation. Policy and regulatory challenges and confined get right of entry to to financial services avert private investment and restrict SMEs. Skills deficiencies lessen group of workers productiveness, at the same time as environmental issues from unsustainable industrialization threaten long-term balance. Additionally, international change limitations and political instability disrupt development efforts and deter foreign investment. Overcoming those demanding situations calls for centered investments, coverage reforms, and sustainable practices.

19 Key Components of SDG 9 and Their Significance for Economic Growth

As shown in Fig. 2 Sustainable Development Goal 9 (SDG 9) focuses on building resilient infrastructure, selling inclusive and sustainable industrialization, and fostering innovation, all of which can be vital drivers of financial increase. Infrastructure Development enhances marketplace get admission to, reduces logistics prices, and fosters regional connectivity, boosting change and business efficiency. Industrialization creates jobs, will increase productiveness, and diversifies the financial system by way of selling advanced manufacturing and fee-introduced sectors. Innovation drives technological improvements, improves performance, and helps new commercial enterprise fashions, fostering sustainable economic enlargement. Sustainable and Inclusive Development guarantees that boom is environmentally pleasant and handy to all, reducing inequality and improving long-term monetary resilience. Access to Financial Services and Markets empowers small and medium Enterprises (SMEs), encouraging entrepreneurship and increasing financial participation. Lastly, International Cooperation promotes worldwide partnerships, facilitating know-how-sharing, generation switch, and economic help, which improve commercial capability and worldwide change systems. Together, those components create a foundation for sustainable and inclusive economic progress.

Fig. 2 Key components of SDG 9

20 Digital Transformation and Industry 4.0

Digital transformation, driven by new technologies, is changing the industrial model and offering new opportunities for economies to grow within the framework of environmental protection. This wave, so-called Industry 4.0, brings into the industrial process advanced tools in the fields of artificial intelligence, robotics, IoT, big data, and blockchain. Such technologies make industries more efficient, with a minimum amount of waste, and thus sustainable for the future.

For example, AI is able to predict when machines need maintenance, hence helping factories economize energy and avoid expensive breakdowns. IoT sensors sing real-time data on energy and resources that help businesses minimize their carbon footprint and waste (Awasthi et al., 2024; Bai et al., 2023). Such smart technologies make industries more productive, greener, and resilient, hence directly connecting to global goals such as SDG 9.

Besides, digital transformation enhances infrastructure. The application of digitization in smart cities contributes to the conservation of energy, reduction in the emission of harmful gases, and better planning of urban systems. Smart grids balance energy supply and demand while integrating renewable sources into the supply, making the overall system of electricity much more effective (Mangla & Ram, 2020).

21 Fostering R&D for Inclusive Economic Progress

Research and Development play an extremely vital role in building up ideas, technologies, and solutions for achieving growth. Targeted towards sustainability, R&D thus helps business and government sectors to make amiable attempts on gigantic challenges involving environmental and social spheres of humanity along with industrial building for people's needs and earth survival (Kaplinsky, 2018).

For instance, R&D enables the establishment of cleaner fuel, such as solar and wind energy, to reduce our dependence on fossil fuels and furthermore make renewable energy more affordable. It can also improve agricultural means with regard to farmers being able to increase production without harming the ecosystem. Such kind of innovation opens up ways to growth that is not harmful to the environment.

Equally important, R&D should be for the benefit of all, especially for those in developing countries or even from any essence of marginalized groups. By addressing specific local needs, R&D can create significant impacts on improving lives—say, creating affordable healthcare devices, distributing renewable energy to off-grid areas, or developing low-cost water filters for clean drinking water. Inclusive R&D ensures progress reaches the people who need it most.

Effective R&D requires collaboration. It requires the collaboration of governments, businesses, and international organizations through the sharing of ideas and resources. Partnerships like these can dramatically accelerate the development of sustainable technologies and ensure that R&D efforts align with global objectives, such as SDG 9, which focuses on promoting sustainable industries, innovation, and infrastructure.

22 Infrastructure Development: The Key to Growth and Opportunity

Infrastructure is the only one that can lead the development of any country. It includes all the essential systems we rely on every day—from highways to railroads, the power grid, fresh water, and the internet. These systems are the connective tissue that knit together people, businesses, and communities, making lives simpler and economies stronger. Inadequate infrastructure makes everything to do with trade and education more difficult, limiting opportunities for expansion and development.

Let me illustrate this with transportation. Good roads and railways enable farmers to move their harvests to the market without delay and enable businesspeople to deliver their products on time (Fei et al., 2021). In the same vein, reliable power at work means that production lines are up and running at full capacity, improving productivity. The internet fosters business through opening up access to new technologies around the world and thereby enabling companies to offer better services.

23 Green and Strong Systems Building

In the twenty-first century, infrastructure construction is more than just building roads or power plants. It's about coming up with systems that are ecologically sustainable and impervious to threats such as climate change or natural disaster. Green infrastructure emphasizes resource sustainability: solar panels reduce dependence on fossil fuels, while green buildings minimize waste and prevent pollution from reaching the environment (Wen et al., 2020). Smart water systems conserve rather than waste water, and recycling systems turn waste into renewables—all in a cost-effective manner.

Resilient infrastructure makes a speciality of getting ready for the unexpected. Systems ought to be designed to endure the developing frequency of challenges like floods, hurricanes, and earthquakes (Khan et al., 2021). Flood-resistant roads withstand heavy rains, while earthquake-resistant buildings guard lives and belongings. By incorporating such resilient designs, groups can better respond to adversities and preserve crucial operations.

23.1 Paying for and Building Big Projects

Building infrastructure is a high-priced enterprise, and most governments lack the sources to shoulder these fees alone. This is wherein partnerships among governments, personal enterprises, and global businesses grow to be invaluable. Such collaborations in infrastructure tasks now not best distribute monetary burdens but also proportion risks and duties.

A key instance is the Public-Private Partnership (PPP) version, wherein governments crew up with private buyers for funding, construction, and management. For instance, a private employer may construct a motorway, earning revenue through toll series, while the authorities ensures the undertaking serves public hobbies. By blending authorities oversight with the performance of the personal quarter, PPPs have emerged as an effective option to infrastructure challenges.

Innovative funding mechanisms, consisting of inexperienced bonds, are getting an increasing number of famous. These bonds fund sustainable initiatives, attracting investors who fee each environmental and financial returns. In addition, loans and offers from development banks and global businesses are essential in helping less advanced international locations to build vital infrastructure.

However, investment on my own is not enough for success. Effective making plans and execution are equally critical. Early engagement with neighborhood groups guarantees that initiatives address their wishes and stable public help. Transparent tactics and green control assist avoid delays, reduce charges, and fight corruption, growing a solid basis for sustainable infrastructure improvement.

24 Policy and Governance for SDG 9

24.1 Governments in Advancing Industry, Innovation, and Infrastructure

Governments play a crucial function in selling SDG 9—Industry, Innovation, and Infrastructure—through serving as system integrators. They accomplish that through strategic measures like country wide commercial techniques, digital transformation roadmaps, and comprehensive infrastructure master plans. Encouraging the adoption of inexperienced technology is equally essential. Initiatives which include awards, R&D funding, and rapid-monitoring approvals for infrastructure tasks help nurture innovation ecosystems which can be crucial for sustainable improvement (Leal Filho et al., 2022).

To foster inclusive industrialization, governments consciousness on bridging gaps in connectivity, electricity access, and shipping networks, specifically in underserved regions. Initiatives like India's Make in India software and South Korea's Innovation Cluster Development Program reveal how centered strategies can stimulate business boom and innovation.

24.2 PPPs: A Collaborative Method for Building Infrastructure and Services

Public-Private Partnerships (PPPs) are crucial for advancing SDG 9, combining the personal area's performance and innovation with the public region's responsibility in carrier shipping and procurement. These partnerships are instrumental in financing and handling big-scale initiatives, in particular in transportation and electricity infrastructure (Marx, 2019).

A amazing example is Istanbul Airport in Turkey, that's extensively diagnosed as a benchmark for large-scale infrastructure advanced beneath the PPP version. Similarly, Brazil's partnerships with personal firms within the electricity area have expedited the transition to renewable electricity.

The fulfillment of such collaborations is based on robust governance, transparent procurement methods, and equitable danger-sharing agreements. When these elements align, PPPs can play a transformative role in advancing sustainable infrastructure improvement (Maslova, 2020).

24.3 International Cooperation and Funding Mechanisms

International collaboration is vital for attaining SDG 9, particularly in developing international locations with limited resources. Organizations just like the World Bank, AIIB, and UNIDO play a crucial function by way of offering investment and technical understanding to help close those gaps (Dominguez Lacasa & Molina Vogelsang, 2024).

At the nearby degree, projects which include the Belt and Road Initiative (BRI) and the African Continental Free Trade Area (AfCFTA) are pivotal in enhancing alternate connectivity and selling infrastructure investment. Furthermore, groups like the Green Climate Fund and the Global Innovation Exchange ensure that infrastructure and innovation projects align with global sustainability dreams, helping long-time period development.

25 Challenges in Implementing SDG 9

25.1 Industrial Deficits

Many growing countries face essential gaps in infrastructure, along with dependable strength systems, efficient transportation networks, and sturdy digital connectivity—key elements essential for using commercial boom.

25.2 Economic Constraints

Insufficient funding for large-scale innovation and infrastructure tasks poses enormous demanding situations, limiting progress and development.

25.3 Regulatory and Governance Issues

Bureaucratic inefficiencies, corruption, and inadequately designed regulatory frameworks avert task execution and deter potential investments.

25.4 Skills Gap

The loss of professional people in era-extensive sectors hampers efforts to modernize industries and foster innovation.

25.5 Environmental Challenges

Balancing business growth with environmental sustainability gives a complex and persistent mission. 15.1 Green Innovation, Sustainable Infrastructure Opportunities (Table 2).

25.6 Renewable Energy

Developing infrastructure for sun, wind, and hydroelectric energy drives industrial increase whilst drastically decreasing carbon emissions.

25.7 Circular Economy

Innovations that convert waste into treasured sources allow industries to decrease their environmental effect and create new business possibilities.

Table 2 Green innovation and its opportunities

2.3 BILLION	40 PERCENT	2.6 BILLION	90 PERCENT	2.3 MILLION	30 PERCENT
Worldwide,2.3 billion people lack access to basic sanitation.	In some low-income African Countries, infrastructure constraunts cut businesses' productivity by around 40 Percent.	2.6 Billion people in developing countries do not have access to constant electricity.	More than 4 billion people still do not have access to the internet; 90 percent of them are in the developing world.	The renewable energy sectors currently employ more than 2.3 million people; the number could reach 20 million by 2030.	In developing countries, barely 30 percent of agriculture products undergo industrial processing, compared to 98 percent high-income countries.

25.8 Smart Cities

Technologies along with IoT-enabled waste control and power-green transportation systems promote sustainable urban industrialization and beautify operational efficiency.

25.9 Sustainable Supply Chains

Embracing green logistics and eco-friendly production strategies enables businesses strengthen their global competitiveness even as meeting the growing call for for sustainable products.

26 Technology Bridges the Gap

26.1 Digital Inclusion

Expanding broadband access and integrating digital gear empower organizations whilst fostering monetary inclusion, specifically in rural and underserved areas.

26.2 Automation and AI

Advanced technology raise productivity across industries and optimize infrastructure utilization, lowering waste and inefficiencies.

26.3 Blockchain for Transparency

Blockchain era strengthens responsibility by using combating corruption in public spending and infrastructure tasks via a obvious framework.

26.4 3D Printing and Local Manufacturing

These improvements decrease fees and democratize access to commercial abilities, making them priceless in useful resource-limited settings.

27 Improving Technology Access to Close the Gap as Outlined by Using SDG 9

Closing the era gap for comprehensive strategies that decorate digital infrastructure and make it reasonably priced for underserved populations. Policymakers ought to spend money on growing broadband networks in rural and distance flung areas, ensuring equitable access to the digital financial device. Subsidizing virtual gadgets and internet services for low-earnings groups can make generation greater low-priced. Public-non-public partnerships can force innovation and amplify technological attain, particularly in underdeveloped areas. Implementing virtual literacy programs guarantees that individuals can effectively use technology for personal and professional increase. Furthermore, open get admission to policies for scientific research and generation-sharing can accelerate understanding dissemination. Governments must additionally adopt regulatory frameworks that foster fair competition and reduce the fees of virtual services. Global cooperation in generation switch and local manufacturing of digital gear is vital to bridge the generation divide.

28 The Role of Innovation in Addressing Environmental Issues While Promoting Industrial Growth

Innovation is crucial for balancing business growth with environmental sustainability underneath SDG 9. Green technology, including renewable strength, sustainable materials, and strength-green systems, reduce environmental impact at the same time as fostering commercial enterprise productivity. Circular economy enhancements, like waste-to-resource structures, sell aid efficiency and restrict waste. Smart manufacturing via automation and digital twins reduces energy intake and complements precision, essential to fewer emissions. Encouraging green R&D can bring about soar ahead answers that mitigate pollutants and enhance business tactics. Innovation additionally allows climate resilience, inclusive of advanced water manage structures and sustainable urban infrastructure. Policymakers can foster innovation via tax incentives, green bonds, and innovation hubs that encourage collaboration amongst enterprise and studies establishments. By prioritizing sustainable innovation, nations can obtain industrial boom at the same time as decreasing environmental damage.

29 Conclusion and Future Research

SDG 9 acts as a foundation for progress, integrating innovation, enterprise, and infrastructure to drive economic growth even as tackling social and financial inequalities. Examples inclusive of virtual technologies, renewable power, and partnerships among governments and companies show off the transformative strength of those efforts.

Emerging and evolved economies alike illustrate how strategic infrastructure investments, mixed with innovation-led regulations, can foster big progress. However, obstacles like inadequate funding, complex guidelines, and a lack of skilled people continue to be important challenges. Achieving inclusive development requires policies that are both robust and adaptable.

30 Recommendations for Policy Makers and Practice

Investment in sustainable infrastructure is essential, and industries should be incentivized to transition to clean technologies. Simple, open, and fair regulations are key to attracting investment and ensuring projects are implemented. Public-private partnerships offer a compelling way to leverage resources and expertise, particularly for large infrastructure projects. Finally, investing in grassroots solutions is vital. Local, human-cantered ideas can effectively address immediate issues and drive inclusive growth.

31 Future Directions for Research and Implementation

In the future, research should explore new methods. To make industrial and infrastructure development more sustainable and inclusive. Technologies such as artificial intelligence, blockchain, and the Internet of Things hold great promise for improving efficiency and transparency. It is important to study how green innovations such as recycling and renewable energy How can this benefit communities and the wider economy? Research should also consider how funding and international cooperation can help countries with fewer resources. Locally, small-scale pilot projects that combine innovation, industry and infrastructure It can serve as an example for scaling up successful strategies. By working together—government, industry and communities—we can make meaningful progress into the future for the benefit of all those aligned with SDG 9.

References

Awasthi, A., Charan Pattnayak, K., Dhiman, G., & Raj Tiwari, P. (Eds.). (2024). *Artificial intelligence for air quality monitoring and prediction* (1st ed.). CRC Press.

Bai, C., Zhou, H., & Sarkis, J. (2023). Evaluating industry 4.0 technology and sustainable development goals—A social perspective. *International Journal of Production Research, 61*(23), 8094–8114.

Denoncourt, J. (2020). Companies and UN 2030 sustainable development goal 9 industry, innovation, and infrastructure. *Journal of Corporate Law Studies, 20*(1), 199–235.

Do Rosário Cabrita, M., Užienė, L., Garcia-Perez, A., & Viedma, J. M. (2023, September). Intellectual capital for enhancing sustainable industrialization: Towards the sustainable development goal (SDG) 9. In *ECKM 2023 24th European conference on knowledge management* (Vol. 1). Academic Conferences and Publishing Limited.

Dominguez Lacasa, I., & Molina Vogelsang, M. (2024). Enhancing international collaboration in science, technology, and innovation to achieve sustainable development goals. *Science and Public Policy, 51*(3), 509–525.

Fei, W., Opoku, A., Agyekum, K., Oppon, J. A., Ahmed, V., Chen, C., & Lok, K. L. (2021). The critical role of the construction industry in achieving the sustainable development goals (SDGs): Delivering projects for the common good. *Sustainability, 13*(16), 9112.

Gera, R., Chadha, P., Banerjee, S. P., Sharma, M., Pandey, A. K., Kampani, S., Dixit, S., Tummala, S. K., & Gatea, M. A. (2023). A narrative review on use of biomaterials in achieving SDG 9: Build resilient infrastructure, promote sustainable industrialization, and foster innovation. In *E3S Web of conferences* (Vol. 391, p. 01180). EDP Sciences.

Hales, R., & Birdthistle, N. (2022). The sustainable development goals—SDG# 9 industry, innovation and infrastructure. In *Attaining the 2030 Sustainable Development Goal of Industry, Innovation and Infrastructure* (pp. 1–8). Emerald Publishing Limited.

Kaplinsky, R. (2018). Fostering inclusive innovation for sustainable development. Pathways for Prosperity Commission Background Paper Series.

Khan, P. A., Johl, S. K., & Akhtar, S. (2021). Firm sustainable development goals and firm financial performance through the lens of green innovation practices and reporting: A proactive approach. *Journal of Risk and Financial Management, 14*(12), 605.

Küfeoğlu, S. (2022). SDG-9: Industry, innovation, and infrastructure. *Emerging technologies: Value creation for sustainable development* (pp. 349–369). Springer International Publishing.

Leal Filho, W., Vidal, D. G., Chen, C., Petrova, M., Dinis, M. A. P., Yang, P., Rogers, S., Álvarez-Castañón, L., Djekic, I., Sharifi, A., & Neiva, S. (2022). An assessment of requirements in investments, new technologies, and infrastructures to achieve the SDGs. *Environmental Sciences Europe, 34*(1), 1–17.

Liu, S. (2020). Interactions between industrial development and environmental protection dimensions of sustainable development goals (SDGs): Evidence from 40 countries with different income levels. *Environmental & Socio-Economic Studies, 8*(3), 60–67.

Mangla, S. K., & Ram, M. (Eds.). (2020). *Supply chain sustainability: Modeling and innovative research frameworks* (Vol. 2). Walter de Gruyter GmbH & Co KG.

Marx, A. (2019). Public-private partnerships for sustainable development: Exploring their design and its impact on effectiveness. *Sustainability, 11*(4), 1087.

Maslova, S. (2020). Achieving sustainable development goals through public-private partnership: Critical review and prospects. *International Journal of Innovation and Sustainable Development, 14*(3), 288–312.

Mavuri, S., Chavali, K., & Kumar, A. (2019, November). A study on imperative innovation ecosystem linkages to map sustainable development goal 9. In *2019 international conference on digitization (ICD)* (pp. 142–147). IEEE.

Opoku, E. E. O., & Yan, I. K. M. (2019). Industrialization as driver of sustainable economic growth in Africa. *The Journal of International Trade & Economic Development, 28*(1), 30–56.

Pradhan, P., Costa, L., Rybski, D., Lucht, W., & Kropp, J. P. (2017). A systematic study of sustainable development goal (SDG) interactions. *Earth's Future, 5*(11), 1169–1179.

SDG, U. (2019). Sustainable development goals. The energy progress report. *Tracking SDG, 7*, 805–814.

Sharma, M., Rawat, S., Kumar, D., Awasthi, A., Sarkar, A., Sidola, A., Choudhury, T., & Kotecha, K. (2024). The state of the Yamuna River: A detailed review of water quality assessment across the entire course in India. *Applied Water Science, 14*(8), 175.

Shukla, P., Sharma, P. K., Pandey, S., & Kunwer, R. (2024). Optimizing municipal solid waste recovery through density-based waste plastic segregation. *Journal of Material Cycles and Waste Management, 26*(4), 2334–2354.

Sinha, A., Sengupta, T., & Alvarado, R. (2020). Interplay between technological innovation and environmental quality: Formulating the SDG policies for next 11 economies. *Journal of Cleaner Production, 242*, Article 118549.

Wen, B., Musa, S. N., Onn, C. C., Ramesh, S., Liang, L., Wang, W., & Ma, K. (2020). The role and contribution of green buildings on sustainable development goals. *Building and Environment, 185*, Article 107091.

Breaking Barriers: The Role of Open Data in Advancing Global Sustainability Innovation

Deeksha, Ranjit Kumar Mukherji, and Amit Uniyal

Abstract

Global connectivity has become the new reality and thus opening up data has become one of the trans formative ways of handling cross-boundary sustainability issues. This chapter discusses how open data is breaking down silos in innovation, providing equal access to data and knowledge and supporting communities to lead local solutions to environmental and social challenges currently obtaining. Real examples included are as diverse as rural businessmen and women who employ water data to develop cheap irrigation or students who use air quality data to campaign for green space in cities. The chapter explores how open data is pointing the way towards overcoming the knowledge divide, energizing local ingenuity and encouraging people to work together worldwide. It also presents obstacles such as access, use and use, values and norms and provides guidance on how to overcome them to effectively and responsibly use open resources. In the future scenario, open data ecosystems are anticipated for being the main pillars that can lead to sustainable development, and can realize the asymptotic human potential for innovative solutions to advance in gauntlet of environmental issues. As a call to arms for governments, organizations, and individuals to fully support and fund open data, this chapter presents a world where knowledge is free from boundary and works together to build a future that is sustainable and fair.

Keywords

Agriculture sustainability • Global sustainability • Innovation • Open data • Water management

Deeksha · R. K. Mukherji (✉) · A. Uniyal
Department of Commerce, Graphic Era Deemed to be University, Dehradun, Uttarakhand, India
e-mail: rkummukh@gmail.com

T. Choudhury et al. (eds.), *Open Access as a Knowledge Platform for Implementing the United Nations Sustainable Development Goals*, World Sustainability Series,
https://doi.org/10.1007/978-3-031-99605-4_12

1 Introduction

Today's technological advancements and growing environmental issues have given birth to open data as a trans-formative force which drives change. Open data describes working data that anyone can access without any limitations to utilize and distribute freely. Open data breaks down information barriers to create a more balanced landscape that stimulates innovation together with global sustainability collaboration and participation for all groups (Open Knowledge Foundation, 2019).

The availability of essential datasets that encompass climate change information alongside public health statistics and natural resource records was previously limited by institutional constraints and both technological capabilities and economic resources. The existing barriers produced disadvantages for developing countries alongside small organizations and underrepresented communities because these groups lacked sufficient funds to buy proprietary datasets and conduct analyses. Through open access to information the Global Partnership for Sustainable Development Data (2021) allows people from all geographic areas and economic levels to work together on worldwide challenges.

Across multiple industries organizations witness transformation through open data initiatives. Through agriculture-related initiatives like the Global Open Data for Agriculture and Nutrition (GODAN) smallholder farmers now gain access to beneficial insights regarding weather forecasting and sustainable soil analysis and crop management Open data resources aid low-income regions through improved food security and crop resilience as well as local economic development (World Bank, 2020a, 2020b, 2020c, 2020d, 2020e). Open data on energy consumption and carbon emissions has turned available information into a potent force that drives innovations in renewable technologies and the optimization of energy efficiency for the sector. Open data initiatives have created faster advances toward die-carbonization and decreased environmental footprints according to OECD (2021a, 2021b).

Moreover, open data fosters transparency and accountability, essential components of sustainable governance. Open data publication initiatives from governments involving real-time pollution data as well as accessible budget information allow citizens to share in governmental decision-making processes. This collaborative approach helps sustainability solutions become response-based to every stakeholder's needs according to UNESCO (2022a, 2022b).

Open data serves as a powerful tool for people because it provides real-life data points which empower marginal communities to participate within the sustainable narrative. Through data accessibility open data solutions enable community groups along with researchers and entrepreneurs in developing nations to join global innovation projects. Community-based projects combining open data have recognized environmental risks while tracking biodiversity loss yet they help minimize natural disasters in vulnerable areas (Global Open Data Index, 2021).

The staggering abundance of open data opportunities remains threatened by various difficulties. Complete adoption of open data requires resolving matters involving data transformation standards as well as digital equity gaps and privacy and security doubts. Through public and private sector collaboration together with civil society institutions open data will establish itself as the backbone for sustainable development (OECD, 2021a, 2021b).

1.1 Background

Digital access to information operates as a key force that drive innovation together with creating economic growth while transforming societies. Open data has transformed into an influential game play charger in the fight against worldwide sustainability issues because it remains freely accessible for modification and sharing purposes. Open data breaks down information barriers to enable businesses researchers and policymakers and communities to base their decisions on sound evidence leading toward environmental and social development.

Through open data access organizations can achieve climate change mitigation alongside responsible resource management while collaborating across international borders with industrial sectors along with different disciplines. The system provides clear governance oversight while granting power to under served communities and speeds up the creation of enduring sustainable solutions. The availability of open climate data enables cities to create infrastructure with resilience while public pollution records pressure industries toward environmental responsibility.

Although open data bring numerous benefits to the table several implementation challenges still exist. The benefits of open data remain inaccessible to various stakeholders because problems with data privacy together with digital inequalities and diverse technological capabilities across populations limit who can actually harness information for progress. Open data must be maintained at peak availability alongside clear accessibility and knowledge of application for all population groups to create world free of barriers.

1.2 Problem Statement

Open data technology maintains great potential for global sustainability standards yet barriers exist to equal data access together with maximum usage. Numerous organizations alongside many regions face difficulties because they do not have the essential digital tools and workforce capabilities or suitable policy standards needed. To maximize their abilities to use open data for sustainability-based decision processes. Protection of personal information together with security worries and misuse concerns block the establishment of a complete inclusive data ecosystem.

The study examines how open data can achieve sustainability innovation equality by analyzing obstacles and recommending methods to advance its effectiveness. The identification of implementation and accessibility hurdles will enable us to develop systems through which open data serves all stakeholders across different regions and economics backgrounds for meaningful sustainable results.

2 Understanding Open Data: A Catalyst for Change

Open data serves as a transformative tool which addresses important global sustainability challenges alongside other worldwide pressing issues. Open data stands as information which the Open Knowledge Foundation (2019) describes as transparent and inclusive data accessible by anyone through no legal transformative, financial or technical barriers or limitations.

2.1 What Is Open Data?

Open data is information that is.

2.1.1 Accessible

Machine-readable formats such as CSV, JSON, and XML let anyone access this data through digital channels. Every user benefits from this simplified usage design.

2.1.2 Free of Restrictions

Open data runs unrestricted because removal of copyright protection clears the way for patent restrictions along with proprietary limitations.

2.1.3 Reusable

Licensing data enables two essential benefits: it helps people adapt the data while also creating links between datasets so innovation develops.

2.1.4 Interoperable

The open data structure produces system compatibility that enables analysis between platforms.

Global meteorological agency open data enables the development of predictive models by citizens and organizations for disaster response planning alongside agricultural venture design.

2.2 How Open Data Is Used in Local Communities

Open data provides power to local communities by enhancing public service quality and encouraging community-based innovations. Real-world examples demonstrate the Power.

2.2.1 Community-Led Air Quality Monitoring (Barcelona, Spain)

Residents from Barcelona joined forces with scientists to develop Luftdaten which is an open-data-based pollution monitoring program. The community used inexpensive air pollution sensors to gather measurement data which became available to the public. Through this approach citizens acquired the tools to defend fair air regulation and decide wisely about going outdoors. The local activist expressed that knowing air pollution was a problem was not enough until the newly gathered evidence enable true change to become achievable (Open Data Institute, 2023).

2.2.2 Fixing Broken Streetlights and Potholes (Chicago, USA)

Chicago citizens now benefit from 311 Service Tracker—an open-data reporting system that helps people file reports about streetlight outages and cracked roads along with garbage collection problems. Users of the platform can see where all submitted complaints have been placed and how their resolutions stand. Residents now check request processing through the new system that shows their information status after contacting the city authorities. Open data practices have enhanced both government accountability while simultaneously improving their response capabilities (City of Chicago Open Data Portal, 2023).

2.2.3 Mapping Food Deserts and Fighting Hunger (New York, USA)

Through its open data system New York City operates Plentiful App which provides users with food pantry location information. Food aid supports low-income families yet these families face challenges finding available resource locations. The use of real-time inventory data and pantry hours information through the Plentiful App enables users to determine pantry availability without waiting in lines or guessing about stock outs as a mother in the Bronx reported. Hundreds of people benefit from this initiative by using NYC Open Data (2022) data to access their food supply quickly.

2.2.4 Open Mapping Serves as a Flood Prevention System in Jakarta (Indonesia)

The Indonesian capital city of Jakarta encounters floods repeatedly while previous government responses toward floods demonstrated sluggishness. PetaBencana utilized open data from social media sources together with sensor information and official government reports to generate operational flood maps. Local residents use this information to choose areas not vulnerable to flooding. The shop owner praised this app because it helped him protect his store inventory against flooding. He expressed gratitude by saying ‘This app saved my business by enabling me to relocate my inventory ahead of rising water levels.’ World Bank (2021) reported this statement.

2.2.5 Transparent Police Accountability (Oakland, USA)

Through the Oakland Police Data platform Oakland California started monitoring police interactions and data about officer use of force together with citizen complaints. The data provided by this system helps activists together with journalists reveal racial inequality while demanding changes to public policies. A community leader fighting for equitable police conduct told Open Oakland that numerical evidence proved their previous experiences (Open Oakland, 2022).

2.3 Challenges Do Communities Face in Accessing and Using Open Data Effectively

Open data utilization at a community level encounters several obstacles which mainly stem from human and social limitations. The following section identifies important barriers with supporting evidence from research studies:

2.3.1 Digital Divide and Technological Barriers

The public through the internet and digital tools and technical learning are not equally available in all communities according to van Schalkwyk et al. (2017). Open data initiatives do not support marginal communities located in deprived rural areas because these groups face challenges in developing digital skills.

2.3.2 Data Complexity and Usability Issues

Open data availability does not guarantee simplicity because organizations present the information in complex unstructured formats that need expert analysis to interpret (Davies, 2019). Data users who lack expertise face challenges while working with datasets because these systems do not provide intuitive visualization systems that would help non-experts develop useful conclusions.

2.3.3 Lack of Awareness and Engagement

The availability of open data remains unknown to communities since they lack information about its local application potential (Zuiderwijk & Janssen, 2014). Open data initiatives maintain little practical connection to societal improvement because government departments and institutions do not engage citizens in creating data-based solutions.

2.3.4 Trust and Data Quality Concerns

Open data usage declines among communities that doubt government and institutional information sources in societies marked by previous data falsification or misinformation (Ruijer et al., 2020). Old or incomplete datasets or inconsistent data can cause negative impacts on decision quality through misinterpretation or wrong decisions.

2.3.5 Language and Localization Barriers

Multiple open data platforms primarily distribute information in major languages thus blocking access to non-English users and local dialect speakers (Gurstein, 2011). Local context representation in data results in reduced practical value for its intended applications.

2.3.6 Privacy and Ethical Dilemmas

People avoid participating with open data platforms because they worry about privacy breaches and improper data use (Taylor, 2016). An improper job of anonymization on sensitive data like health or personal information creates ethical challenges about data ownership protection and ownership rights.

3 Global Sustainability: A Shared Responsibility

Sustainable resource management and environmental issue mitigation operate as core attributes of global sustainability which protects current standards alongside future opportunities. Universal environmental protection demands worldwide collaboration to establish links between national borders and their ecosystems with global societies. Multiple global challenges such as climate change and biodiversity loss along with water scarcity and socio-economic inequality have become international problems requiring global industrial teamwork since the previous two decades.

Various sustainability issues reinforce one another necessitating deep planning to achieve sustainable goals. Interdisciplinary whole solutions require capabilities that autonomous nations together with independent organizations and regional units cannot achieve individually. MODERN CLIMATE CHANGES IN LOCAL ENVIRONMENTS CONTINUE PROPELLING FORENSIC OUTCOMES ACROSS ADJACENT ENVIRONMENTS THROUGH ROOM-TO-ROOM WEATHER PATTERNS AND HARMS IN HABITAT AND FOOD CHAINS. A globalized world requires international heritage programs which bind scientific information sources with pooled resources for unified response efforts.

The successful execution of international sustainability collaborations requires both complete access to data and its effective use. The fundamental role of open data in stakeholder collaboration stems from its ability to offer governments businesses and NGOs and civil society members precise real-time data at large scales. Transparency along with enhanced accountability and better decision-making results from data accessibility. Data-driven monitoring fundamentally supports assessment of SDG progress together with the identification of gaps in sustainability efforts throughout the world (United Nations, 2020).

Organizations together with countries use data as a tool to focus alike directions and monitor environmental results and create approaches that use evidence-based methods. The integration of innovative solutions happens through practical solutions while public-private partnerships form and sustainability initiatives work effectively for all populations. Digital transformation through data analytics drives operational efficiencies allowing organizations to reduce waste and advance sustainable resource usage.

3.1 Data as a Catalyst for Collaboration in Sustainable Development

The collective work of sustainable development depends heavily on open data which enables governments to unite with businesses to collaborate with researchers and communities. Data openness with transparent access helps groups form better decisions along with creating trust among stakeholders and enables innovation throughout communities. The following points explain specific methods through which open data helps collaborative partnerships.

3.1.1 Breaking Silos and Enabling Cross-Sector Partnerships

The solution of multiple sustainability problems which include climate change as well as pollution and resource management demands united participation between different groups. Open data functions as a gap-bridging solution because it enables organizations to exchange their findings for collaborative solution development. Environmental data in the public domain allows NGOs together with policymakers and businesses to coordinate their sustainability plans (Janssen et al., 2012).

3.1.2 Empowering Communities and Local Government

Open data access to communities allows them direct involvement in administrative decision-making processes. The combination of government partnerships with citizens produces an approach to sustainability that operates from the ground up (Davies, 2019). Open climate data serves as an illustration which enables farmers to make better agricultural choices for sustainable farming practices that protect both natural resources and achieve food security.

3.1.3 Enhancing Transparency and Trust

Open data contributes to trust-based collaboration because stakeholders can check information and monitor progress to ensure accountability. Business transparency is achieved through open availability of sustainability reports and emission data which allows consumers together with investors to measure environmental dedication (Ruijer et al., 2020). The open view into business practices and policies helps drive responsible management choices and policy agreement.

3.1.4 Driving Innovation Through Open Knowledge Sharing

The sharing of open data enables startups researchers and entrepreneurs to access beneficial data which they use for creating sustainable solutions. Open energy consumption data accessibility enabled researchers to generate smart grid technologies which optimize energy optimization (Zuiderwijk & Janssen, 2014).

3.1.5 Strengthening Global Cooperation for Sustainability Goals

International cooperation requiring data-intensive methods will help achieve the United Nations Sustainable Development Goals (SDGs). The Global Partnership for Sustainable Development Data enables nations to exchange both successful techniques and development metrics regarding poverty reduction and climate action through its open data platform (Smith et al., 2021).

4 Democratizing Innovation Through Open Access

Open access democratizes innovation by producing fundamental changes in the way knowledge sharing and resource accessibility distribute across communities. Mainstream innovation activity previously focused exclusively on regions and institutions that hold special access to advanced research elements and ample funding resources. The unrestricted availability of data combined with information is dismantling previous boundaries thus enabling worldwide users to innovate with complete freedom.

4.1 Empowering Individuals

Open data systems now provide infinite access to restricted data once exclusively held by educational facilities academic departments and large corporations. Through open scientific repositories together with publicly available datasets and open-source software individual researchers can address local issues and create new products. Through analysis of open data entrepreneurs in remote areas develop localized solutions such as agricultural technology applications that forecast local weather conditions and improve farming results (OECD, 2021a, 2021b).

Open access facilities access to innovation opportunities for people from various socio-economic environments to become involved in launching new creative concepts. Freely accessible materials offer students and independent researchers a path to participate in scientific and technological research despite their lack of funding for premium content. The democratization of innovation occurs through this system while working to bring essential diverse viewpoints required to solve international scale difficulties.

4.2 Empowering Communities

Open access delivers broad benefits to neighborhoods that experience limited resources especially in remote communities and marginalized areas. Open data provides local groups with necessary information so they can tackle their critical issues no matter if the matters concern healthcare or education or urban development. Geospatial data combined with demographic information allows communities to ask for improved infrastructure development as well as better disaster response plans. Open data initiatives throughout Kenya enable citizens to monitor government projects while holding their authorities accountable according to World Bank (2020a, 2020b, 2020c, 2020d, 2020e).

Through communal intelligence city residents can stimulate neighborhood-based innovations. Citizen science continues to grow as more people join teams to both gather data and examine its implications through collaborative efforts. Community members can participate in environmental and well-being research through these initiatives creating a sense of ownership which builds trust in innovation processes.

4.3 Empowering Organizations

Equal competitive opportunities now present themselves to organizations through open access systems primarily serving small businesses and non-profit entities which usually operate with limited financial resources of large corporations. Open data allows small businesses and non-profit organizations to receive critical information without incurring high costs so they can maintain competitiveness while they innovate for advancement. Through open access organizations within the renewable energy sector can quicken sustainable solution development through utilizing freely accessible research about energy-efficient technologies.

The free dissemination of data permits various stakeholders to team up with organizations from different sectors. Communities of organizations can develop collaborative solutions to difficulties through unfettered information exchange which tackles environmental challenges as well as medical innovations. The swift development of COVID-19 vaccines occurred through historic levels of data collaboration between governments and pharmaceutical companies and research institutes (Peters et al., 2021).

4.4 Removing Barriers to Innovation

The elimination of traditional impediments to innovation through open access surrounds researchers with unrestricted access that removes cost issues and geographical factors and exclusive barriers. Open access removes barriers to knowledge access which enables innovation to remain unrestricted from privileged groups located anywhere. The rate of technological advancement accelerates because open access enables new depths of collaborative problem-solving abilities leading to fairer economic opportunities worldwide.

4.5 Open Data as a Catalyst for Local Entrepreneurs and Researchers in Sustainability

Open data enables researchers and local entrepreneurs to create purpose-fit solutions that tackle community sustainability issues through directly accessible information. Free access to valuable information through open data stimulates advancements in innovation and improves decision quality and eliminates barriers between scientific disciplines and business sectors and community involvement.

4.5.1 Open Data Accelerates Innovative Processes of Both Green Startup Venture and Business Projects

Business owners from developing areas encounter challenges because they lack sufficient data about their markets and environmental conditions. Through access to open data entrepreneurs develop sustainable business approaches thanks to market trend and natural resource data and environmental consequence information. Open climate and agricultural databases enable start-ups to build precise farming technologies which maximize water and fertilizer efficiency while minimizing waste and strengthening food security nationwide (Davies, 2019).

The HiLo Tractor program in Africa demonstrates how open agricultural data can unite tractor operators with cultivating farmers to enhance agricultural productive along with lowering agricultural carbon emissions (Omidyar Network, 2018).

4.5.2 Empowering Researchers to Drive Local Sustainability Solutions

Through open data access researchers can identify local environmental and socioeconomic patterns which lead to evidence-based problem solutions. Open air quality platforms facilitate scientists'; development of pollution forecasting models which prompts changes within urban planning policies (Zuiderwijk & Janssen, 2014).

Indian scientists applied open deforestation data to pinpoint areas of illegal logging so they could lobby officials for stricter environmental regulations (Smith et al., 2021). Data-driven advocacy lead to sustainability solutions which produce concrete benefits that solve issues at specific locations.

4.5.3 The Removal of Market Entry Obstacles Helps Social Enterprises Participate More Easily in Business Activities

Small sustainable businesses face major financial challenges when they need to acquire proprietary data. Through open data everyone in society achieves vital knowledge without needing to expend financial resources. Entrepreneurs in renewable energy solution development gain access to free meteorological and solar radiation data to find their ideal locations for solar farms (Janssen et al., 2012).

Through open energy data Solar Sister has created a social enterprise to deliver affordable solar lamps in African communities which decreases kerosene usage and decreases carbon emissions (World Bank, 2020a, 2020b, 2020c, 2020d, 2020e).

4.5.4 Strengthening Community-Led Sustainability Initiatives

Open data enables grassroots organizations to receive empowerment beyond academic institutions and businesses. Locals across communities employ open mapping solutions to monitor natural resource deficiencies together with waste management problems and biodiversity reduction patterns. The indigenous communities in Brazil use open satellite data to track illegal Amazon mining operations as part of their environmental conservation efforts (Ruijer et al., 2020).

5 Case Studies: Real-World Impact of Open Data on Sustainability

5.1 Resource Efficiency: Open Energy Data in India

Resource efficiency within India has advanced rapidly through the implementation of open energy data systems. Through its National Power Portal the Ministry of Power makes energy consumption and power generation and renewable energy source data publicly available. The available energy data has enabled public and private organizations to maximize energy consumption efficiency while enhancing their operational performance levels.

5.2 Conservation and Biodiversity: Global Forest Watch (GFW)

Through GFW the World Resources Institute offers free monitoring tools that use satellite data to track forest erosion and land management changes. The initiative stands as the global benchmark contributing monumental changes to worldwide conservation work. Brazilian authorities used GFW data to identify illegal logging in the Amazon rainforest which helped them perform swift intervention. GFW enables Indonesian local communities to protect endangered habitats through its real-time distinctions of deforestation activities. GFW provides transparent data which supports governments and NGOs in both enforcing conservation policy along with protecting biodiversity because of its easily implementable format (Hansen et al., 2013).

5.3 Water Management: Open Data in the Netherlands

The Dutch water management approaches depends heavily on open data functionality to manage flood avoidance and water supply challenges. Users can access real-time water data on water levels as well as rainfall and river flow information through Rijkswaterstaat's government platform. Real-time water analytics help municipalities along with engineers and researchers optimize their water management while building better flood defense systems.

Smarter dike systems function through prediction analytics which leverage current data to detect breaches. The integration of open water data allows farmers to find better irrigation methods which optimize water conservation alongside productive agricultural outcomes. Excellent examples from Rijkswaterstaat demonstrate that open data solutions help resolve environmental concerns together with generating economic benefits (Rijkswaterstaat, 2021).

5.4 Agricultural Sustainability: Africa's Open Data Initiative

Through open data programs across sub-Saharan Africa smallholder farmers gained the know-how to introduce sustainable farming techniques. Through the Africa Open Data for Environment Agriculture and Land Initiative farmers obtain access to soil quality information along with weather forecasts and information about suitable growing areas.

Soil map data together with weather information enables farmers to take better decisions through choice of appropriate crops and best planting dates. Data-driven mobile applications driven by open data help Kenyan farmers boost their crop yields and minimize pesticide intake.

6 The Impact of Open Data on Sustainability Innovation

Numbers demonstrate open data capability to extend global GDP by $4.6 trillion because it enhances operational efficiency while fueling innovation and building market clarity (McKinsey Global Institute, 2023). The strategic implementation of open data into public services enables a 10–30% reduction in operational expenses leading to more efficient governance (OECD, 2022) (Table 1).

Open data serves as a transformation force for environmental impact by motivating 75% of cities throughout the world to direct climate action through air quality tracking and energy optimization (World Resources Institute, 2023). Direct evidence shows that climate change reception received a 15% reduction through open data-supported policy initiatives (United Nations Report, 2021).

Open data proves to have dramatic effects in advancing both research activities and inventive practices. Open datasets have become available throughout the scientific community leading to scientific productivity growth of 40% and quickening innovations in renewable energy and biodiversity conservation and pollution control (Nature, 2022).

To fully realize its trans-formative potential open data faces substantial hindrances to adoption because 68% of organizations recognize data privacy as

Table 1 Sustainable Innovation outcomes

Category	Metric	Value	Source/year
Economic impact	Estimated GDP boost via open data	\$2.8–\$4.6 trillion	McKinsey Global Institute (2023)
	Increase in efficiency in public services	10–30% cost reduction	OECD (2022)
Environmental impact	CO_2 emissions reduced via open data-driven policies	15% reduction	UN Report (2021)
	% of cities using open data for climate action	75%	WRI (2023)
Innovation and research	Increase in research output due to open data availability	40% growth	Nature (2022)
	Number of sustainability startups leveraging open data	2500+	Open Data Institute (2023)
Challenges and barriers	% of organizations citing data privacy as a major concern	68%	World Bank (2023)
	% of developing countries with limited open data infrastructure	55%	UNDP (2022)

a significant issue (World Bank, 2023). An open data infrastructure benefits developed nations but 55% of developing countries lack sufficient access which generates a digital gap that blocks equal sustainability participation (UNDP, 2022).

The implementation of open data initiatives delivers multiple sustainability benefit through their functions in enhancing transparency and citizen empowerment alongside their role in promoting informed choices.

6.1 Brazil's Open Budget Transparency Portal

Brazil's Office of the Comptroller General established the Transparency Portal during 2004 as a platform to enhance public budget transparency through unrestricted data disclosure. The tool functions as the main weapon against corruption by attracting between 900,000 different users every month. Open data initiatives in local Brazil public finances and in three more Latin American countries have been inspired by this successful project.

6.2 Mexico's "Mejora Tu Escuela"

The Mexican Institute for Competitiveness (IMCO) built Mejora Tu Escuela which presents school performance information online for citizens to use. Through this platform parents discover appropriate schooling options while gaining authority to obtain better education services and gather information about unsatisfactory areas which drives educational advancement at the national level in Mexico.

6.3 Uruguay's "A Tu Servicio"

The organization DATA in Uruguay worked with the Ministry of Health to establish A Tu Servicio which shows health government data through simple website interfaces. Users can leverage this portal to examine healthcare service providers using different evaluation standards which ultimately helps citizens make more educated choices about their treatment.

6.4 Kering's Environmental Profit and Loss (EP&L) Initiative

Through the EP&L Environmental Profit and Loss method the luxury group Kering monitors and evaluates environmental effects stemming from its operations. Kering tracks CO_2 emissions along with water consumption and pollution and land use and waste generation across its complete supply chain so it can direct its sustainable development strategy and share information directly with its stakeholders.

6.5 City of Orléans' Open Data Portal

Orlèans in France ensures open data accessibility through a portal to present environmental programs to the public. Users have access to environment-related datasets about radio wave emissions together with pollutant concentration data and information regarding regional biodiversity through this portal. The portal improves public transparency while it empowers citizens to learn more about their environment which directly assists regional development alongside environmental protection agendas.

7 Bridging the Knowledge Gap: Open Data for Inclusivity

Under the paradigm of open data, individuals from underserved populations can now obtain vital information in order to bridge the knowledge gap and achieve information parity with more developed communities. Throughout time, access to essential information about education, healthcare and economic possibilities has demonstrated unequal distribution across populations. Numerous areas face limited sustainable development because this disparity between knowledge distribution continues to fuel persistent cycles of poverty. Through open data initiatives, we establish a more uniform global society that permits everyone to actively impact their prospective scenarios.

7.1 Equal Access to Opportunity

Valuable precise data enables a transformative shift for marginalized communities. Communities experiencing marginalization transform when they gain access to reliable data in real time. Open data systems remove financial restrictions in regions with insufficient infrastructure, so these areas can obtain essential data they couldn't afford before. Medical providers in developing nations obtain essential research from PubMed's open-access medical database so they can deliver improved care through informed clinical decisions.

7.2 Empowering Marginalized Communities

Decisions platforms exclude marginalized communities from their structure because of inadequate representation which results in failed perspective recognition. Open data equips marginalized communities through accessible tools which help them defend their rights and share their ideas effectively. The Open Data for Development (OD4D) initiative enables grassroots organizations to utilize data in their efforts to resolve local problems. Open data in Nepal supports the identification of relief distribution imbalances which allow aid to reach marginalized communities suffering from natural disasters according to World Bank's 2020 findings.

Open data projects bring equality to governance through universal access to information. Through Open Government Data platforms citizens exercise the ability to monitor public spending while tracking projects and exercising oversight of public officials. Transparent systems broker trust between citizens and governments and establish equitable resource distribution which targets vulnerable sectors.

7.3 Addressing Global Inequities

Through its data accessibility open data helps overcome worldwide economic differences in the fields of agriculture and climate resilience management. The availability of open data about weather patterns and soil conditions enabled African smallholder farmers to implement sustainable agricultural practices which both raised their crop yields and decreased their production expenditures. Through open data initiatives such as GODAN (2021a, 2021b) farmers can break through system limitations to safeguard their economic welfare.

Open data helps vulnerable communities adapt through developing resilience in situations of climate change. By providing real-time information about sea level rises and extreme weather events together with data about deforestation at-risk citizens gain knowledge which helps them prepare climate-risk mitigation strategies. Southeast Asian coastal communities developed local early tornado and flood safety systems using freely available climate information which has saved residents and property from harm (Asian Development Bank, 2021).

7.4 The Role of Technology in Bridging the Gap

Open data capabilities expanded through digital transformation now extend to the remote areas throughout the globe. Mobile applications combined with open-source software along with low-cost digital devices allow widespread effective distribution of information. Project Loon utilizes balloon technology to offer internet connectivity throughout rural domains and therefore functions to bridge digital gaps and deliver open data service accessibility independent of geographic location or infrastructure condition constraints.

7.5 Barriers to Open Data Access and Use for Marginalized Communities

Through access to open data minorities gain essential information needed for planning decisions and campaigning purposes and resource allocation needs. The communities struggle to obtain and utilize openly available data despite these potential benefits. Open data faces substantial hurdles that combine technical elements with social and economic and political requirements.

7.5.1 Digital Divide and Limited Technological Access

All the necessary fundamental tools represent essential blockages in reaching the benefits of open data (Davies, 2019). Environmental and agricultural data becomes accessible online even in Sub-Saharan African regions yet smallholder farmers cannot use it because they lack digital literacy and suffer from poor network connectivity (Smith et al., 2021).

7.5.2 Data Illiteracy and Lack of Capacity-Building

Open data accessibility does not guarantee proper utilization by marginalized groups who need skills to interpret and make effective use of it. Complex datasets need specific analytical or visualization skills from users to effectively interpret them. The purpose of data stays in the realm of theory instead of functional use since training programs and community support are absent (Janssen et al., 2012).

The lack of localization efforts for data literacy prevents Indigenous communities in Latin America from maximizing the public availability of deforestation data (Ruijer et al., 2020).

7.5.3 Language and Cultural Barriers

The majority of open data platforms and government portals use dominant languages that block participation of communities who talk native or minority languages. The existing plan format of data systems often fails to match regional information organization methods. Scientific climate data formats fail to match the traditional forms of knowledge about ecology leading to reduced effectiveness for local adaptation methods (Smith et al., 2021).

7.5.4 Mistrust and Lack of Inclusion in Decision-Making

As a matter of history marginalized communities received no opportunity to participate in the development of data governance frameworks. Open data initiatives receive skepticism because doubts exist about the fundamental reasons for their implementation. When community observations differ from the data it indicates that the information seems irrelevant or tainted to those who experienced it. Data privacy and exploitation problems become major trust concerns when corporations use open land-use data for commercial gain (Davies, 2019).

7.5.5 Data Relevance and Contextualization Issues

Open data initiatives typically distribute generic high-level indicators which fail to provide solutions for marginalized community needs. Local farming communities and urban slum dwellers need detailed information that global climate datasets cannot properly deliver. The scarcity of indigenous surface data generated by the community hinders accessibility of the information (Zuiderwijk & Janssen, 2014).

7.5.6 Power Imbalances in Data Governance

Because open data usually rests under governmental and international organizational and large corporate authority marginalized communities face the risk of having their data needs ignored. The attempt of communities to create their own data through local activist water quality monitoring usually fails due to their lack of institutional backing and limited resources (Smith et al., 2021).

7.6 Affected the Accessibility and Utilization of Open Data, Particularly in Rural or Underserved Areas

Open data accessibility together with its utilization has experienced significant positive effects due to digital transformation initiatives in remote and under served regions. Open data benefits from technological advancements but these improvements have not solved the many obstacles that prevent successful community utilization of open data in rural areas.

7.6.1 Infrastructure Limitations

Strategies to increase digital connectivity across nations have not resolved the problem of broadband connectivity and essential technology infrastructure gaps in rural communities. The digital gap curbs local populations from accessing online open data platforms which reduces their capacity to participate in data-based initiatives. The digital equity challenge continues to affect South African students because rural students lack equal access to internet and computer resources when compared to urban populations.

7.6.2 Digital Literacy Gaps

The incomplete digital literacy skills create a major impediment to overcome. The people in underserved areas who gained internet connectivity often lack skills needed for proper open data interpretation and navigation. Numerous studies show internet access alone does not lead to meaningful use so these communities need structured digital literacy education programs to get real benefits from data-sharing resources.

7.6.3 Relevance and Standardization of Data

Open data becomes effective through successful application of local context information. The absence of standardized data formats together with unclear ownership issues leads to unequal distribution of resources because big corporations control the digital agricultural sphere thus pushing out small farmers and locally based agripreneurs. Open data initiatives need proper evaluation of rural contexts to create successful programs for all communities.

7.6.4 Socio-Cultural Factors

Digital tool adoption and utilization gets influenced by socio-cultural dynamics which include language barriers along with traditional practices. Digital inclusion programs such as the Digital India Program perform differently in India depending on how older the population gets and what educational status the country maintains and what rates of technology adoption exist. Successful digital initiatives need to conform with socio-cultural elements because it ensures both their successful implementation and effectiveness.

7.6.5 Community-Led Data Initiatives

Open data initiatives serve rural communities better when they establish systems which allow community members to handle and regulate their data autonomously. Data model combinations between data pools and data cooperatives allow communities to create their own local datasets which promotes fair digital participation.

7.6.6 Policy and Regulatory Support

Digital transformation needs supportive policies with regulations as fundamental elements to succeed in rural areas. Open data implementation becomes more efficient when digital infrastructure collects investment while programs build capacity and data governance practices welcome inclusivity. The European Commission stresses how digital technologies can improve rural areas by making them more sustainable and attractive and intelligent simultaneously.

8 The Role of Technology: Open Data Reveals Connections Through Technology Application

Open data interactions become more powerful through artificial intelligence (AI) combined with blockchain technology and Internet of Things (IoT) emerging technologies. This combination converts simple information into usable knowledge. Through technological breakthroughs we can analyze substantial amounts of data to uncover patterns while manufacturing actionable solutions for vital world problems. These technological tools create relationships between data points to create actionable insights which drive both innovation and sustainable development goals.

8.1 Artificial Intelligence: Unlocking Insights from Complex Data

The transformation of open data into perceptible insights relies heavily on artificial intelligence systems. The rapid processing power and analysis of enormous datasets by these technologies enables researchers and organizations as well as governments torecognize patterns so they can base their decisions on data. The use of AI algorithms allows organizations to perform weather pattern prediction while tracking tree loss conditions and identifying disease outbreak signs.

Climate modeling represents one of the chief ways AI systems become operational. Both open climate data used with predictive AI systems reveals probable pathways of extreme weather patterns to enable vulnerable populations with better disaster readiness. Through open patient databases medical AI tools produce quick diagnostic results then generate specific treatment strategies in areas with limited medical services according to UNEP (2022a, 2022b).

Massive climate simulations regularly benefit from Artificial Intelligence systems as an evident example. AI systems analyze public climate datasets to forecast extreme weather tracks thus helping populations in danger prepare for emergencies. Artificial intelligence systems currently assess healthcare data stored in open repositories to detect conditions early while tailoring individual treatment plans toward vulnerable populations (UNEP, 2022a, 2022b).

8.2 Block-Chain: Ensuring Transparency and Trust

The decentralized character of block-chain together with its tamper-proof foundations enables full transparency of open data. Dependable information regarding product origins and delivery routes in supply chains requires a centralized approach from stakeholders.

Sustainable sourcing operations are tracked using block-chain systems that serve cocoa and coffee industries and others. Through block-chain technology users can track supply chain progress through encrypted ledger systems to verify product integrity including ethical standards and environmental regulations.

8.3 Internet of Things: Real-Time Data Collection and Monitoring

Real-time data collection becomes possible when physical devices, including sensors, cameras, and smart appliances, connect to internet networks through the IoT. The devices continuously create open data streams that can produce valuable insights from their analysis.

Soil moisture sensors, alongside temperature and crop health sensors deployed in agricultural areas, help farmers monitor their fields correctly, leading to enhanced irrigation efficiency and decreased water wastage through feedback data collection. IoT devices across urban areas continuously measure traffic flow and air quality while tracking energy use to help cities create more efficient, sustainable frameworks. The Barcelona smart city project shows IoT devices together with open data channels improving waste removal efficiency while decreasing energy requirements in the city (European Commission, 2021a, 2021b, 2021c).

8.4 Synergy: When Technologies Work Together

Comprehensive benefits occur from integrating multiple technologies. Through this combination, AI processes real-time IoT data, but block-chain provides data security together with complete transparency. The fight against illegal fishing combines IoT equipment with fishing vessels to acquire location data and activity logs, while AI detection tools find irregular patterns, and block-chain establishes transparent documentation of all information. Through uniting these technologies, authorities can both protect marine biodiversity while effectively enforcing regulations (FAO, 2022).

Renewable energy management benefits from AI, IoT, and block-chain technology united in a transformative fashion. The integration of IoT into smart grids allows for consumption data collection, and AI uses the data to achieve optimized distribution, while block-chain powers the transaction system for peer-to-peer renewable power sharing between sustainable community members.

8.5 How Do They Enhance the Effectiveness of Open Data?

Open data initiatives receive substantial improvement from the emerging technologies Artificial Intelligence (AI) and Block-chain and the Internet of Things (IoT). Here's how each contributes:

8.5.1 Artificial Intelligence (AI)

The analytical strength of AI allows extensive data processing to find important patterns while supporting better decisions. AI processes real-time information collected from multiple data sources to detect environmental pollution patterns which results in improved targeted interventional strategies.

8.5.2 Blockchain Technology

With block-chain technology organizations can maintain transparent data systems through decentralized record storage which also prevents modifications. The use of blockchain technology together with IoT devices creates an enhanced system for tracing supply chain movements because it confirms product authenticity and fight counterfeiting operations.

8.5.3 Internet of Things (IoT)

The IoT technology helps devices record physical world data in real-time thus adding current data to open data collections. The nonstop data stream enables smart city applications by enabling infrastructure sensors to track both urban health along with traffic flows and environmental variables thus enabling preventive upkeep and better urban design.

8.5.4 Convergence of Technologies

When these three technologies work together they enhance all their separate capabilities. These technological approaches generate improved outcomes when implemented together in industrial environments to create better data protection and operational organization and predictive equipment monitoring systems.

9 Challenges on the Path to Openness

Fully accessible data ecosystems face multiple hurdles on the path to becoming open and available to everyone despite their potential trans-formative power. A variety of obstacles involving data privacy along with standardization difficulties and digital access disparities hinder the path toward effective open data solutions that also promote security and fair distribution of its benefits. The pathway to balanced openness requires protocols which safeguard individual human rights behind the transparent access to information.

9.1 Data Privacy: Protecting Individual and Organizational Rights

Open data faces an urgent difficulty in the area of maintaining data privacy. The disclosure of information for public access promotes innovation and partnership yet exposes vital data to potential exploitation. The improper anonymization of healthcare information makes it possible for patient-sensitive details to leak thus causing privacy invasion. Organizations operating with big data systems face potential consequences when they uncover proprietary information due to disclosures.

The implementation of advanced privacy frameworks remains crucial because of this challenge. The European Union's General Data Protection Regulation

(GDPR) provides specific rules about data sharing along with protocols to maintain individual rights protection. Construction of global data protection systems requires international jurisdictions to work together alongside investments in modern encryption strategies and anonymization methods (European Data Protection Board, 2021).

9.2 Lack of Standardization: Bridging Fragmented Data Systems

Multiple organizations including public agencies and private sector entities and research institutions supply open data but different presentation methods between these sources complicate both analysis and integration. Diverse measurement standards utilized by nations when collecting climate data create conflicts which prevent successful comparison between datasets for global climate modeling.

Standardization initiatives represent the key solution to overcome this challenge. The Open Data Institute leads efforts to promote standard data formats and interoperability methods while establishing metadata standards for maximal organizational connectivity. Acquiring worldwide agreement about these standards proves difficult because different participating entities possess different technological expertise and prioritization approaches (ODI, 2022).

9.3 Digital Inequity: Ensuring Access for All

The value of available data depends entirely on its accessibility for people and communities intending to use it. Digital inequity emerges from unequal internet access conditions combined with varying digital skills and inadequate technology infrastructure that restricts the geographic spread of open data projects. Internet connectivity failure in rural spaces that serve under served areas blocks access to online data repositories and prevents communities from utilizing open data benefits.

The digital divide shows its clearest manifestation in developing countries through rural–urban differences since World Bank records show 32% internet access among rural populations versus 89% in urban territories (World Bank, 2021). Open data benefits require affordable internet infrastructure alongside programs to build local capacity for effectively using data and enhancements to digital literacy. Global expansion of Project Loon's remote area internet connectivity platform shows promise in bridging the accessibility gap as long as suitable implementation methods are found.

9.4 Balancing Openness with Security

The disclosure of public datasets might unintentionally reveal system weaknesses, especially when dealing with essential domains such the energy sector, transportation, or finance. Open data platforms experience cyberattacks, which result in the loss of critical sensitive data and interruption of fundamental operational services.

Securing data-sharing infrastructure requires constant government support together with dedication from organizations. Block-chain technology runs data sharing operations on a decentralized network that permits secure access while creating an unalterable document system. Only authorized users can access or modify shared information on this platform. To minimize potential risks, organizations must maintain active contingency plans while performing periodic cybersecurity audits (World Economic Forum, 2021).

9.5 Cultural and Organizational Resistance

Organizations and governments refrain from accepting open data because of apprehension regarding losing oversight of their information. Stakeholders often shun open data initiatives because they worry about being responsible for their information, and fearing damage to their reputation along with possible misuse of shared data. Various governments suppress open data because their operations may get exposed due to insufficient efficiency or corruption, while private sector organizations maintain concerns about market advantage losses.

To establish an open culture, organizations must show concrete advantages that come from open data, such as better public services combined with economic increases. Collaboration between public institutions and private entities, combined with recognition of implementation successes, works to defeat reluctance and construct trust.

10 Collaborative Networks: Building a Global Data Ecosystem

Open data maintains its ongoing value within our interconnected world because successful partnerships act as the core foundation that enables its deployment. A global data ecosystem powered by government-to-government, organization-to-organizational, and individual-to-individual partnerships is necessary because it enables transparency, innovation, and broad inclusivity. Open data initiatives maintain their robust, productive character through partnerships that enable diverse stakeholders to unite resources while exchanging expertise to handle common challenges, so these initiatives remain impactful and equitable.

10.1 The Role of Governments: Enabling Infrastructure and Policy

Open data ecosystems function as the primary support structure constructed by government institutions. Through funding distribution and policy development with data system maintenance these entities support collaborative tasks. Through the Open Government Partnership (OGP) about 75 nations actively join forces to ensure people get access to public data while producing inviting spaces for civic involvement.

Data.gov.uk operated by the United Kingdom brings together more than thousand datasets on its platform so citizens and researchers and businesses can access public information. The practice of transparent data management creates an innovative atmosphere while sustaining loyal relationships between public institutions and their residents (OGP, 2021). The effective expansion of these strategies requires government intervention together with essential stakeholder partnerships between non-governmental organizations and communities.

10.2 Organizations as Catalysts for Innovation

Open data programs receive funding alongside specialized skills along with innovative solutions from private organizations and nonprofit entities. Modern technology companies establish market leadership by creating frameworks for accessing and implementing data properties. Through collaboration with the United Nations Google implemented Global Forest Watch which uses satellite data to track deforestation instantly. These partnerships enable advanced technological use while expanding the open data impact on major global issues such as climate change (WRI, 2022).

The important function of non-profits consists of filling gaps particularly in regions that receive inadequate attention. The Open Data for Development Network (OD4D) enables organizations to connect with local communities thus assuring relevant data solutions can be effectively used at the community level. These initiatives demonstrate the essential role which partnerships play in delivering global information into local implementations.

10.3 Individuals as Data Stewards and Advocates

When individuals maintain authority and access to participate open data ecosystems flourish. Open data engagements are driven by citizen scientists together with researchers while activists work towards transparency through advocacy and analysis. OpenStreetMap depends on global volunteers who collaborate to produce detailed and accurate maps of local neighborhoods for disaster relief preparation and urban development and human relief efforts.

The citizens of Kenya employed open data to create healthcare facility maps that revealed service gaps throughout the COVID-19 pandemic. Through grassroots engagement people strengthened government operations while proving that citizens possess the ability to develop data-centric solutions (World Bank, 2020a, 2020b, 2020c, 2020d, 2020e).

10.4 Fostering Multi-stakeholder Partnerships

Success with open data projects requires powerful connections between governments and organizations and people who work together to leverage individual strengths. The Open Data Charter provides an operational basis for stakeholder partnerships through guidelines which support collaborative practices.

These strategic collaborations help maintain inclusive data initiatives which address key issues surrounding privacy needs and data standards and digital access barriers. The European Data Strategy leads the European Union and tech companies to jointly share data between industries safely (European Commission, 2021a, 2021b, 2021c).

10.5 Addressing Global Challenges Together

The resolution of worldwide complications including climate change and public health emergencies and resource challenges depends heavily on collaborative network participation. When international organizations unite with national governments and local communities through open data initiatives they produce salable solutions more effectively.

During the COVID-19 pandemic the joint effort of the World Health Organization (WHO) with data providers delivered real-time global health information to support prompt decision-making processes. These joint efforts illustrated the ability of a worldwide interlinked data environment to handle crises both speedily and with fairness (WHO, 2021).

11 Future Horizons: Open Data for a Resilient Planet

The growing concept of open data presents opportunities to defeat escalating worldwide environmental risks. Open data transforms the world through collective partnerships which produce novel solutions and support accurate decisions. Further, imagining this future necessitates an attempt to utilize open data positively. Scaling, inclusivity, and sustainability are goals that must be pursued together. With such approaches, solutions to protect the environment and enable future generations to thrive can be achieved.

11.1 Scaling Solutions Through Data-Driven Innovation

The future has the potential for open data to play an even more significant role. With its ability to drive constant innovation, open data has already aided the world in achieving feats never thought to be possible. The continued advances in data quantity alongside improved data quality will enable us to tackle resource scarcities along with climate change and biodiversity decline.

The open data platform GODAN Global Open Data for Agriculture and Nutrition helps farmers use resources effectively for better crop yield optimization. Maps enabled by advanced data analytics solutions lead precision farming techniques to minimize waste while improving global food security for billions of people (GODAN, 2021a, 2021b). Also, in the case of renewable energy, the established open data on solar irradiance and wind patterns makes energy systems clean and can transform the focus into a much cleaner new world for energy. This helps in speeding up the transitions people are looking for at a global level.

11.2 Encouraging Broad Reach Cooperation

Maintaining the vision about the planet being resilient is a collaboration opportunity at global front. Open data bridges the gap between governments, organizations, and citizens enabling coordinated takedown across borders. For example like the Paris Agreement uses open climate data for climate tracking, monitoring, and stakeholders reporting.

The expectation from future Global collaborations would be heavily invested in carbon capture technologies, circular economy and sustainable cities planning. Open data makes it possible for developing countries to actively participate in the effort to democratize access to cutting edge research and data (UNEP, 2022a, 2022b).

11.3 Enabling Improved Decision-Making

The decision to work and change policies are done on base of concrete data, which is available due to open data. Resilient focuses heavily on guiding towards fully informed decision making.

For instance, disaster-prone areas might make use of weather and geospatial data to construct predictive models that strengthen their ability to withstand floods, hurricanes, or fire. Open data infused early warning systems have initiated the saving of immeasurable lives and the next step is broadening access to these systems alongside making them more intricate (World Bank, 2021).

11.4 Empowering Grassroots Action

Future advancements in open data will not involve powerful organizations only, these steps will also need a bottom up approach. Already, communities across the globe are leveraging open data to address local issues, including but not limited to monitoring deforestation or enhancing waste management systems. It is not unimaginable to picture a world where every community possesses the tools and skills to think globally while acting locally.

For example, OpenStreetMap might one day develop into an advanced hyper-local aid which will allow individuals to manage resources and counter environmental dangers. Such community participation will boost the planetary resilience (OSM Foundation, 2022).

11.5 Overcoming Problems for a Beneficial Future

Despite having immense potential, open data comes with its own set of concerns including privacy, standardization, and accessibility. There is a need for investment in digital infrastructure together with the embracing of global standards to make open data a tool everyone can use. Co-working will greatly decrease the chances of open data inflicting harm and instead make it more likely to be beneficial—since for data to inflict harm there is always a need to hide something.

12 Conclusion

Open data is essential because we live in a period where crucial global challenges such as environmental threats together with social disparities, require immediate collaborative approaches and sustainable solutions. The open data framework establishes a cross-boundary environment that enables teams from every nationality, culture and financial level to work together on solving complex modern issues.

Open data distribution enables marginalized groups to gain power through knowledge-sharing goals that both inspire new ideas and build international alliances. It serves as a bridge, connecting diverse stakeholders and enabling them to align their goals with a higher purpose: we need to use data as a tool to construct both equitable and lasting structures for society. Real-world stories demonstrate how collective endeavors between data-driven climate solutions adoption and industry resource transformation yield concrete benefits.

We have yet to achieve our ultimate destination. Open data's complete potential requires perpetual investment in technology platforms alongside continued policy inclusiveness and the development of transparent environments. Open data must maintain collaborative relationships through which it can function as a trans-formative force rather than remain an unused collection of data.

When we unite global understanding with purposeful data sharing, humans strengthen connections while embracing their common humanity along with their mutual responsibility to each other. Open data provides both the technical resources for developing new solutions while granting society a platform for redefining our advancement path centered around human dignity, together with environmental stewardship and economic abundance. Modern technology and open data will become our pathway to build a sustainable future shared by all humanity (Ministry of Power, 2022).

References

Asian Development Bank. (2021). *Open data for climate resilience.* https://www.adb.org

City of Chicago. (2023). *Chicago Data Portal [Open data portal]. City of Chicago Data Portal.* https://data.cityofchicago.org

City of New York. (2022). *NYC Open Data [Open data portal].* https://opendata.cityofnewyork.us

Davies, T. (2019). *The state of open data: Histories and horizons.* African Minds.

European Commission. (2021a). *Smart cities and communities.* https://ec.europa.eu

European Commission. (2021b). *Copernicus climate change service.* https://climate.copernicus.eu

European Commission. (2021c). *European data strategy: Building a data-driven Europe.* https://ec.europa.eu

European Data Protection Board. (2021). *Guidelines 07/2020 on the concepts of controller and processor in the GDPR (Version 2.1).* https://edpb.europa.eu/our-work-tools/documents/public-consultations

FAO. (2022). *Technology and innovation in fisheries and aquaculture.* https://www.fao.org

Global Open Data Index. (2021). *Empowering communities through open data.* https://index.okfn.org.

Global Partnership for Sustainable Development Data. (2021). *The role of data in achieving the SDGs.* https://data4sdgs.org

GODAN. (2021a). *Global open data for agriculture and nutrition.* https://www.godan.info

GODAN. (2021b). *Open data for sustainable agriculture.* Retrieved from https://www.godan.info

Gurstein, M. (2011). Open data: Empowering the empowered or effective data use for everyone? *First Monday, 16*(2).

Hansen, M. C., et al. (2013). High-resolution global maps of 21st-century forest cover change. *Science, 342*(6160), 850–853.

Janssen, M., Charalabidis, Y., & Zuiderwijk, A. (2012). Benefits, adoption barriers, and myths of open data and open government. *Information Systems Management, 29*(4), 258–268.

McKinsey Global Institute. (2023). *The economic value of open data.*

Ministry of Power, India. (2022). *National power portal.* https://npp.gov.in

Nature. (2022). *The rise of open data in scientific research.*

OECD. (2021b). *The role of open data in promoting innovation.* www.oecd.org

OECD. (2021a). *Harnessing open data for sustainable development.* https://oecd.org

OECD. (2022). *The role of open data in public service efficiency.*

Omidyar Network. (2018). *Open data's impact: Open data initiatives driving change.*

Open Data Institute. (2023). *Startups leveraging open data for sustainability.*

Open Government Partnership (OGP). (2021). *Open government partnership annual report.* https://www.opengovpartnership.org

Open Knowledge Foundation. (2019). *What is open data?* https://okfn.org

Open Oakland. (2022). *Budget Transparency Project [Dataset]. Open Oakland.*https://www.openoakland.org/projects/budget

OpenStreetMap Foundation (OSM). (2022). *Open data for localized solutions*. https://www.openstreetmap.org
Overseas Development Institute. (2022). *Financing a just transition*. https://odi.org/en/publications
Peters, M. D., Godfrey, C. M., McInerney, P., et al. (2021). Open access and collaborative innovation: Lessons from the COVID-19 pandemic. *Journal of Global Innovation, 12*(4), 45–52.
Rijkswaterstaat. (2021). *Water management data*. https://rijkswaterstaat.nl
Ruijer, E., Grimmelikhuijsen, S., & Meijer, A. (2020). Open data for democracy: Developing a theoretical framework. *Government Information Quarterly, 37*(3), Article 101485.
Smith, M. L., Reilly, K. M. A., & Scholte, J. A. (2021). Open data for sustainable development: Theory, practice, and impact. *Sustainability, 13*(4), 2021.
Taylor, L. (2016). The ethics of big data as a public good: Promises and challenges. *Philosophy Technology, 29*(1), 67–83.
UNEP. (2022a). *Harnessing artificial intelligence for environmental sustainability*. https://www.unep.org
UNEP. (2022b). *Open data and global collaboration for climate resilience*. https://www.unep.org
UNESCO. (2022a). *Open science for sustainable development*. https://unesco.org
UNESCO. (2022b). *Global open educational resources (OER)*. https://www.unesco.org
United Nations. (2020). *The 17 goals*. https://sdgs.un.org/goals
United Nations (UN) Report. (2021). *Open data's role in reducing global* CO_2 *emissions*.
United Nations Development Programme (UNDP). (2022). Bridging the digital divide in open data access.
Van Schalkwyk, F., Willmers, M., & McNaughton, M. (2017). Open data in developing countries: Toward building an evidence base on what works and how. *Information Technology for Development, 23*(3), 518–534.
World Bank. (2020a). *Open data in agriculture and nutrition*. https://worldbank.org
World Bank. (2020b). *Open data for development*. www.worldbank.org
World Bank. (2020c). *Africa open data for environment, agriculture, and land initiative*. https://www.worldbank.org
World Bank. (2020d). *Open data in action: Mapping healthcare facilities in Kenya*. https://www.worldbank.org
World Bank. (2020e). *Open data for sustainable development: Unlocking solutions for the future*.
World Bank. (2021). *Harnessing open data for disaster preparedness*. https://www.worldbank.org
World Bank. (2023). *Data privacy concerns in open data adoption*.
World Economic Forum. (2021). *Blockchain for social impact*. https://www.weforum.org
World Health Organization (WHO). (2021). *Data sharing for global health crises*. https://www.who.int
World Resources Institute (WRI). (2022). *Global forest watch: Real-time data on deforestation*. https://www.globalforestwatch.org
World Resources Institute (WRI). (2023). *Data-driven climate action in cities*.
Zuiderwijk, A., & Janssen, M. (2014). Open data policies, their implementation and impact. *Government Information Quarterly, 31*(1), 17–29.

Smart City and Intelligent Transportation System (ITS): Ahead Towards Greener and Sustainable Future

Riddhima Jaiswal, Shipra Agarwal, Malika Pahwa, Tanya Rawal, and Jaya

Abstract

A Smart City uses new technologies in adapting and sustaining everything in the buildings, including the infrastructure of Intelligent lighting, energy management, Artificial Intelligence, and IoT implemented in public transport systems. In the broad view of an intelligent transport system: it is one of the top elements of the intelligent transportation system integrated with the usage of technology into building adaptability and security in the flow of traffic. The concept of Smart Cities recently gained tremendous momentum because it came to realize that the creation of efficient, sustainable, and livable urban environments would require the integration of new technologies. Using an Intelligent Transportation System is one of the most transformational elements a Smart City can have for improving the flow of traffic, reducing congestion, and lessening the environmental impact. This manuscript reviews the development and effectiveness of Smart Cities, using ITS in driving economic progress with environmental sustainability. It examines; therefore, how technological transportation development serves as a critical building block toward a more intelligent or efficient ecosystem. Key areas derived from the research will describe how Smart Cities evolve to be part of urban life with great impact, how economies develop and become sustainable for intelligent transport systems, determination of components that configure smart mobility, and then study its contribution to attaining a low-carbon environment. It concludes that this paper is important in forming the integration of Smart Cities and ITS, which can represent a crucial step towards greener and more sustainable urban futures, wherein transport can play a vital role in mitigation strategies related to climate change and enhancing the quality of life. Intelligence transportation systems can change or make alterations to

R. Jaiswal · S. Agarwal (✉) · M. Pahwa · T. Rawal · Jaya
Department of Commerce, Graphic Era Deemed to be University, Dehradun, Uttarakhand, India
e-mail: shipra.agg26@gmail.com

T. Choudhury et al. (eds.), *Open Access as a Knowledge Platform for Implementing the United Nations Sustainable Development Goals*, World Sustainability Series,
https://doi.org/10.1007/978-3-031-99605-4_13

the way people travel to and from work in metropolitan regions. ITS offers a broad range of approaches to provide numerous advanced infrastructures, types of transportation and solutions of traffic mobility management.

Keywords

Growth • Smart-city • Sustainable • Technology • Future • Intelligent transportation

1 Introduction

With the goal of transforming public housing district into more equitable and sustainable environments, India launched National Smart Cities Mission (SCM) in 2015 (Bibri & Krogstie, 2017). The way cities compete is changing—they're shifting from traditional rivalries to what's being called "smart" competition. In this new era, questions arise about how smart cities are owned, organized, and evolve over time. At its heart, a smart city is a community that harnesses digital technology to make life better for its residents. These cities capture information from every possible source that comes their way: be it citizens, devices, buildings, or cameras and put the data into important usages such as transportation, utilities, water management, waste disposal, safety, and even schools and hospitals.

Smart cities take it a step further by sharing this data with businesses, citizens, and other groups that can put this to good use. They integrate ICT and IoT-attached devices to optimize their services and strengthen their linkages with residents. Their aim is to make life more convenient, efficient, and enjoyable (Harmon et al., 2015).

The essence of a smart city is how the leadership, or the local government, monitors, analyses, plans, and governs their communities. As of 2022, the largest investments in smart cities have been in areas like visual surveillance, public transit, and outdoor lighting. However, creating a truly smart city isn't just about technology; it's about collaboration (Barns, 2018). Public and private sectors, along with residents, need to work together to bring meaningful change.

Of course, smart cities face challenges. Critics point to issues like surveillance, inequality, limited citizen participation, and environmental concerns. Others highlight the difficulty of creating vibrant, culturally rich communities that attract and retain residents (Belaïd & Arora, 2024).

Despite these hurdles, the Smart Cities Mission remains focused on driving innovation and improving quality of life. By fostering economic growth and leveraging cutting-edge technologies, these cities aim to build a brighter, more inclusive future for everyone who lives there (Kumar & Dahiya, 2017). This mission is a Centrally Sponsored Scheme, with the centrally government providing financial support and the state and the local bodies contributing equally, private sector participation is also involved through Public-Private Partnership (PPP).

A smart city operates intelligently also known as Urban Intelligence. The cities are structured using the 3-C concept which includes competence, convenience and cleverness. A concise light has also been thrown on the concept of smart cities in India which defines the adequacy and selection criteria which is a necessary condition for availing the benefits of this plan. There are some schemes and policies like Atal Mission for Rejuvenation and Urban Transforming (AMRUT) have also been mentioned. The government is offering financial aid and investments in the smart city projects but if it breaks a certain amount then funds have to be raised by the state government (Ahmed & Ali, 2021).

In the twenty-first century, smart cities are leading the way in transforming urban living. These cities are embracing cutting-edge technologies and data analytics to improve the livability of citizens, further enhance services, and accommodate sustainable economic growth (Shaaban et al., 2021). Smart Cities integrate key sectors such as transportation, energy, health, education, and governing into an interconnected intelligent environment that meets the needs and requirements of modern living and lifestyles.

This systematic review sets the modern boundary for smart cities by exploring what defines smart cities, how budgets are allocated and spent, the tools and metrics in use, the real-world applications of such cities, the benefits, and the challenges that come along. This review also looks into reasons why cities struggle with the adoption of smart solutions and the strategies that were used to overcome them (Winkowska et al., 2019).

Even as smart cities offer several advantages, there are indeed challenges. Additionally, there are still concerns being under discussion when it comes to data protection, the digital divide based on unequal access to certain facilities, and the need felt for close collaboration between government agencies, public institutions, and the private sector itself.

Despite these obstacles, the vision of smart cities is clear: to create sustainable, inclusive, and efficient spaces where people can live, work, and thrive in harmony with technological innovation (Silva et al., 2018).

In spite of the growing trend in combining (ITS) with smart city frame work, there is a noteworthy gap in systematically to give an address to the crucial elements of smart city frame work that can add to the effectiveness of ITS, the specific approaches for merging ITS with smart city frame work to improve pedestrian and cyclic safety, traffic management, public transportation, the recognition and determination of challenges and hurdles related with accumulation process and the true qualification and analyzation of the benefits derived from merging ITS with smart city framework (Mora et al., 2023).

2 Methodology

This research employs a mixed-methods approach that combines qualitative research methods to investigate the definition and technological development aspects of smart cities. Relevant literature was reviewed to identify concepts and gaps in research. It was secondary research done by extracting information from databases, reports and cases studies of smart cities across the world. Data relevant to the set objectives were selected using purposive sampling. Qualitative data were subjected to thematic analysis and quantitative data to statistical analysis. Ethical guidelines were adhered to and limitations acknowledged including bias and sample size. The current methodology intends to present this approach as comprehensive in the knowledge of how smart city development is utilized and its implications.

3 Necessities of Smart Cities

The need for smart cities in India is driven by the quest to enhance urban living while addressing environmental challenges. Smart cities improve sustainability to environment, decent quality of life, full utilization of resources, clean environment, reduce pollution for reducing health issues and effective transportation systems. By integrating AI and IoT, we can design cities that optimize resources, reduce waste, and improve the quality of life. Smart cities promote growth of economy and digital changes make cities smart (Qian et al., 2019). This vision of smart, sustainable cities isn't just about technology—it's about building a future where urban spaces nurture and sustain their inhabitants, making daily life easier and more enriching.

The Smart Cities Mission is about building better and livable cities (Dai et al., 2024). It aspires to rejuvenate the local economy while improving daily lives through thoughtful planning intertwined with technology. The strategy involves a three-pronged approach: retrofits of existing neighbourhoods, rebuilding or adding new sections in areas, and offering smart city solutions pan-city to solve the core issues a city faces. At the core, this is a people-first mission to better connect cities for all citizens with efficiency and joy.

4 Objective of the Study

This manuscript therefore embarks on this exciting journey into the smart city revolution and its effects on human sustainable urban environments. Amidst the present world complexities, ITS plays a profound role in improving how we think of moving in our town. These systems not only make living in our cities easier with everyday chokes but boost economic factors through reducing traffic congestion and improving general mobility situations. Through this research, we aim to shed light on how smart cities and their transportation systems are paving the way toward a greener, more sustainable future that benefits everyone. Together, let's envision a world where technology not only transforms urban spaces but also enhances the quality of life for all its residents.

5 Evolution of Smart City

Earlier "smart" word was termed to be as smart than it names changed by Smart. People did not know about the smart city concept or smart development (Yin et al., 2015). In 1992, IBM launched Simon Personal communicator which was first smart phone than by this invention people started getting to know about the smart development. But they did not think that much about the meaning and importance of smart cities. Then in 1994, Netherland city Amsterdam created an effective digital city of itself which was known to be as "First SMART CITY of Present Times". And in 2007, there was an iPhone launched which was played a very important role in smart city concept. This concept of smart city was found in 2011 which was initiated by Smart City Expo World Congress held in Barcelona. This event was known to be as annual event charting course of smart city development.

The Smart City Mission in India was officially launched on June 25, 2015, with the ambitious goal of transforming 100 cities across the country into smart, sustainable, and citizen-friendly urban spaces. The Smart Cities Challenge picked 20 cities in the very first round of selection to invite cities for submission of proposals for urban transformation in a number of areas, including but not limited to affordable housing, efficient urban mobility, clean energy, water management, and better sanitation. The launch ushered in an era whereby the transformation of urban landscapes began to create a quality of life all over the country.

In 2016, the work entered the planning stage when cities began to do their SCPs. The initial list of 20 announced in January came up with roadmaps for transformation. This phase was all about comprehensive urban planning, broad-based citizen participation, and the preparation of a master plan that assured a green, smart, and citizen-friendly city. By August 2016, an expanded mission covered 13 more cities, taking the total to 33 as the enthusiasm for the ambitious scheme grew.

The implementation phase really took off in 2017. A third batch of 27 cities was picked to take the total to 60. The mission was also financially strengthened, with the budget of the central government costing Rs. 48,000 crore, approximately $7 billion. Major infrastructure projects—smart roads, traffic management systems, and digitalized public services—had the funds allocated to them, thus solidifying the infrastructure development mantle of the mission.

By 2018, tangible changes started to appear: cities integrated digital technologies for service management and delivery; initiatives such as smart meters for utilities, advanced waste management, and surveillance systems were undertaken for better functionality and security (Shukla et al., 2024). Governance also became smarter through support from digital platforms on e-governance and engagement with citizens. In addition, renewable energy, including solar power and smart grids, was integrated into infrastructure to make it sustainable.

In 2019, the cities prioritized smart transportation and sustainability. Smart buses, electric car charging stations, and synchronized traffic signals enhanced urban mobility and reduced congestion. Other initiatives included projects of waste-to-energy, rainwater harvesting, and more green spaces that enhanced the environmental friendliness of such cities. Telemedicine and mobile health services improved healthcare access, marking a shift toward holistic smart city development.

The COVID-19 pandemic in 2020 accelerated digital transformation. Cities with pre-implemented digital solutions were better equipped to handle the crisis through remote services, virus tracking, and public health safety measures. Contactless services, digital health checks, and real-time monitoring became essential. The crisis reshaped priorities, emphasizing public health infrastructure, digital governance, and remote work solutions for future resilience.

In 2021, cities focused on resilience and smart technologies. Green transport solutions were encouraged by integrating electric vehicle infrastructure with renewable energy. Public engagement increased through mobile apps allowing citizens to report issues and engage with local governments. AI-based disaster resilience systems, predicting floods and heatwaves, further enhanced adaptability to environmental challenges.

By 2022, the mission deepened its commitment to sustainability. Initiatives ranged from energy-efficient street lighting to sustainable drainage systems, smart grids, and water conservation technologies. Smart governance expanded and became more transparent and efficient. Funding for state-of-the-art solutions increased, improving the sustainability of urban areas and enhancing quality of life.

In 2023, AI, ML, and IoT continued to enhance urban operations: smart surveillance, advanced traffic management, and public safety created efficient and safe environments. Cities shared best practices on an international level, while public-private partnerships made sure that smart infrastructure was sustainable and scalable.

In 2024, it gave its nod for 12 new smart cities under NICDP and expanded its horizon. Other cities that became part of the initiative include Khurpia, Rajpura-Patiala, and Agra. This speaks volumes about the broadening scope of the mission. This extension up to March 2025 reflects the resolve of the government to transform urban spaces into sustainable, citizen-friendly areas with the help of technology. It was one more evolution that showed the adaptability of the mission and its long-term vision for changing urban living in India.

6 Components of Smart

6.1 Smart Mobility

There are different important component required for the smart city as shown in Fig. 1. Smart mobility concentrates on the optimization of transport systems through smart traffic management, electric vehicles, cars that can run autonomously, and shared mobility (Sagaris & Tiznado-Aitken, 2020). These technologies will reduce congestion and carbon emissions, while also improving

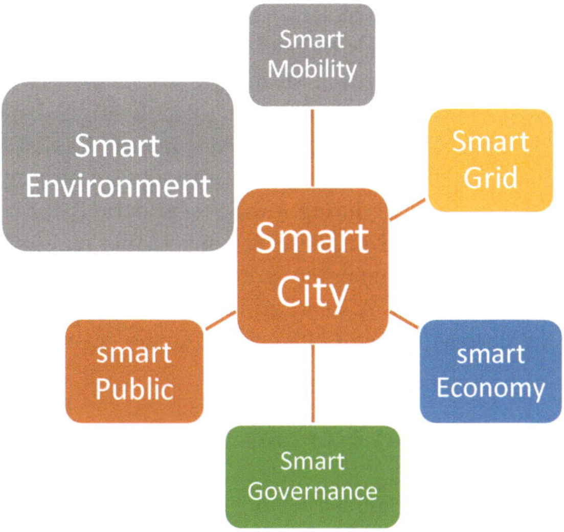

Fig. 1 Components of smart cities

access. Smart mobility will help in enhancing the quality of life and increasing productivity.

Cities with integrated advanced technologies such as ITS can monitor and control the flow of traffic in real-time and dynamically adjust routes to maximize smooth flow and safety. The use of intelligent traffic lights, sensors inserted into roadways, and adaptive signal systems can all be controlled dynamically to manage traffic patterns based on actual demand. They also encourage the use of electric vehicles by providing EV charging infrastructure that is convenient and widely available. The self-driving cars and smart buses central to smart mobility increase transportation safety and efficiency by reducing human error. All the various transportation modes of bike-sharing systems, electric scooters, etc., greatly reduce the need for private cars and therefore bring down traffic and lower pollution levels. It also makes public transport better by real-time tracking of arrival and dynamic scheduling. The convenience will increase with less waiting time for the passengers. All these factors combined go on to make the transportation network in a smart city more sustainable, accessible, and efficient.

6.2 Smart Grids

Contribute to increasing energy efficiency by integrating renewable sources into the grid, deploying smart meters, and using intelligent networks for the management of electricity distribution in real time. They can minimize energy consumption more effectively and thus help cities adapt better in response to changing demands. Unlike traditional electrical grids that rely on one-way communication of electricity distribution, smart grids utilize two-way digital communication technologies to monitor, control, and optimize electricity distribution in real time. These grids incorporate smart meters for detailed information on energy usage by both consumers and utilities to cater to efficient use and dynamic pricing for demand. A principal concept of a smart grid comes out as the incorporation of renewable energy, easily accommodating solar, wind, and other renewable sources within this grid. This decreases dependence on fossil fuels and lowers carbon traces within the city. Energy storage systems are also crucial for the storage of excess energy generated during low-demand periods for use when demand peaks. Additionally, demand response systems enable real-time adjustment in energy use to incentivize consumers to reduce consumption during peak hours and avoid overloading the grid. A smart grid enhances energy reliability, promotes sustainability, and improves overall efficiency in urban environments.

6.3 Smart Economy

Entails digital infrastructure, innovation hubs, smart manufacturing, economic growth, entrepreneurship, and job creation. The smart economy also incorporates various digital payment systems into e-commerce and the gradual transition to a digital workforce. Its measures for a smart economy in creating a sustainable environment, effectiveness, fuller resource utilization, and the welfare of mankind are as follows.

A smart economy and good city will rest upon digital technology, innovation, and sustainability. Thus, it should reference digital systems, from robust internet access, to creating a 5G environment, to cloud computing, to support businesses and entrepreneurs. Resources and access to network ventures will benefit innovativeness and entrepreneurial development through innovation hubs, tech parks, and startup incubators. Industry 4.0 would be supported in smart cities, meaning that modalities such as IoT, robotics, and AI would be employed within manufacturing, and resource use services would become more efficient and sustainable. With e-commerce and digital payments, transactions can be made much simpler and easier without too much reliance on cash and enable more people to take part in the economy. Smart manufacturing systems automate much of the production, enable predictive maintenance, leverage data analytics, therefore have the potential to cut costs and raise productivity. It also creates employment and workforce development, as it prepares its citizens with the necessary digital skills to take on an increasingly digitized job market.

6.4 Smart Governance

Enhances public administration through e-governance services, data-driven decision making. and citizen engagement platforms To ensure transparency responsibility and improved service delivery. This component of a smart city Creative ideas for government services… and use technology It also creates a sustainable environment and a liberal and inclusive society. Smart governance means using technology and data to improve the efficiency, transparency and responsiveness of government services and public administration. Citizens can access essential services online through e-government services, such as paying taxes, applying for permits. and reporting problems such as potholes or outdated streetlights. And this facility increases citizen participation and satisfaction with public services. Data-driven decision-making is another hallmark of smart governance. Government agencies analyze data to make decisions using comprehensive information regarding urban planning. Resource allocation and policy setting Real-time data analysis makes governance more responsive and adaptive. Civic engagement platforms handle emerging issues, and the platforms let residents participate in decision-making: providing feedback and voting on all types of initiatives for increased responsibility and inclusion. That makes public policy answer the needs and wants of the community. Digital identity also provides easy and secure access to government services, hence enabling all citizens to be at an advantage when engaging with their government.

6.5 Smart Environment

Focus on creating sustainable urban ecosystems using technology to monitor and manage air and water quality. Increase efficiency in waste management and promote energy efficiency in buildings It also supports efforts to mitigate climate change through the adoption of green spaces and renewable energy sources. Smart environments focus on using technology to address environmental challenges and create sustainable and healthy urban ecosystems (Hannan et al., 2020). Air and water quality monitoring systems use sensors along with AI techniques to detect pollution (Awasthi et al., 2024). It provides real-time information to officials and residents. This information helps manage resources and ensure good living conditions by reducing pollution levels. Waste management is also optimized with smart bins that alert full-time waste collectors. Helps to collect garbage efficiently and in a timely manner. Reduce the use of landfills Improve recycling efforts Energy efficiency is the cornerstone of a smart environment. Where smart buildings and energy systems optimize energy consumption, cities that use sensors to reduce waste can use green roofs city park and public parks Improve air quality Increase biodiversity and allows residents to access nature in urban areas. Climate change mitigation Smart cities using low, renewable energy sources support carbon transport options and sustainable urban planning with the goal of reducing overall carbon emissions and promote environmental sustainability.

6.6 Smart Public

Addresses citizens' well-being by improving healthcare, education, social services. Public safety and access Smart cities aim to ensure equal access to essential services through telemedicine. Smart classroom and comprehensive public services to ensure that all residents will have a better quality of life. Together, these components make smart cities more efficient, sustainable, and responsive to citizens' needs. The smart public component of a smart city focuses on improving the well-being and quality of life of citizens through technology and innovation. Healthcare services in smart cities are improved through telemedicine and IoT-based health monitoring, enabling remote consultations (Lim et al., 2024). Collect health data in real time and more accessible care this is especially important in underserved areas. Where access to medical care may be limited Smart education system integrates e-learning platform interactive classroom and digital library to provide residents with flexible and personalized learning opportunities. This approach also facilitates lifelong learning. This is to ensure that citizens continue to develop their skills to adapt to the rapidly changing job market. Social services are delivered more efficiently through digital platforms. Improving the welfare access process Housing assistance and unemployment benefits strengthening public safety through surveillance systems AI-powered crime prediction and emergency response networks to improve overall safety Respond to incidents faster. Ultimately, accessibility features ensure that every citizen including people with disabilities Can visit the city and access various services easily Promote integration and equality.

7 Significance of Smart City

7.1 Better Quality of Life

Smart cities efficiently manage public services, simplifying daily living with personalized experiences. Real-time traffic monitoring ensures smooth travel, while smart apps assist with parking and reporting issues like faulty streetlights. Integration across healthcare, education, and transportation streamlines daily tasks. Additionally, event notifications, public safety updates, and optimized routes enhance city functionality, tailored to individual needs.

7.2 Sustainability and Environmental Impact

Smart cities promote sustainability through energy efficiency and waste reduction. Automated systems manage lighting and temperature to conserve energy, while smart grids optimize energy distribution, lowering carbon footprints. Water sensors prevent wastage, and smart trash collection maximizes resource utilization. Green transport systems and eco-friendly public transit ensure sustainable commuting, preserving the environment for future generations.

7.3 Economic Growth and Innovation

With advanced infrastructure, smart cities foster economic growth, encouraging startups and entrepreneurship. High-speed internet and cutting-edge technology make these cities innovation hubs. They create diverse job opportunities in fields like software development and urban planning, boosting local economies and supporting sustainable development.

7.4 Greater Efficiency and Cost Savings

Smart cities enhance efficiency through automated resource management. Sensor-driven waste systems optimize collection routes and schedules, reducing fuel consumption and landfill usage. These advancements lower operational costs, enabling effective resource allocation without compromising service quality. Automation leads to substantial cost savings.

7.5 Improved Safety and Security

Safety is prioritized in smart cities. Integrated systems ensure rapid emergency response by instantly notifying responders. Data analysis and smart technologies like automated lighting enhance neighborhood safety, fostering a secure environment and greater peace of mind for residents.

7.6 Improved Governance and Public Involvement

Data-driven governance enables proactive problem-solving, improving urban living. Predictive analytics allow city officials to anticipate and address challenges. Digital tools provide real-time updates on projects and budgets, fostering transparency and accountability. Open engagement builds trust between governments and citizens.

7.7 Mobility and Connectivity

Smart cities revolutionize transportation with sustainable and connected solutions. Electric vehicles, bike-sharing programs, and intelligent traffic systems reduce reliance on private cars, making travel greener and more efficient. Universal high-speed internet bridges the digital divide, keeping residents and businesses connected and fostering innovation.

7.8 Health and Well-Being

Smart city technologies enhance health by improving healthcare access and environmental quality. Telemedicine services enable remote doctor consultations, while air quality sensors monitor pollution, creating healthier living conditions for everyone.

7.9 Resilience and Disaster Management

Smart cities are equipped to handle emergencies and adapt to climate challenges. Real-time data systems predict and respond to disasters like floods or earthquakes, ensuring quicker recovery and minimizing damage. Resilient infrastructure supports adaptation to changing climates and ensures a sustainable future.

7.10 Global Competitiveness

Adopting advanced technologies, smart cities attract top talent, entrepreneurs, and investors, boosting their global appeal. These cities set benchmarks in urban innovation, collaborating with others to share knowledge and accelerate progress. Such international partnerships enhance competitiveness and shape the future of urban development.

8 Challenges Faced by Smart City

A smart city has the potential to remarkably transform urban living but also invokes a number of challenges to be thoughtfully addressed, including privacy concerns, barriers in cost, social inequities, and the integration of technology. A long list of such issues calls for thoughtful planning with careful consideration given to open processes and inclusiveness regarding what will truly make such a city thrive and meet the promise of the digital era for all its citizens.

8.1 Privacy and Information Security

Smart cities are based on voluminous data collection, where everything—from the flow of traffic to air quality to even individual behaviors—is under observation. That raises many serious concerns about privacy, as people are afraid of their personal information being used in a wrong manner or misused by unauthorized hands. Further, the interconnectedness of smart cities links them to the risk of cyberattacks. A security breach in city infrastructure and smart systems may lead to disruption in basic services, safety, and violation of privacy.

8.2 High Implementation Costs

It involves immense investment in infrastructures like sensors, IoT devices, and data centers. This especially has presented a challenge to most cities, especially those in developing regions, in raising such funds. Besides, a smart city is expensive in the long run to upkeep or upgrade the technology. A lack of financial planning will see the infrastructures deteriorate or running on obsolete technology, hence minimizing its effectiveness in the function.

8.3 Digital Divide and Accessibility

At the same time, smart city technologies can be a risk factor in widening the digital divide if they are not available for everybody. A minimal possibility to access the Internet or missing electronic equipment may mean exclusion in poorer areas. The lack of sufficient digital literacy some people have would create barriers in using such technologies for living in smart cities to a full degree and amplify social disparities.

8.4 Integration of Integrated Systems

Fragmented technology is perhaps the most common stumbling block to the smart city: Systems developed by different companies or departments often won't communicate well with other systems, causing functionality and compatibility issues. Without standardization, cities may invest in technologies that become obsolete or incompatible with future upgrades, hindering long-term success.

8.5 People's Resistance to Change

People might be resistant to new technologies out of fear of Big Brother, loss of personal freedoms, or the complexity of smart systems. Public trust is crucial in ensuring the success of smart city initiatives. If citizens feel these projects value profit or surveillance over improving their quality of life, they will lose trust in city authorities and limit adoption and engagement.

8.6 Environmental Concerns and Sustainability

Whereas smart cities are to be energy-efficient, most of the involved technologies, which include data centers, sensors, and communication networks, are incredibly power-hungry. Such systems, unless sustainably powered, will be contributory to environmental problems. Because of rapid technological changes, the world is facing electronic waste resulting from devices becoming outdated; this calls for prudent e-waste management to be environmentally responsible.

8.7 Social and Cultural Consequences

For any given solution in technology to click, it has to correspond to the local customs and infrastructures. What ultimately applies in one city for all reasons may not relate perfectly to another because there lies a difference in the cultures. Furthermore, automation substituting traditional jobs becomes a headache in many instances, along with job displacement. Such adjustment to new roles naturally demands reskilling labor resources, and the social aspect poses a challenge in adjustment as well.

8.8 Moral Issues

With all that said, it does spur ethical issues: increased surveillance and data collection do point to a fine balance between safety and personal freedom. Over-surveillance may even infringe on individual rights, thus giving birth to ethical debates in the process. Algorithms within smart city systems might be biased, too. Predictive policing algorithms can be skewed against communities for unfair targeting and perpetuate inequity in society through such policing.

8.9 Issues of Governance and Coordination

Inefficient coordination between the departments, agencies, and private companies involved in the smart city projects could yield redundancies and create much confusion. Resources could go to waste, and for citizens, finding one's way through fragmented systems is not easy. Also, complicated legal and regulatory frameworks contribute to slowing down development. Governments have to adapt rapidly to technological changes, ensuring that regulations move at the same pace as emerging trends.

8.10 Long-Term Sustainability and Scalability

Smart city technologies need to be adaptable to the requirements of growing populations and changing urban landscapes. Systems that work for smaller cities may not scale effectively to larger metropolitan areas. Long-term planning with a focus on sustainability is necessary to avoid the obsolescence of technologies. A forward-thinking approach is required to ensure the viability and effectiveness of smart city projects over time.

9 Policies of Smart City

9.1 Infrastructure Development

It is the process of improving certain infrastructural components vital to a countries or companies' growth. The mission focuses on providing core infrastructure, such as adequate water supply, electricity, sanitation, and efficient public transportation.

9.2 Sustainable Environment

Sustainable development means that development and advancement of human which aims to reach the necessity of present without adjusting the ability of future generation. So, sustainable development in smart cities uses technology to improve life of people by making sure that future needs of future generation are fulfilled. The mission aims to create a sustainable environment, including clean energy technology deployment.

9.3 Inclusive Development

It is a development approach which aims to ensure that all society members including the poor and marginalized share the benefits of the economic growth.

9.4 Safety and Security

Safety refers to being protected from harm caused by faults, while security is referred to the state of being protected from harm caused by intentional human actions. This mission is basically focus upon the safety purpose and security of peoples like women, children and the elder citizens.

9.5 Education and Health Development

Smart city helps in preventing people from disease and disability. Also helps in securing and reducing premature death. It involves health knowledge, behavior, intelligence and viewpoint of peoples.

9.6 Public-Private Partnerships

Public private help to combine the strength of public and government sector which helps in bringing the best expertise and best resources to tackle the challenges. Public private partnership also helps to use the technology to improve the services delivery. In short, it uses develop the infrastructure development, financing, technology and service delivery. The mission uses public-private partnerships to create smart outcomes.

9.7 Area-Based Development

Area based development means regional approach that focuses on particular area with a challenges of common development challenges. It's aims to address were issues like creating job, seeking gender equality, promoting youth in that particular area.

10 Master Plans of Smart Cities Mission

10.1 Reconstruct

It is the process of changing and improving the existing structure of buildings, vehicles and bridges. It can used to reform damage, safety improvement, livable and make structure efficient of more energy.

10.2 Renewal

It helps in replacing old structure to create and develop a new design with advance infrastructure, use mixed land and rise density.

10.3 Development of Greenfield

It is especially for poor peoples to whom housing facility is given. Greenfield development mainly happened in vacant area with finance planning, planning of innovations and tools of plan implementation for sustainable housing.

10.4 Development of Pan-City

It is a type of development which involve with traffic management, utility management, improvement of communication among citizens and electronic device technologies.

11 Government Initiatives for Promoting Smart Cities Mission

11.1 Atal Mission for Rejuvenation and Urban Transformation (AMRUT)

AMRUT, launched in June 2015 by the Ministry of Housing and Urban Affairs, is a transformative initiative focused on urban renewal and sustainable development. Uttarakhand and Himachal Pradesh have been particularly recognized for their progress under this mission.

The primary objective of AMRUT is to conserve and utilize every drop of water effectively, incorporating rainwater harvesting across urban structures. The mission prioritizes ensuring adequate water supply, improving sewerage systems, and reducing pollution by promoting public transport and creating non-motorized transport facilities like walking and cycling pathways.

Reforms under AMRUT aim to enhance citizen services, reduce costs, improve financial sustainability, increase resource efficiency, and foster transparency. Initiatives such as replacing conventional streetlights with energy-efficient LED lights have been instrumental in achieving these goals.

Initially budgeted at ₹50,000 crores for five years (2015–2020), the mission was extended to December 2021 after challenges in meeting urban renewal targets across 500 cities.

11.2 Pradhan Mantri Awas Yojana—Urban (PMAY-U)

Launched on June 25, 2015, PMAY-U embodies the vision of "Housing for All" by 2022. This mission focuses on providing **pucca houses** with basic amenities such as toilets, water supply, electricity, and kitchens for families without homes or living in temporary housing.

The scheme includes a restructured rural housing program, **Pradhan Mantri Awas Yojana-Gramin (PMAY-G)**, under the Ministry of Rural Development, which caters to rural housing needs. The urban component of PMAY covers all statutory towns, planning areas, development authorities, and special area authorities, ensuring comprehensive reach.

PMAY-U promotes inclusivity and empowerment by mandating house ownership in the name of women or jointly with male members. Preferences are extended to vulnerable groups such as differently abled individuals, senior citizens, SCs, STs, OBCs, minorities, single women, and transgenders.

By January 2021, 55% of the construction target (1.27 crore out of 2.28 crore houses) had been completed, and funds were sanctioned to 85% of the identified beneficiaries. Despite these achievements, meeting the full target remains a challenge.

11.3 Climate Smart Cities Assessment Framework (CSCAF) 2.0

The CSCAF 2.0 initiative provides cities with a strategic roadmap to combat climate change through sustainable urban planning and investments. Supported by the Climate Centre for Cities under the National Institute of Urban Affairs (NIUA), the framework fosters a climate-sensitive approach in urban development.

The framework is based on **28 indicators** across five key categories:

1. Energy and Green Buildings
2. Urban Planning, Green Cover, and Biodiversity
3. Mobility and Air Quality
4. Water Management
5. Waste Management.

CSCAF aims to inspire cities to create walkable, vibrant streets and revitalized spaces, particularly in areas with high footfall. Initiatives include reimagining under-flyover spaces, transforming dead neighborhood areas, and building pedestrian-friendly links through parks and institutional zones.

All cities with populations exceeding 5 lakhs, along with state capitals, are encouraged to participate in this initiative, ensuring a robust and collaborative effort to address climate challenges while fostering urban resilience.

12 Achievements of Smart City Mission

12.1 Integrated Command and Control Centers (ICCCs)

All 100 Smart Cities now have operational ICCCs, serving as the brain of city management. These centers use data-driven insights and advanced technologies like AI, IoT, and data analytics to make informed decisions, enhancing services such as transportation, water supply, and waste management. During the COVID-19 pandemic, ICCCs proved invaluable, functioning as war rooms to manage the crisis effectively as shown in Table 1.

12.2 Public Safety and Security

To ensure safer cities, over 83,000 CCTV cameras have been installed for real-time crime monitoring. Additionally, cities have implemented 1884 emergency call boxes, 3000 public address systems, and advanced traffic systems for detecting red-light violations and automatic number plate recognition, significantly enhancing public safety measures.

Table 1 Smart city achievements

Sector	Completed		Work in progress	
	Projects	Cost (₹ crore)	Projects	Cost (₹ crore)
ICCC	100	11,775*	–	–
Smart mobility	1546	36,449	164	5013
Smart energy	667	13,712	27	527
WASH	1412	46,203	128	3587
PPP	195	9193	3	26
Vibrant public spaces	1284	10,120	111	2188
Economic infrastructure	522	10,792	114	2192
Social infrastructure	829	11,847	61	1281
Smart governance	636	16,035	52	1615
Environment	146	2279	6	280

Integrated Command and Control Centres (ICCC) at all 100 smart cities, 84,000 CCTV surveillance cameras, 1,884 emergency call boxes, more than 3,000 public address systems, more than 1,740 kms of smart roads, 713 kms of cycle tracks, 17,026 kms of water supply system being monitored through Supervisory Control and Data Acquisition (SCADA) system, more than 66 cities are managing Solid Waste Management with increased usage of technology, around 9,194 vehicles have been Radio Frequency Identification (RFID) enabled for Automatic Vehicle Location (AVL), more than 9,433 smart classrooms and 41 Digital Libraries have been developed, 172 e-health centers and clinics developed and 152 health ATMs have been installed

12.3 Water Supply

Smart Cities have developed a 9900-km water supply network monitored by SCADA systems, enabling efficient detection and repair of leaks while reducing non-revenue water. This innovation ensures more reliable and sustainable water management.

12.4 Solid Waste Management

Advanced technologies are transforming waste management in over 50 cities. With 4400 vehicles equipped with RFID-based Automatic Vehicle Location (AVL) systems, waste collection has become more streamlined and efficient, ensuring cleaner cities.

12.5 Streetlights

More than 52 lakh solar and LED streetlights have been installed, reducing energy consumption and improving sustainability. Additionally, 86,000 km of underground electricity cabling have been completed, ensuring safer and more reliable energy distribution.

12.6 Mobility

To promote sustainable transportation, 4700 km of smart roads and 580 km of cycle tracks have been constructed or upgraded. The Intelligent Transport Management System (ITMS), monitored through ICCCs, has optimized traffic operations, reduced travel time, and improved rule enforcement.

12.7 Affordable Housing and Shelter

The mission has provided 49,300 affordable housing units and 1562 community shelter rooms, including facilities like Rain Basera, hostels, and night shelters, ensuring secure living spaces for those in need.

12.8 Vibrant Public Spaces

Urban spaces are being revitalized with over 1300 parks, green spaces, and waterfront promenades either completed or under development. These projects enhance the quality of life by providing residents with serene and engaging recreational areas.

12.9 Education

Education is being transformed with the creation of 7654 smart classrooms and 40 digital libraries. These modern facilities are designed to provide students with an enriched learning experience and better access to resources.

12.10 Health

Healthcare access has been significantly improved with the establishment of 172 e-health centers and clinics, along with 155 health ATMs. These initiatives make basic healthcare more accessible and convenient for urban residents.

12.11 Economic Hubs

The Smart City Mission is fostering economic growth through 21 incubation and skill development centers and the completion of 56 market redevelopment projects. These hubs provide opportunities for entrepreneurship and local business expansion.

12.12 Public-Private Partnerships (PPP)

Collaboration is key to the mission's success. Over 50 cities are driving progress through 199 projects worth ₹9200 crore, implemented via Public-Private Partnerships. These initiatives demonstrate the power of collective effort in creating smarter cities.

13 Smart Mobility in Smart City

It refers to using a different way of transportation alongside or whether instead of buying a vehicle which is gas powered. The need of smart mobility is arisen due to increasing heavy traffic and its side effects includes time wasting, pollution from gases and accidents. While there is some advance technology that is contributed upgrade the sources in more effective and efficient manner such as Intelligent Transportation Systems (ITS) which are a key component of smart cities. The transportation sector in India is witnessing significant advancements, driven by initiatives from both the Government of India and various state governments. A key focus in this area is the development of Intelligent Transportation Systems (ITS), a transformative approach aimed at enhancing efficiency, safety, and sustainability in surface transportation. ITS integrates electronic devices, communication technologies, and advanced management strategies to address critical aspects such as real-time parking management, smart cards for seamless transactions, advanced driving assistance systems, and efficient incident response mechanisms.

The ITS sector is poised for substantial growth, supported by policy reforms that have fostered an investor-friendly taxation system and promoted innovation. These measures aim to accelerate the adoption of smart transportation technologies and create a robust ecosystem for their development and implementation. Despite these promising developments, India faces several challenges that hinder the full realization of ITS potential. Issues such as inadequate infrastructure, poor air quality, and inconsistently implemented intelligent transportation systems persist. Additionally, the relatively slow pace of development threatens to create delays and widen the technological gap with other countries advancing rapidly in smart transportation solutions. Nevertheless, initiatives like "Make in India" have opened new avenues for manufacturing components essential for future transportation systems, offering a platform to address these challenges. By leveraging these opportunities, India can align itself with global advancements and pave the way for a more efficient and sustainable transportation network. Safe and trouble-free transportation for computers is raised by transportation authorities. Urban traffic, under developed infrastructure of road, road safety, declining infrastructure and inefficient public transport are some of the ITS that are restricted by adopting the advanced technology. ITS can better control and manage transportation at actual time and ITS can play a vital role in solving these problems. Some easy-to-understand technologies such as smart parking systems, advance driver assistance systems, traffic management systems, autonomous driving, telematics etc. are integrated with intelligence transportation system to enable transport system for smart cities. Intelligent Transportation Systems (ITS) can use resources and infrastructure effectively, manage and control traffic of urban cities and make conveyance safer and more efficient reachable, manageable and eco-friendly.

Smart Cities are transforming urban mobility with over 4700 km of upgraded or newly constructed smart roads and 580 km of dedicated cycle tracks. These developments promote safer, more sustainable travel options for residents. Additionally, an Intelligent Transport Management System (ITMS), monitored through Integrated Command and Control Centers (ICCCs), is revolutionizing traffic management. By optimizing traffic flow, enforcing regulations, and reducing travel times, the ITMS ensures smoother and more efficient journeys for everyone.

14 Components of Intelligent Transportation System

Management of Traffic

Intelligent transportation system helps in managing traffic system as sometime there is rush arise on travelling ways like on road due to poor traffic management. So, ITS helps in removing this conflict.

14.1 System of Toll

ITS develops a system of toll by which peoples are paid for travelling on paved roads. It also helps to manage traffic on roads and speed up the journey of drivers.

14.2 Public Transport

Intelligent transportation system provides public transport system which has the unique capacity to regulate different ways of transportation such as metros, auto, electronic vehicles and CNG vehicles. This system focuses on the sustainability of environment and less time consumption of peoples.

14.3 Vehicle-to-Everything Communication

ITS provides Vehicle-to-vehicle, vehicle-to-infrastructure, vehicle-to-pedestrian, vehicle-to-cloud communication are making connecting environment which helps in safety from traffic, reports of accidents, avoidance of accidents and many other uses.

15 Significances of Intelligent Transportation System

15.1 Improved Safety

Intelligent Transportation Systems (ITS) enhance traffic safety by using sensors, cameras, and real-time data to monitor roads and warn drivers about potential hazards. These systems can detect possible collisions, alert drivers to dangerous conditions like ice or fog, and even assist autonomous vehicles in avoiding accidents.

15.2 Reduced Traffic Congestion

With real-time traffic data, ITS can adjust traffic signals, manage traffic flow, and suggest alternative routes to help drivers avoid congestion. This reduces travel time, stress, and fuel consumption (Cheng et al., 2020).

15.3 More Efficient Traffic Flow

ITS optimizes transportation by using data to fine-tune traffic light patterns and adjust traffic flow, reducing delays. This means spending less time stuck in traffic, cutting emissions, and enjoying smoother rides.

15.4 Environmental Benefits

By reducing congestion and optimizing routes, ITS helps lower fuel consumption and air pollution. Cleaner air and a smaller carbon footprint contribute to healthier, more sustainable cities.

15.5 Smarter Public Transport

ITS improves public transport systems through real-time tracking and communication tools, ensuring buses and trains run on time. Passengers benefit from updated arrival information and seamless connections between modes of transport.

15.6 Smart Parking

ITS simplifies parking by providing real-time information about available spaces. Apps and sensors guide drivers to open spots quickly, reducing unnecessary driving and easing frustration.

15.7 Traveler Convenience

Travelers gain access to up-to-date information on traffic conditions, accidents, weather, and road closures. Tools like mobile apps and in-car displays enable better decision-making, saving time and reducing stress.

15.8 Support for Autonomous Vehicles

ITS provides the necessary data for the safe integration of autonomous vehicles, including traffic light statuses, road conditions, and vehicle-to-vehicle communication, ensuring smooth navigation.

15.9 Cost Savings

By reducing congestion, accidents, and travel times, ITS helps individuals and businesses save money. For example, delivery services benefit from faster routes, cutting operational costs significantly.

16 Developments Performed Under Intelligent Transportation System

A well-developed transportation system plays a pivotal role in improving connectivity and driving a country's progress. In India, the transportation sector has seen steady advancements in recent years, reflecting the nation's commitment to enhancing mobility and accessibility. Significant improvements in public transportation across the country have been achieved, driven by various government initiatives focused on boosting nationwide connectivity. These efforts aim not only to make transportation more efficient and accessible but also to support India's broader development goals by fostering economic growth, reducing regional disparities, and improving the quality of life for its citizens.

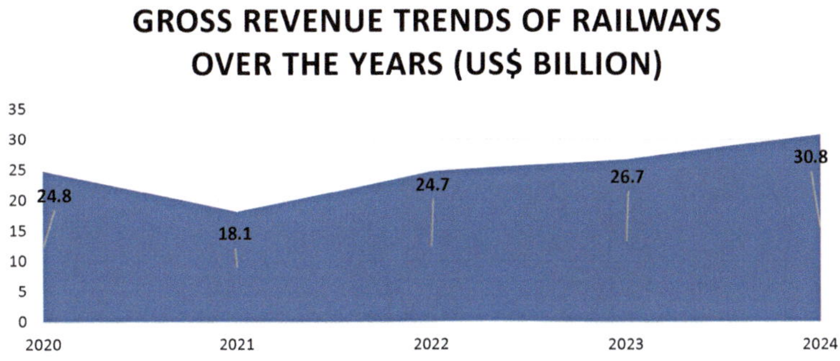

Fig. 2 Revenue of railways. *Source* (India Brand Equity Foundation [IBEF], 2024a)

16.1 Railway Networks

India's railway network is one of the largest and busiest in the world, playing a pivotal role in the country's transportation and logistics infrastructure. Operated by Indian Railways, it connects remote regions to urban centres, facilitating passenger travel and freight movement. With its vast reach, the railways contribute significantly to economic growth and regional connectivity. Over the years, modernization initiatives such as high-speed rail projects, electrification, and digitalization have enhanced the efficiency and sustainability of the network.

As shown in Figs. 2 and 3, the gross revenue of Indian Railways showed a fluctuating but upward trend. In 2020, revenue stood at $24.8 billion, dropped to $18.1 billion in 2021, likely due to pandemic-related disruptions, and then steadily recovered to $24.7 billion in 2022, and reached to $30.8 billion in 2024. This indicates a strong post-pandemic recovery and enhanced operational efficiency. Figure 3 shows the Revenue by Segment, Freight accounts to the majority of revenue at 67.1%, reflecting the railway's crucial role in transporting goods. Passenger services contribute 26.2%, highlighting the network's importance for mobility. This segmental distribution emphasizes the critical dependence on freight operations for financial sustainability (India Brand Equity Foundation [IBEF], 2024a).

16.2 Road Networks

Smart cities play a pivotal role in enhancing roadways networks by leveraging advanced technologies and innovative infrastructure planning. The integration of Internet of Things (IoT) devices, real-time data analytics, and intelligent traffic management systems ensures smoother and more efficient transportation networks (Rani & Sharma, 2023). In smart cities, roadway networks are designed to accommodate sustainable transportation options, reduce congestion, and improve connectivity. Additionally, smart city frameworks emphasize the maintenance

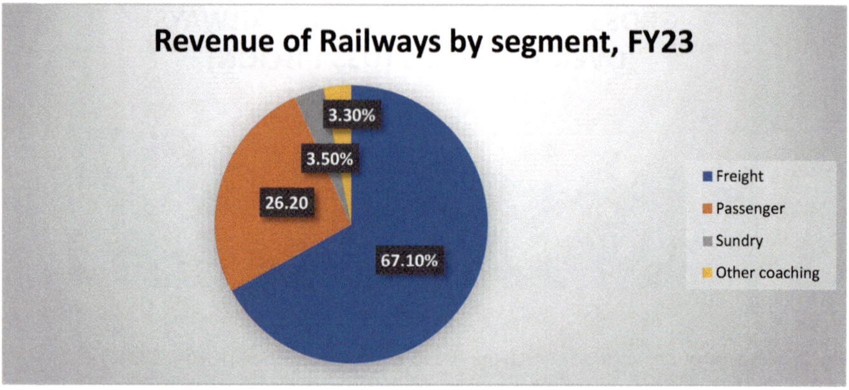

Fig. 3 Revenue of railways by sector. *Source* (India Brand Equity Foundation [IBEF], 2024a)

and expansion of national highways (NH), ensuring the infrastructure meets the growing demands of urbanization and economic growth.

The total length of National Highways (NH) in India has steadily increased from 2020 to 2024. In 2020, the network was recorded at 132,995 km, which rose to 138,376 km in 2021, marking a significant growth of over 5000 km. The upward trend continued, with the length reaching 141,345 km in 2022 and 145,240 km in 2023. By 2024, the network extended to 146,145 km. This consistent growth highlights the government's commitment to expanding the roadways infrastructure, crucial for facilitating trade, improving mobility, and supporting the development of smart cities (Fig. 4).

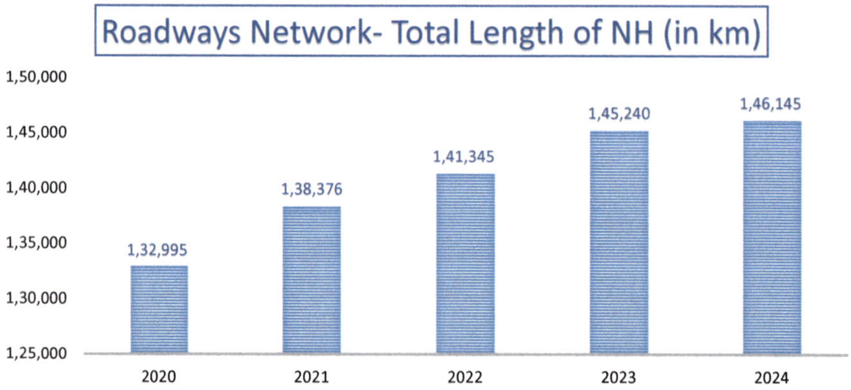

Fig. 4 Length of highways in roadways network. *Source* IBEF (2024b)

16.3 Tolling Management

It is a system in which peoples are payable for travel on specific roads, tunnels, bridges and facility of transportation to infrastructure of finance. The advance structure of road mean needs for ITS like system of toll management and systems of new traffic management. India is guaranteed about the advancement of tolling systems that they are successful, systematic and clear. Tolling system is established by Indian Highway Management Company Limited (IHMCL) and commandment of National Electronic Toll Collection (NETC) which makes payment easier by tolls digitally.

The year 2020–21 saw a relatively low growth rate in toll collections, growing by just 1.04%. This can largely be attributed to the impact of the COVID-19 pandemic, which caused significant disruptions in travel. During this period, many highways experienced reduced traffic as lockdowns and travel restrictions limited movement. However, post-pandemic recovery has been robust. From 2021 onward, toll collections have seen rapid growth, fueled by several key factors. As the economy rebounded, travel increased, and more vehicles began using highways, boosting toll revenues. Additionally, the mandatory adoption of **FASTag** since 2021 has streamlined toll collection, making it more efficient by reducing congestion at toll plazas and improving overall compliance. Another significant contributor has been the continued expansion and improvement of India's road infrastructure, particularly national highways and expressways, which has likely encouraged more drivers to use these routes, thereby increasing toll collection.

As a result, the total toll revenue has seen a remarkable rise, from ₹27,637.64 crores in 2019–20 to ₹64,809.86 crores in 2023–24 as shown in Fig. 5. This sharp increase signals that toll collection systems have become far more effective, with more people using toll roads, likely due to better infrastructure, convenience, and connectivity. The growth also reflects an overall recovery in the economy, which has translated into greater mobility and higher toll receipts. Looking ahead, if the current growth trend persists, toll collections in the coming years are likely to remain strong. With ongoing investments in highway infrastructure and the increasing adoption of digital tolling systems like **FASTag**, the growth in toll revenue could stabilize at a high level, further supporting the development of India's road network.

16.4 Metro Networks

It is a type of network that creates linkage within an area through metro rails that is known to be as metropolitan area. Metro networks in India have emerged as a vital component of urban transportation, addressing the challenges of congestion and pollution in rapidly growing cities. These networks provide efficient, reliable, and eco-friendly transit options for millions of commuters daily. Modern metro systems incorporate advanced technologies such as automated operations, smart ticketing, and energy-efficient trains, contributing to sustainable urban mobility.

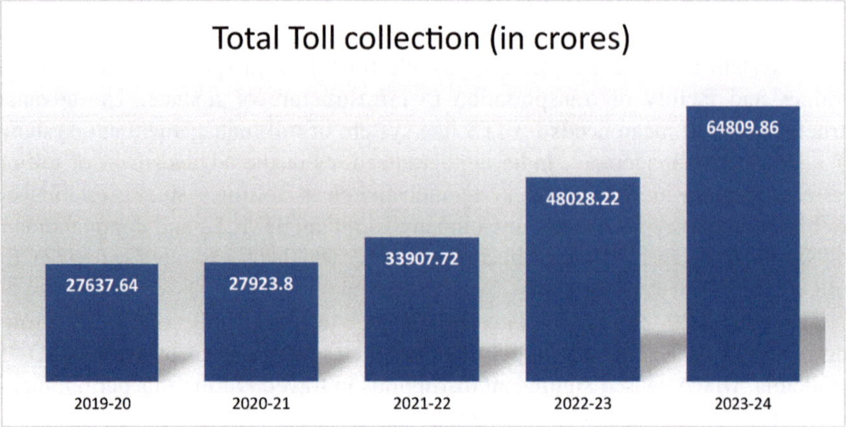

Fig. 5 Toll collection in India. *Source* Moneycontrol (2024)

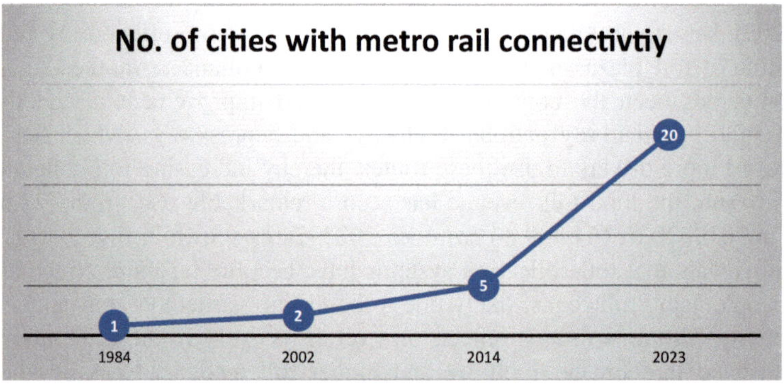

Fig. 6 No. of metro cities in India. *Source* IBEF (2024b)

There was only 1 connectivity of metro rail in 1984 than in 2002 it increased to 2 connectivity. In 2014, there were 3 more metro rail connectivity were developed. Shri Hardeep Singh Puri, who is a Union Minister, informed about that in 2023, there are 20 cities in which metro rail is operational by 874 km and in some cities 986 km is in process as shown in Fig. 6. And after this he stated that India moves close towards the achievement of having second largest network of metro in world.

16.5 E-Bus Sewa

The Prime Minister-eBus Sewa was launched in August 2023, as an initiative to start electric buses in city bus operations. The estimated outlay under the scheme is US$ 6.9 billion (Rs. 57,613 crore). Eligible cities that can benefit from the scheme are cities having a population of 3 lakh and above as per the 2011 census, which sums upto 169 cities. The aim of the initiative is to bring 10,000 e-buses under the umbrella of city operations through the PPP model. This plan also supports the development of the e-buses ecosystem toward enhancing last-mile connectivity.

17 Future of Intelligent Transportation System in India

The future of Intelligent transportation System in India is poised for a transformative shift, driven by advancements in technology, sustainability, and urban planning. As cities become more congested and environmental concerns intensify, India is embracing smart transportation solutions to improve mobility and reduce pollution. Electric vehicles (EVs) are expected to play a crucial role, with the government pushing for EV adoption through incentives, charging infrastructure, and policies to reduce reliance on fossil fuels. The rise of ride-sharing platforms, autonomous vehicles, and the integration of ITS will help streamline traffic management, making travel more efficient and safer. Public transportation is also undergoing a revamp, with metro networks expanding in major cities and bus services becoming more connected and technologically advanced. Additionally, the development of hyperloop and high-speed rail systems promises to revolutionize long-distance travel, reducing travel time significantly. In rural areas, the focus will be on improving road connectivity and integrating digital technologies for better last-mile connectivity. With a growing focus on sustainability, inclusivity, and convenience, the future of transport in India is set to be smarter, cleaner, and more accessible for everyone.

18 Smart Cities and Their Connection to Sustainable Development Goals

SDG 3: Public fitness and safety are valuable to SDG 3 (Good Health and Well-Being), and smart towns contribute through enhancing air nice, avenue safety, and emergency response structures. Smart air pollution tracking systems, along with London's Ultra Low Emission Zone (ULEZ), have effectively reduced nitrogen dioxide ranges with the aid of 36%, thereby decreasing respiratory illnesses. AI-powered street protection analytics and emergency reaction coordination in addition beautify urban nicely-being with the aid of reducing injuries and enhancing healthcare get admission to as shown in Table 2.

SDG 4: Incorporating AI and IoT into town planning also blessings SDG 4 (Quality Education) with the useful resource of improving gaining knowledge of environments and accessibility. Smart gaining knowledge of systems, at the side of South Korea's AI-powered school rooms, use virtual fact (VR) and AI tutors to provide a more interactive instructional enjoy. Additionally, smart faculty delivery monitoring ensures protection and overall performance in scholar mobility.

Table 2 Linked SDGs with smart city (Sharifi et al., 2024)

SDGs	Linked SDGs	How smart cities and ITS contribute
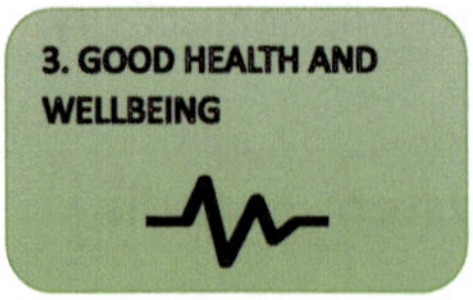	SDG 3: good health and well-being	Improved air quality (clean energy) and safer transportation systems enhance public health
	SDG 4: quality education	AI and IoT in Smart Cities improve access to digital learning tools and educational resources
	SDG 5: gender equality	Inclusive urban planning ensures equal access to clean energy, transport, and opportunities for all
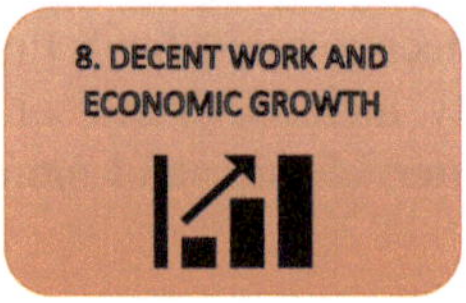	SDG 8: decent work and economic growth	Transition to clean energy and sustainable transport creates jobs and fosters economic development
	SDG 9: industry, innovation, and infrastructure	Smart Cities use renewable energy (solar, wind) and energy-efficient systems (smart grids, EVs) to build resilient infrastructure
	SDG 10: reduced inequalities	Affordable clean energy and accessible public transport reduce urban disparities
	SDG 11: sustainable cities and communities	Clean energy and ITS promote sustainable urban development, reducing slums and improving public transport

(continued)

Table 2 (continued)

SDGs	Linked SDGs	How smart cities and ITS contribute
	SDG 12: responsible consumption and production	Reduces waste through IoT and AI and Optimize energy & water use with real-time data. Helps in promoting circular economy via smart manufacturing and recycling
	SDG 13: climate action	ITS reduces carbon emissions by optimizing traffic, promoting EVs, and minimizing energy consumption
	SDG 17: partnerships for the goals	Public-private partnerships (PPPs) drive innovation, funding, and scalability of Smart City projects

SDG 5: Gender-responsive city planning allows SDG 5 (Gender Equality) via prioritizing stable transport alternatives, AI-powered surveillance, and inclusive public areas, as seen in New Delhi's Women's Safety Smart Initiative.

SDG 8: Sustainable public transport systems play a important function in network financial development, helping SDG 8 (Decent Work and Economic Growth). Curitiba's Bus Rapid Transit (BRT) tool is a top instance, because it has boosted business enterprise employer hobby along transit corridors and created new gadget possibilities in inexperienced mobility sectors which incorporates EV production and public shipping control.

SDG 9: Smart metropolis technologies also boost SDG 9 (Industry, Innovation, and Infrastructure) via growing resilient, technology-driven urban infrastructure. The implementation of IoT-enabled clever grids, AI-driven logistics, and 5G-powered connectivity enhances the efficiency and resilience of city services. Barcelona's clever infrastructure, which includes sensor-primarily based waste control and adaptive street lighting fixtures, is an example of the way generation can optimize urban structures at the same time as decreasing environmental impact.

SDG 10: Smart cities address metropolis disparities, aligning with SDG 10 (Reduced Inequalities) through making sure handy public transportation, digital inclusion, and much less high-priced housing. Initiatives together with Berlin's Smart Accessibility Plan, which integrates wheelchair-first-rate routes into ITS, help bridge mobility gaps.

Despite those upgrades, clever towns face large demanding conditions, specifically in scalability, fee, and records privateness problems. High implementation expenses can restrict adoption in developing regions, even as technological disparities may furthermore widen socio-financial gaps, impacting SDG 10. Additionally, problems over records protection and surveillance ethics pose risks to virtual governance. Without inclusive rules, the sustainability effect of clever city era can be erratically distributed, probably hindering improvement in the route of more than one SDG.

SDG 11: Smart Cities and Intelligent Transportation Systems (ITS) play a essential position in advancing the United Nations' Sustainable Development Goals (SDGs) through integrating generation and sustainability into urban planning. The Smart Cities Mission aligns with SDG eleven (Sustainable Cities and Communities) by promoting sustainable city development via efficient land use, clever governance, digital infrastructure, and advanced public services. Cities like Singapore have efficiently applied real-time visitors monitoring, green building technologies, and IoT-based waste management systems to decorate urban sustainability. These efforts make a contribution to lowering pollution, enhancing mobility, and making towns more liveable.

SDG 12: By integrating IoT and AI, these cities significantly reduce waste through smart waste management systems, automated sorting, and data-driven recycling initiatives. Real-time data analytics help optimize energy and water consumption, ensuring efficient resource utilization and minimizing environmental impact. Additionally, smart cities foster a circular economy by incorporating intelligent manufacturing processes, sustainable supply chains, and innovative recycling methods to reduce material wastage. Digital platforms further encourage sustainable consumer habits by providing real-time insights, eco-friendly recommendations, and incentives for responsible consumption. Through these technology-driven strategies, smart cities contribute to a more sustainable and resource-efficient future.

SDG 13: Intelligent Transportation Systems (ITS) directly make contributions to SDG 13 (Climate Action) by means of decreasing carbon emissions through smart traffic management, electric powered car (EV) adoption, and congestion manage. Cities like Stockholm have applied congestion pricing, main to a 14% reduction in emissions, whilst AI-pushed traffic go with the flow optimization reduces gas wastage from idling motors. Additionally, the combination of renewable power assets, together with solar-powered EV charging stations, further helps climate motion.

SDG 17: Public-non-public partnerships (PPPs) are fundamental in scaling sustainability projects, as they mobilize investments, drive technological innovation, and make sure the long-time period achievement of clever metropolis initiatives. The collaboration between governments and personal enterprises, as seen in Google's Sidewalk Labs in Toronto, has brought about facts-pushed urban improvement and revolutionary solutions for transportation and infrastructure. These partnerships boost up studies and development while ensuring sustainable boom.

19 Conclusion and Discussion

Smart cities promote different advance education policies and health policies which make peoples safe from diseases and enhance knowledge and also play a vital role in growth of economy. There are various master plans of smart cities mission to make cities smart such as reconstruct, renewal, development of greenfield and development of pan-city which are discussed above in details. Intelligent transportation systems give traffic management system, toll management, travel and transport vehicles advancement. By this system, urban issue reduces and progress in effective working of peoples. Traffic management solve problems of rush on roads and helps in decreasing of travel times. For the achievement of growth of economy, various policies states above play an important role. And some key components of intelligent transportation system discussed above which helps in making transport systems best and nature friendly like public transport which reduce pollution and helps peoples to travel anywhere easily and vehicle-to-everything communication.

The concept of the Smart City Mission has evolved considerably since its inception, reflecting the changing needs and technological advancements of urban environments. The primary goal of the Smart City Mission was initially to enhance basic urban services like water supply, electricity, and sanitation utilizing technology. Advances in the Internet of Things (IoT) and data analytical tools led to the integration of real-time data management and citizen engagement through digital platforms. Artificial intelligence (AI) has further revolutionized urban planning and resource management by providing smart solutions for traffic management, waste disposal, and energy conservation. The Smart City Mission has evolved to prioritize sustainability, inclusivity, and resilience, with an aim to construct cities that are intelligent, adaptive to future challenges, and cater to the needs of every resident.

20 Future Scope

In the forthcoming developments, we anticipate the integration of cutting-edge technologies like 5G, edge computing, and quantum computing into the enhancement of smart transportation systems. This vision encompasses a futuristic city where autonomous vehicles and drones play a central role in optimizing public transport and delivery services, leading to less congested and cleaner streets. We foresee the ability to anticipate and address current urban challenges such as traffic management and energy efficiency in real time. Leveraging AI and machine learning for smart grid systems powered by renewable energy sources moves us towards creating self-sustainable cities. It is our goal that these advancements are not only advanced technologically but also widely accessible and beneficial to all. Imagine urban landscapes that adapt to our needs and foster innovation to promote inclusive, vibrant living experiences in our daily routines.

References

Ahmed, A., & Ali, S. (2021). Smart cities in India: Practices, policies, current status and gaps. In *ICIDSSD 2020* (p. 399). Springer. https://doi.org/10.1007/978-981-33-6801-3_30

Awasthi, A., Charan Pattnayak, K., Dhiman, G., & Raj Tiwari, P. (Eds.). (2024). *Artificial intelligence for air quality monitoring and prediction* (1st ed.). CRC Press.

Barns, S. (2018). Smart cities and urban data platforms: Designing interfaces for smart governance. *City, Culture and Society, 12*, 5–12. https://doi.org/10.1016/j.ccs.2017.09.006

Belaïd, F., & Arora, A. (2024). Smart cities: Social and environmental challenges and opportunities for local authorities. *Sustainable Cities and Society*.

Bibri, S. E., & Krogstie, J. (2017). Smart sustainable cities of the future: An extensive interdisciplinary literature review. *Sustainable Cities and Society, 31*, 183–212. https://doi.org/10.1016/j.scs.2017.02.016

Cheng, Z., Pang, M. S., & Pavlou, P. A. (2020). Mitigating traffic congestion: The role of intelligent transportation systems. *Information Systems Research, 31*(3), 653–674. https://doi.org/10.1287/isre.2020.0921

Dai, Y., Hasanefendic, S., & Bossink, B. (2024). A systematic literature review of the smart city transformation process: The role and interaction of stakeholders and technology. *Sustainable Cities and Society, 101*, Article 105112. https://doi.org/10.1016/j.scs.2024.105112

Hannan, M. A., Begum, R. A., Al-Shetwi, A. Q., Ker, P. J., Al Mamun, M. A., Hussain, A., Basri, H., & Mahlia, T. M. I. (2020). Waste collection route optimisation model for linking cost saving and emission reduction to achieve sustainable development goals. *Sustainable Cities and Society, 62*, Article 102393. https://doi.org/10.1016/j.scs.2020.102393

Harmon, R. R., Castro-Leon, E. G., & Bhide, S. (2015, August). Smart cities and the Internet of Things. In *2015 Portland International Conference on Management of Engineering and Technology (PICMET)* (pp. 485–494). IEEE. https://doi.org/10.1109/PICMET.2015.7273140

India Brand Equity Foundation. (2024a, December 17). *Enhancing public transport: Steps taken by the Indian government to strengthen nationwide connectivity*. India Brand Equity Foundation. https://www.ibef.org/blogs/enhancing-public-transport-steps-taken-by-the-indian-government-to-strengthen-nationwide-connectivity

India Brand Equity Foundation. (2024b, December 17). *Indian Railways: Infographic*. India Brand Equity Foundation. https://www.ibef.org/industry/indian-railways/infographic

Kumar, T. V., & Dahiya, B. (2017). Smart economy in smart cities. In T. V. Kumar & B. Dahiya (Eds.), *Smart economy in smart cities* (pp. 3–76). Springer. https://doi.org/10.1007/978-981-10-1610-3_1

Lim, Y., Edelenbos, J., & Gianoli, A. (2024). What is the impact of smart city development? Empirical evidence from a Smart City Impact Index. *Urban Governance, 4*(1), 47–55. https://doi.org/10.1016/j.ugj.2023.12.002

Moneycontrol. (2024, December 17). India's total toll collections rise 35% on year to Rs 64,810 cr in FY24; target for FY25 at Rs 70,000 cr. *Moneycontrol*. https://www.moneycontrol.com/news/business/indias-total-toll-collections-rise-35-on-year-to-rs-64810-cr-in-fy24-target-for-fy25-at-rs-70000-cr-12626331.html

Mora, L., Gerli, P., Ardito, L., & Petruzzelli, A. M. (2023). Smart city governance from an innovation management perspective: Theoretical framing, review of current practices, and future research agenda. *Technovation, 123*, Article 102717. https://doi.org/10.1016/j.technovation.2022.102717

Qian, Y., Wu, D., Bao, W., & Lorenz, P. (2019). The Internet of Things for smart cities: Technologies and applications. *IEEE Network, 33*(2), 4–5. https://doi.org/10.1109/MNET.2019.1800287

Rani, P., & Sharma, R. (2023). Intelligent transportation system for internet of vehicles based vehicular networks for smart cities. *Computers and Electrical Engineering, 105*, Article 108543. https://doi.org/10.1016/j.compeleceng.2023.108543

Sagaris, L., & Tiznado-Aitken, I. (2020). Sustainable transport and gender equity: Insights from Santiago, Chile. In *Urban mobility and social equity in Latin America: Evidence, concepts, methods* (pp. 103–134). Emerald Publishing Limited. https://doi.org/10.1108/9781839091146-007

Shaaban, K., Elamin, M., & Alsoub, M. (2021). Intelligent transportation systems in a developing country: Benefits and challenges of implementation. *Transportation Research Procedia, 55*, 1373–1380. https://doi.org/10.1016/j.trpro.2021.07.123

Sharifi, A., Allam, Z., Bibri, S. E., & Khavarian-Garmsir, A. R. (2024). Smart cities and sustainable development goals (SDGs): A systematic literature review of co-benefits and trade-offs. *Cities, 146*, Article 104659. https://doi.org/10.1016/j.cities.2023.104659

Shukla, P., Sharma, P. K., Pandey, S., & Kunwer, R. (2024). Optimizing municipal solid waste recovery through density-based waste plastic segregation. *Journal of Material Cycles and Waste Management, 26*(4), 2334–2354.

Silva, B. N., Khan, M., & Han, K. (2018). Towards sustainable smart cities: A review of trends, architectures, components, and open challenges in smart cities. *Sustainable Cities and Society, 38*, 697–713. https://doi.org/10.1016/j.scs.2018.01.053

Winkowska, J., Szpilko, D., & Pejić, S. (2019). Smart city concept in the light of the literature review. *Engineering Management in Production and Services, 11*(2), 70–86. https://doi.org/10.2478/emj-2019-0019

Yin, C., Xiong, Z., Chen, H., Wang, J., Cooper, D., & David, B. (2015). A literature survey on smart cities. *Science China Information Sciences, 58*(10), 1–18. https://doi.org/10.1007/s11432-015-5397-4

Law and the 2030 Agenda: Strengthening Sustainable Development Through Legal Accountability

Parineeta Goswami and Jorge Isaac Torres Manrique

Abstract

The 2030 Agenda for Sustainable Development aims to create a more equitable and sustainable world through its 17 Sustainable Development Goals (SDGs). While widely endorsed, its implementation faces significant challenges, particularly in regions like Latin America and the Caribbean, where disparities persist. This study explores the extent to which the 2030 Agenda aligns with fundamental rights and examines the role of regional mechanisms in its execution. The key research question is: How effectively does the 2030 Agenda integrate fundamental rights, and what are the barriers to its implementation in Latin America and the Caribbean? Using a qualitative approach, the study analyses report from the Economic Commission for Latin America and the Caribbean (ECLAC) and the UN's quadrennial progress assessments. It also reviews scholarly literature on governance, environmental sustainability, and technological advancements concerning the SDGs. Findings indicate that while the agenda promotes sustainability, its voluntary commitments limit enforcement. Key concerns include inadequate attention to environmental degradation, artificial intelligence, and governance. The COVID-19 pandemic has further hindered progress, deepening inequalities. The study concludes that stronger political will, mandatory compliance mechanisms, and broader stakeholder participation are essential for meaningful progression. Strengthening regional cooperation and addressing emerging challenges will be crucial to achieving the SDGs by 2030.

P. Goswami (✉)
School of Law, UPES, Dehradun, India
e-mail: Parineeta.Goswami@ddn.upes.ac.in

J. I. T. Manrique
Universidad Católica de Santa María, Arequipa, Peru

T. Choudhury et al. (eds.), *Open Access as a Knowledge Platform for Implementing the United Nations Sustainable Development Goals*, World Sustainability Series,
https://doi.org/10.1007/978-3-031-99605-4_14

Keywords

Sustainable development goals • Fundamental rights • Latin America • Governance • Environmental sustainability • Regional mechanisms

1 Introduction

A sluggish global economy coexists with ongoing social disparities and deteriorating environmental conditions, which hinder sustainable long-term development paths (United Nations, 2018). Current production systems, together with existing methods of energy use and economic development patterns, have resulted in unsustainable patterns, so a transition to an inclusive, sustainable framework becomes necessary. Such change represents essential requirements rather than options for the regions having the most pronounced inequalities. Latin America and the Caribbean (LAC) stands out as a region that shows the worst distribution of wealth yet holds the lowest rank regarding poverty (United Nations, 2019). The international community made its landmark move towards sustainable development when it adopted the 2030 Agenda for Sustainable Development through a global adoption in September 2015. The United Nations member states who endorsed the agenda established 17 SDGs and 169 targets to construct a resilient and equitable global society. The 2030 Agenda stands out against preceding development plans through its integrated methodology, which acknowledges that environmental protection requires partnership with societal development and economic expansion (Sharma et al., 2025). Every nation across the world must embrace this initiative through obligatory participation because of its universal scope (United Nations Environment Programme [UNEP], 2021).

The execution of the 2030 Agenda tasks proves essential for Latin American countries throughout the Caribbean. Sustainable development remains difficult to achieve because the region has historically experienced economic instabilities alongside structural inequality issues, poor governance, and environmental challenges. The agenda allows Latin American nations to establish new development paths through enhanced social equity, along with improved governance systems and strengthened environmental resistance.

Resolution 700 (XXXVI) led to the creation of the Forum of Latin American and Caribbean Countries on Sustainable Development to facilitate regional implementation and progress tracking, which serves as a regional monitoring mechanism. This forum plays a crucial role in ensuring accountability, fostering cooperation among nations, and aligning regional policies with the global SDG framework. Additionally, it provides a platform for evaluating national policies, sharing best practices, and addressing common regional challenges (United Nations, 2019).

Despite the ambitious nature of the 2030 Agenda, significant challenges continue to hinder its effective implementation in Latin America and the Caribbean.

While the agenda envisions a transformative shift toward sustainable development, its voluntary nature and lack of legally binding enforcement mechanisms present critical barriers to progress. Several persistent challenges stand in the way of achieving the SDGs.

Inequity and social exclusion are ongoing challenges to the implementation of the SDGs, especially in LAC, where inequality is among the highest on the planet. Economic disparities disproportionately impact vulnerable groups, limiting their access to quality education, health care, and jobs (United Nations, 2018). To overcome these challenges, far-reaching education system reform is needed. Curriculum revision to include sustainability and climate change can enhance awareness and promote responsible citizenship. In addition, teacher education investment to advance sustainable development capabilities will ensure the delivery of pertinent knowledge and skills effectively. Also, enhanced access to quality education among marginalized groups utilizing special policies and fiscal incentives, can abate educational disparities and support inclusive growth. By adopting these practices, nations can establish a more inclusive society and quicken the pace of achieving the SDGs (International Labour Organization [ILO], 2022).

The success of the 2030 Agenda largely depends on strong governance, transparent institutions, and policy coherence. However, many countries in the region face institutional weaknesses, political instability, and corruption, which undermine the effective implementation of sustainable policies (United Nations, 2019).

Climate change and environmental vulnerabilities pose significant threats to sustainable development, particularly in areas vulnerable to deforestation, biodiversity loss, and climate-related extreme weather events. These environmental degradations are risks to both social welfare and economic stability. They are paramount to the goal for the attainment of various SDGs, some of which include Clean Water and Sanitation (SDG 6), Climate Action (SDG 13), and Life on Land (SDG 15) (United Nations, 2018). Inadequate implementation of good environmental policies may erode achievements in different sectors, worsening poverty and inequality. To avoid such risks, policymakers must prioritize environmental sustainability by initiating tighter regulations and incentives for sustainable practices. Also, incorporating climate resilience strategies into all development activities can improve adaptive capacity and vulnerability reduction. Effective coordination and anticipatory policymaking are necessary to protect natural ecosystems and achieve long-term sustainable development (Universal Rights Group, 2018).

While technological advancements offer opportunities for economic growth and innovation, they also raise concerns related to job displacement, ethical considerations, and digital inequality. The 2030 Agenda does not fully address the implications of artificial intelligence, data governance, and the digital divide, which have become pressing issues in today's global economy.

The COVID-19 pandemic severely impacted Latin American economies, exacerbating existing inequalities and reversing progress toward the SDGs. The quadrennial progress report highlights that the pandemic increased poverty levels, weakened healthcare systems, and disrupted labour markets, further delaying the region's sustainable development efforts (United Nations, 2019).

2 Research Objective and Significance

This study aims to evaluate the effectiveness of the 2030 Agenda in addressing fundamental rights and to assess the role of regional mechanisms in overcoming barriers to implementation. The key research question guiding this analysis is:

To What Extent Does the 2030 Agenda Integrate Fundamental Rights, and What Are the Primary Challenges in Achieving the SDG in Latin America and the Caribbean?

By answering this question, the study contributes to the academic and policy discourse on sustainable development by:

- Identifying key implementation gaps in the Latin American context.
- Analysing the effectiveness of regional institutions such as ECLAC and the Forum of Latin American and Caribbean Countries on Sustainable Development in facilitating SDG progress (Economic Commission for Latin America & the Caribbean [ECLAC], 2023)
- Assessing the impact of global crises, such as the COVID-19 pandemic and climate change, on the achievement of SDGs.
- Proposing recommendations for stronger governance, improved stakeholder engagement, and more effective policy implementation.

Enhancing stakeholder engagement is crucial for the successful implementation of the SDGs. Developing inclusive mechanisms that facilitate multi-stakeholder participation, including governments, non-governmental organizations, and the private sector, fosters collaboration and shared responsibility. Transparent and participatory decision-making processes ensure that diverse perspectives are considered, leading to more effective and equitable policies. Furthermore, encouraging corporate social responsibility initiatives aligned with SDG targets can leverage private sector resources and innovation to drive sustainable development. By creating an environment where all stakeholders actively contribute, nations can accelerate progress towards achieving the SDGs and build resilient and inclusive societies (International Energy Agency [IEA], 2022).

3 ECLAC's Priorities to Support the Implementation and Follow-Up of the 2030 Agenda for Sustainable Development in Latin American and Caribbean Countries

The 2030 Agenda for Sustainable Development, which was adopted in 2015 by the United Nations, is a worldwide pledge to tackle urgent issues like poverty, inequality, climate change, and environmental degradation. It has 17 SDGs and 169 targets that seek to promote social inclusion, economic growth, and environmental sustainability. For LAC nations, the implementation of this ambitious agenda calls for strategies adapted to their specific socioeconomic and environmental challenges and regional coordination (Feiring & König-Reis, 2020).

The ECLAC has emerged as a central partner in promoting the 2030 Agenda in the region. ECLAC's integrated strategy is centred on overcoming structural obstacles, promoting policy coherence, and strengthening institutional and statistical capacities to facilitate sustainable development. To realize these goals, ECLAC has outlined four main priorities that act as a roadmap for accelerating progress in the LAC region (United Nations, 2019). A strong regional institutional framework is critical to facilitate collaboration, tackle transboundary issues, and encourage joint action towards attaining the SDGs. ECLAC insists on the reinforcement of current institutions and the creation of regional cooperation mechanisms that bring nations together in their quest for sustainable development (Economic Commission for Latin America & the Caribbean [ECLAC], 2023).

ECLAC promotes the creation of inclusive spaces that integrate governments, private sectors, civil society, and universities. These spaces facilitate dialogue, exchange of knowledge, and collective action to address shared issues like climate change, natural resource management, and economic inequality. The Forum of the Countries of LAC on Sustainable Development is one such initiative, offering a regional forum for reporting on progress, challenges, and good practices in the implementation of the 2030 Agenda (International Monetary Fund [IMF], 2021).

ECLAC collaborates intimately with countries' national governments to strengthen institutional capacities. It does this through the training of public servants, technical assistance, and open and accountable governance arrangements. This helps ensure that institutions can manage properly the multi-dimensional requirements of the SDGs and respond to the development gaps in the region. Policy consistency at national and regional levels is vital in the attainment of the SDGs. ECLAC promotes conformity of national policies with regional and international policy frameworks, allowing for consistent and coherent approaches to responding to common development issues. Harmonization of policies enables nations to leverage synergies and avoid trade-offs, which ensures that regional actions reinforce the greater agenda of the 2030 Agenda. The implementation means include the financial, technological, and institutional resources to attain the SDGs. Disparities in access to such resources are high in the LAC region and present major challenges. ECLAC is committed to examining and closing these gaps to facilitate balanced and effective 2030 Agenda implementation. Access to sufficient financial resources is an urgent issue for most LAC nations, especially middle-income ones, which find it difficult to obtain concessional funding. ECLAC encourages the implementation of new financing mechanisms like blended finance, public–private partnerships, and green bonds. These strategies harness various sources of funds and channel investments into priorities for sustainable development (Inter-American Development Bank [IDB], 2019).

Technological innovation plays an important role of spurring sustainable development by propelling productivity, enhancing service delivery, and supporting environmental sustainability. The ECLAC assists nations in the region in the utilization of digital technologies to realize these goals. Closing the digital divide continues to be an important priority to ensure the marginal population also gains from technological innovation. Increased data gathering and surveillance using artificial intelligence and big data can be beneficial in generating insight for evidence-based decision-making and successful policy application. Further, enhanced access to digital public services and strong infrastructure can ensure wider inclusivity and connectivity. Bridge the digital divide using focused interventions and equal access to technology to ensure the benefits of innovation reach all walks of society to speed up efforts towards the SDGs (United Nations, 2020a, 2020b).

Regional cooperation is critical to respond to transboundary challenges and to take advantage of economies of scale. ECLAC promotes cooperation between countries, international organizations, and private actors to join forces, exchange knowledge, and address together issues of development such as climate resilience, disaster preparedness, and regional trade.

ECLAC offers technical cooperation and training to strengthen national capabilities in resource mobilization and management. These programs aim at improving financial governance, raising tax revenues, and curtailing illicit financial flows, which all help improve the means of implementation. The incorporation of the SDGs into mainstream national development plans is necessary for achieving policy consistency and resource prioritization. ECLAC proactively encourages LAC countries to harmonize their development strategies, plans, and budgets with the 2030 Agenda. ECLAC supports governments in formulating national priorities and incorporating them into development plans. This entails linking SDG targets to existing policies, finding gaps, and establishing measurable goals. Through harmonizing national plans with the SDGs, nations can achieve a strategic and coordinated effort towards sustainable development. One of the most important features of integration is connecting resources to SDG results. ECLAC advocates for result-based budgeting, whereby budget allocations are linked to definite targets and indicators. This strengthens accountability, ensures effective resource use, and makes it possible to monitor progress toward the SDGs. Participatory and inclusive strategies are key to the 2030 Agenda's success. ECLAC highlights the necessity of involving different stakeholders, such as indigenous peoples, women, and young people, in the planning and implementation process. This ensures that policies are sensitive to the needs of all sectors of society and help in lessening inequalities. To facilitate the incorporation of SDGs in national budgets, ECLAC helps nations develop financing strategies that mobilize public as well as private investments. These strategies maximize domestic resources and attract foreign funding to meet development needs. Reliable and timely data are the cornerstone of successful SDG monitoring and decision-making. ECLAC gives high priority to enhancing the capacity of national statistical systems and assisting in the integration of measurement procedures into statistical development plans. ECLAC offers technical support to the national statistical offices to improve the skills of their data collection, analysis, and dissemination. These include the improvement of methodologies, data quality, and compliance with international statistical standards. Applying big data, geospatial technologies, and digital platforms can change the manner of data collection and analysis. It is recommended that countries embrace such technologies to help produce more accurate and disaggregated data, hence better-targeting policies and programs (World Bank, 2018).

Regional cooperation among LAC nations is needed to fill statistical capacity gaps. ECLAC promotes regional activities that enhance knowledge sharing, standardize methodologies, and construct integrated databases. This promotes uniformity in reporting data and allows for comparative analysis throughout the region. ECLAC emphasizes the need to generate data that reflects the realities of all groups in the population, especially the poor and vulnerable groups. This includes creating disaggregated data according to income, gender, ethnicity, and geography, which is critical for tracking progress on selected indicators and confronting inequalities (Intergovernmental Panel on Climate Change [IPCC], 2022).

The 2030 Agenda has not been without difficulties in the LAC region. Enduring inequalities, scarce financial resources, political unrest, and external shocks like the COVID-19 pandemic have constrained advancement. But the difficulties also bring challenges to innovate and transform. Many studies have been done to study the effect of COVID-19 and its variation with the different variables. The pandemic, for example, underscored the need for resilient health systems, digital infrastructure, and social protection. By responding to these systemic weaknesses, LAC nations can construct more inclusive and sustainable societies. ECLAC's initiatives to enhance regional integration, drive innovation, and encourage capacity-building offer a firm basis for addressing these challenges. Its focus on equity, sustainability, and partnership guarantees that the region is well on its way to meeting the 2030 Agenda. ECLAC's agenda priorities for the facilitation of the implementation and follow-up of the 2030 Agenda in LAC demonstrate a holistic vision of sustainable development. Through the consolidation of regional institutions, examination of means of implementation, incorporation of SDGs into national plans, and strengthening of statistical systems, ECLAC prepares the region to confront complicated development issues. With Latin American and Caribbean nations moving toward the 2030 Agenda, ECLAC's strategic actions continue to be central. Its focus on cooperation, innovation, and inclusiveness guarantees that the values of sustainable development lead the way for the region, shaping a future in which no one is left behind (Economic Commission for Latin America and the Caribbean (CLAC), 2023).

4 Quintessence of the 2030 Agenda for Sustainable Development

The United Nations General Assembly adopted the 2030 Agenda for Sustainable Development in September 2015, a game-changing international pledge to secure economic, social, and environmental sustainability. This far-reaching plan calls for 17 SDGs and 169 targets based on an integrated approach to development that does not leave anyone behind. Involving 193 Member States, the agenda establishes a strong platform for a more inclusive, equitable, and sustainable future in the next 15 years after adoption (United Nations Human Rights Council, 2018).

The 2030 Agenda holds special importance for LAC. The region has its own distinctive challenges like pervasive inequality, extensive poverty, climate change vulnerability, and pressures of urbanization. These are intricately interwoven in the texture of the SDGs that focus on eliminating systemic obstacles to development, encouraging social equity, and protecting the environment (Food & Agriculture Organization [FAO], 2020).

The 2030 Agenda is not just a development blueprint: it's a blueprint for civilization. It is rooted in the promise to maintain human dignity, end poverty, and limit inequalities. It acknowledges that development is multidimensional and interdependent and needs an integrated plan based on social, economic, and environmental dimensions. For Latin America and the Caribbean, the agenda offers a historic moment to face deep-rooted structural challenges and design inclusive sustainable development pathways.

The high-priority issues addressed in the 2030 Agenda align closely with the region's needs, including:

- Despite notable progress in reducing poverty, significant disparities persist. The SDGs call for eliminating poverty in all forms, emphasizing targeted strategies for vulnerable populations.
- Inequality remains one of the most pressing challenges in LAC, spanning income, gender, ethnicity, and access to resources. The agenda's emphasis on reducing inequalities across dimensions is particularly relevant.
- Growth and Decent Work: The SDGs advocate for sustained economic growth that generates decent jobs and economic opportunities for all, a critical need in a region characterized by high unemployment and informal labour markets.
- Urbanization in LAC has created challenges related to housing, transportation, and basic services. The agenda promotes sustainable urban development that prioritizes inclusivity and environmental sustainability.
- As one of the regions most vulnerable to climate change, LAC faces challenges such as rising sea levels, extreme weather events, and deforestation. The agenda underscores the importance of resilience and climate action.

The 2030 Agenda puts human dignity and equality at the forefront. This vision acknowledges that development is not just about economic growth but also has to deal with the root causes of inequality, exclusion, and environmental degradation.

The fulfilment of these objectives requires the active engagement of all sections of society, including governments, civil society, academia, and the private sector (United Nations Development Programme [UNDP], 2021).

The agenda emphasizes inclusivity in policymaking and policy implementation. It calls for the active participation of marginalized groups such as women, indigenous peoples, and youth to ensure that their voices are heard and their concerns addressed. Participatory approach is critical in developing inclusive societies that promote human rights and social justice. The 17 SDGs represent a bold and visionary vision for the future. These objectives are interlinked and respond to urgent global issues, ranging from eliminating poverty (Objective 1) and promoting gender equality (Objective 5) to addressing climate change (Objective 13) and encouraging partnerships (Objective 17). Their ambitious nature calls for creative solutions, collective action, and long-term commitment from all stakeholders.

The 2030 Agenda underlines the accountability of today's generations to ensure a sustainable future for posterity. Intergenerational equity is the central principle here, which points towards meeting current development needs without harming the ability of future generations to meet their own needs. The fulfilment of the 2030 Agenda calls for a multisectoral and multidimensional response. Governments, civil society, academia, and the private sector need to collaborate to establish an enabling environment that supports sustainable development. Governments have a central role in shaping the implementation of the SDGs. This entails incorporating the goals into national development strategies, policies, and budgets. Effective implementation and monitoring require strong governance, transparency, and accountability.

Civil society organizations (CSOs) contribute to the agenda through representing marginalized communities, keeping governments in check, and providing services addressing development problems. Their grass-roots linkages allow them to engage with vulnerable groups and give them a voice. Academia contributes to research, innovation, and capacity development. Universities and research centres offer evidence-based answers to intricate development challenges and assist in developing the knowledge and expertise necessary to reach the SDGs (Organisation for Economic Co-operation and Development [OECD], 2021) The private sector is an essential partner in funding and delivering the 2030 Agenda. Companies can integrate their strategies with the SDGs, embrace sustainable practices, and invest in activities that generate social and environmental value (Fig. 1).

Adapted from *Progress Across Selected SDGs: A Regional Analysis*, by United Nations Development Programme (n.d.). © UNDP. Retrieved from https://www.undp.org. While detailed progress analysis varies across countries and regions, some key trends and achievements can be highlighted:

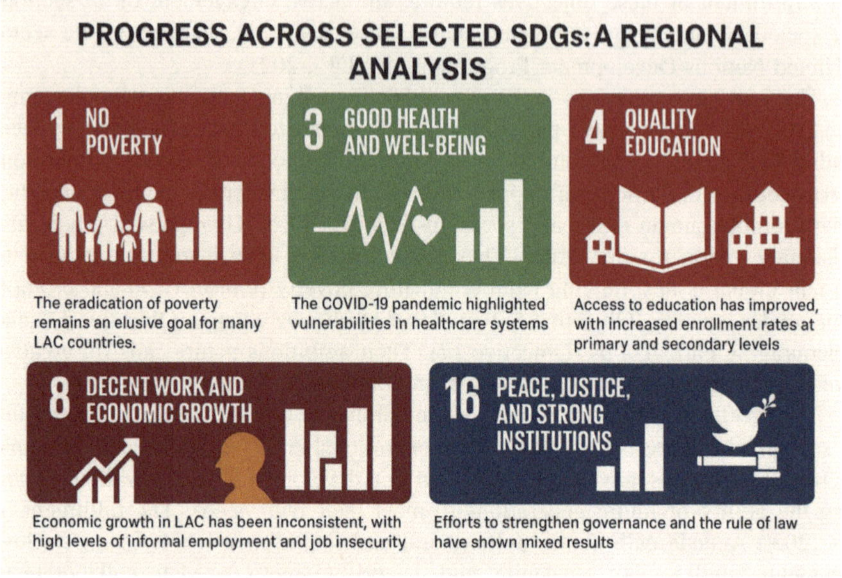

Fig. 1 Progress across selected Sustainable Development Goals (SDGs) in the Latin America and Caribbean (LAC) region. The infographic highlights ongoing challenges such as poverty eradication (SDG 1), healthcare vulnerabilities post-COVID-19 (SDG 3), improvements in education access (SDG 4), inconsistent economic growth (SDG 8), and mixed progress in governance and institutional strength (SDG 16)

4.1 Goal 1: No Poverty

LAC has made strides in reducing poverty, but economic shocks, pandemics, and inequality have slowed progress. Targeted social protection programs and inclusive economic policies are vital for achieving this goal.

4.2 Goal 3: Good Health and Well-being

Investments in healthcare systems have improved access to essential services in many LAC countries. However, disparities in healthcare quality and access remain significant challenges.

4.3 Goal 5: Gender Equality

Gender equality has seen progress in areas such as education and political participation. However, gender-based violence and economic disparities persist, requiring intensified efforts.

Designing gender-sensitive policies is essential for ensuring inclusive progress toward the SDGs. Implementing empowerment programs that enhance women's participation in decision-making processes can foster equitable representation and more inclusive governance. Addressing socio-economic disparities that disproportionately affect women, such as unequal access to education, healthcare, and economic opportunities, is crucial for achieving gender equality. Additionally, strengthening legal protections against gender-based violence and discrimination can create safer and more supportive environments for women and girls. By integrating gender perspectives into policy frameworks and promoting targeted initiatives, governments can accelerate progress towards gender equality and contribute significantly to the overall success of the SDGs.

4.4 Goal 7: Affordable and Clean Energy

The region has advanced in adopting renewable energy sources, with countries like Costa Rica and Uruguay leading the way. Expanding access to affordable and clean energy remains a priority.

4.5 Goal 13: Climate Action

LAC countries have committed to ambitious climate targets, including reducing greenhouse gas emissions and increasing resilience to climate impacts. Regional cooperation and international support are essential to meeting these commitments. Despite its transformative potential, the 2030 Agenda faces several implementation challenges in Latin America and the Caribbean. Inequalities in income, gender, ethnicity, and geographic location continue to hinder progress. Addressing these disparities requires targeted policies and inclusive growth strategies. Limited financial resources, coupled with high levels of debt, constrain the capacity of many LAC countries to invest in sustainable development. Innovative financing mechanisms, such as green bonds and blended finance, are needed to bridge the funding gap. Economic volatility, natural disasters, and health crises such as the COVID-19 pandemic have underscored the region's vulnerability to external shocks. Building resilience and diversifying economies are critical for mitigating these risks. Weak institutional capacities and governance challenges undermine the effective implementation of the SDGs. Strengthening institutions, promoting transparency, and combating corruption are essential for building trust and ensuring accountability.

To reach the 2030 Agenda, LAC needs to pursue strategies that speed progress and overcome system obstacles: Regional action is crucial to work on cross-boundary challenges like climate change, commerce, and migration. Efforts such as the Forum of the Countries of LAC on Sustainable Development build dialogue, learning, and joint action. Technology and innovation are catalysts for achieving the SDGs. Digital transformation can enhance the delivery of services, support better data collection, and drive economic development, especially in isolated and underserved communities. Policy alignment across sectors and governments at all levels guarantees that measures in one area do not undermine achievements in another. Comprehensive approaches to policymaking and planning are necessary for achieving synergies and reducing trade-offs. Involving a wide array of stakeholders, such as women, youth, indigenous peoples, and the business community, is vital to building inclusive and equitable development trajectories. Good and timely data are key to tracking progress and guiding policy interventions. Improved national statistical systems and new approaches to data collection enhance the region's capacity to monitor SDG implementation.

The 2030 Agenda for Sustainable Development is a visionary plan for a more equitable, inclusive, and sustainable world. For Latin America and the Caribbean, the agenda presents a special opportunity to overcome structural obstacles, close inequalities, and construct resilient societies. Although challenges remain, the region has shown a firm commitment to the SDGs through innovative policies, regional coordination, and stakeholder participation. Through the adoption of the values of inclusivity, equity, and sustainability, LAC can fast-track progress and deliver a more prosperous future to both current and future generations. The essence of the 2030 Agenda is that it understands the interlinkages of development issues and has made a pledge to leave no one behind. It invites all sections of society to collaborate, pool their strengths, and make the vision of a world where dignity, equality, and sustainability are universal realities a reality.

5 Analysis of the Progress of the 2030 Agenda for Sustainable Development. Eight Years Later and Six Years from the Day D and Hour H

The United Nations General Assembly, in 2015, adopted the 2030 Agenda for Sustainable Development as a comprehensive, far-reaching guide to resolving interconnected challenges to poverty, inequality, and the sustainability of the environment. Since the world is halfway toward fulfilling these targets, there is a need to measure progress, identify challenges by region, and steer forward. Eight years into operation and with six years left, the agenda's progress is uneven, especially in LAC, an area that represents global opportunities as well as challenges. Structural inequality and social exclusion persist, constraining equal access to education, health, and job opportunities. Fragile institutional frameworks and governance problems further hamper policy implementation and enforcement. In addition, LAC is also beset by acute environmental risks, such as climate change and loss of biodiversity, which require urgent and concerted action. Yet, these are often undermined by a lack of policy support. Adding to these challenges, the voluntary nature of the SDGs weakens legal enforcement, constraining accountability among countries. Moreover, although new technologies such as artificial intelligence hold great promise for sustainable development, their potential is underleveraged, widening gaps in digital inclusion and governance. Overcoming these challenges calls for bolstered institutional capacity, increased technological integration, and more international collaboration to secure substantial progress towards the 2030 Agenda.

5.1 Quadrennial Progress Report: Regional Progress and Challenges

The Quadrennial Progress Report on the 2030 Agenda provides an overall picture of the LAC's progress toward the attainment of the SDGs, noting both successes and setbacks. Although the goals are global, LAC's unique regional dynamics have influenced their implementation (United Nations, 2024). The global political economy has experienced considerable changes since the adoption of the agenda in 2015. By 2019, unforeseen geopolitical tensions, disruptions in trade, growing populism, and the disastrous effects of the COVID-19 pandemic had unsettled global cooperation and influenced global stability. These issues have gravely hampered LAC's ability to achieve SDG targets, especially in eliminating poverty, promoting economic growth, and driving climate action. The pandemic significantly rolled back poverty reduction gains, plunging tens of millions into extreme poverty, overwhelming health systems, and limiting access to basic medical care. School closures also further exacerbated educational inequalities, most adversely affecting the poor and marginalized. The economic slowdown also led to increased unemployment and financial uncertainty, especially in emerging economies. Adding to these are setbacks in climate action and environmental sustainability efforts with national priorities focusing elsewhere.

To reverse these challenges and restore momentum towards the SDGs, increasing political commitment is necessary. The approach must be multi-faceted, encompassing the conclusion of legally binding agreements to advance policy frameworks and provide long-term accountability. Entrenching strong monitoring systems via independent scrutiny and promoting regional coordination can increase transparency and enable efficient policy implementation. In addition, global partnerships are instrumental in bringing national policies into harmony with international SDG goals, enhancing mutual responsibility, and facilitating cross-border cooperation. By embracing these steps, UN member states can reaffirm their political will and expedite progress toward realizing the SDGs by 2030.

As highlighted in CEPAL (2016), the development path of the region has been long hampered by structural imbalances, such as economic reliance on commodity exports, fragile industrial bases, and high inequality. These structural imbalances have continued to hinder progress toward the transformative goals of the 2030 Agenda. Furthermore, the pandemic deepened existing vulnerabilities, driving millions into poverty and increasing socio-economic.

Despite the challenges, LAC has made notable strides in several areas of the SDGs. These achievements highlight the region's resilience and commitment to sustainable development.

- Many countries implemented social protection programs and cash transfer schemes to mitigate poverty. Notable examples include Brazil's Bolsa Família and Mexico's Prospera programs, which have contributed to reducing poverty and inequality. However, the pandemic-induced economic slowdown reversed some gains, underscoring the fragility of progress.
- The region has made significant advancements in renewable energy adoption. Countries such as Uruguay, Costa Rica, and Chile have emerged as global leaders in renewable energy, achieving substantial progress toward Goal 7: Affordable and Clean Energy.
- Progress toward Goal 5: Gender Equality has been visible in increased political participation of women and gender-focused policies in education and employment. However, gender-based violence remains a pervasive issue.
- Initiatives like the Forum of the Countries of LAC on Sustainable Development have fostered regional dialogue and collaboration, enabling knowledge sharing and joint problem-solving.

While achievements are notable, significant challenges continue to hinder progress across the SDGs in Latin America and the Caribbean. These challenges highlight the complexities of implementing a universal agenda in a diverse and unequal region. The economic impacts of the pandemic further exacerbated these existing inequalities and development gaps, placing additional strain on efforts to achieve the 2030 Agenda.

- Economic disparities remain stark, with wealth concentrated among a small elite while a significant portion of the population experiences poverty and deprivation. Inequalities are also evident across rural and urban areas, gender, and ethnic groups, creating barriers to achieving inclusive development.
- The region is particularly vulnerable to climate change impacts, including hurricanes, droughts, and rising sea levels. Countries like Haiti, Honduras, and Nicaragua are among the most affected by extreme weather events. Efforts to achieve Goal 13: Climate Action are hampered by insufficient resources and capacity for climate adaptation and mitigation.
- Reliance on commodity exports has left many LAC economies vulnerable to global market fluctuations. The economic impacts of the pandemic further exacerbated this
- Governance challenges, including corruption, political instability, and weak institutions, undermine the implementation of the SDGs. Strengthening institutional capacities is essential for building trust and ensuring accountability.
- Reliable and disaggregated data are crucial for tracking progress and identifying gaps. Many countries face challenges in building robust statistical systems, hindering effective monitoring of the SDGs.

5.2 Progress Across Selected SDGs: A Regional Analysis

Goal 1: No Poverty: The eradication of poverty remains an elusive goal for many LAC countries. While extreme poverty rates declined in the years following 2015, economic disruptions caused by the pandemic and inflationary pressures have reversed gains.

Goal 3: Good Health and Well-Being: The COVID-19 pandemic highlighted vulnerabilities in healthcare systems. Countries like Chile and Uruguay have made strides in expanding healthcare access, but disparities in rural and marginalized communities persist.

Goal 4: Quality Education: Access to education has improved, with increased enrolment rates at primary and secondary levels. However, learning outcomes remain unequal, particularly for indigenous and rural populations.

Goal 8: Decent Work and Economic Growth: Economic growth in LAC has been inconsistent, with high levels of informal employment and job insecurity. The pandemic exacerbated unemployment, particularly among women and youth.

Goal 16: Peace, Justice, and Strong Institutions: Efforts to strengthen governance and the rule of law have shown mixed results. Corruption and political instability continue to hinder progress in many countries. With six years left to 2030, LAC needs to implement transformative measures to break down obstacles and speed up progress towards the SDGs. Scaling up and enhancing social protection programs can act as a safety net for the poor and vulnerable, alleviate poverty, and support post-pandemic recovery plans needed to ensure sustainability and inclusivity. Investments in green infrastructure, renewable energy, and sustainable agriculture can spur economic growth while mitigating climate change. Regional cooperation is necessary to respond to transboundary issues like migration, climate change, and trade. Reinforcing efforts such as the Forum of the Countries of LAC on Sustainable Development can improve joint action. Enhancing the quality and access to education, especially for vulnerable groups, is important to develop human capital. Focusing on digital literacy and vocational training can equip the workforce for the future. Improving governance, reducing corruption, and developing public sector capacity are essential in establishing an environment conducive to sustainable development. Focusing on investments in statistical capacity and embracing innovative data collection will enhance monitoring and decision-making mechanisms (Lemos, 2024).

The implementation of the 2030 Agenda for Sustainable Development in LAC is a complex interplay of successes and setbacks. The resilience and commitment of the region to the SDGs are apparent, but systemic impediments and external shocks have highlighted the necessity of transformative action. The SDGs can only be achieved with the active engagement of all stakeholders, including governments, civil society, academia, and the private sector. Joint actions can spur breakthrough solutions and make development gains available to the most vulnerable. Harnessing technology and innovation have the power to improve service delivery, increase data collection, and sustain economic growth. Digital transformation holds the potential to solve development problems in far-flung and underserved places. International and regional collaborations are vital to mobilize resources, share knowledge, and manage transboundary challenges. Bolstering connections with international agencies, financial institutions, and the countries of the neighbourhood can build cooperative action. As we look at the implementation progress of the 2030 Agenda for Sustainable Development, LAC has achieved remarkable progress while struggling with ongoing challenges. The mid-point represents a significant moment to rebalance approaches, capitalize on achievements, and fill gaps. The area needs to give top priority to inclusivity, resilience, and sustainability in development. Through adopting transformative change and cooperation, LAC can make the 2030 Agenda a reality for everyone without leaving anyone behind. The next six years are decisive to drive the SDGs forward and achieve a future of dignity, equity, and prosperity for current and future generations (Mantelero & Esposito, 2024).

5.3 Sustainable Development Goals report

The 2020 SDGs Report is a through assessment of the implementation of the SDGs globally, with special emphasis on the effect of the COVID-19 pandemic. Although much progress had been achieved in most of the goals, the pandemic interrupted decades of progress, revealing weaknesses in global health, economic systems, and social structures. This analysis looks at the main results of the report, discusses disaggregated outcomes according to specific goals, and makes policy suggestions aimed at confronting challenges exacerbated by the pandemic.

The SDGs, agreed by all UN member states in 2015, seek to respond to pressing global issues including poverty, hunger, inequality, and environmental degradation. The 2020 report picked out some major conclusions, the COVID-19 pandemic being the principal disruptor to progress.

The world poverty rate rose for the first time in decades, with about 71 million individuals projected to fall back into extreme poverty. The rise was fueled by losses in income, the absence of social protection, and increased living costs, which exacerbated exposure to poverty and hunger. Informal workers, comprising a large percentage of the world's workforce, had their earnings decline by as much as 60% in the early period of the pandemic. More than a billion slum residents lived with increased vulnerability due to poor housing, overcrowding, and poor sanitation. The pandemic affected critical health services, which in turn raised mortality risks for women and children. Domestic violence similarly spiked, increasing gender disparities. School closures impacted 90% of students globally, depriving millions of accesses to meals and learning materials, especially in poor areas (Nagda and Sharma, 2024). The overall setbacks across multiple Sustainable Development Goals (SDGs) due to the pandemic are summarized in a global snapshot (Fig. 2).

Adapted from *The 2020 SDGs Report: Key Findings*, by United Nations, 2020a, 2020b. © United Nations. Retrieved from https://www.un.org/sustainabledevelopment

Objective 1: End Poverty in All Its Forms Everywhere: The pandemic reversed progress in poverty reduction, especially in Least Developed Countries, where social protection gaps exposed millions to extreme vulnerabilities. Economic losses due to the pandemic were estimated at $23.6 billion in 63 countries in 2018 alone. The global goal of eradicating extreme poverty by 2030 is now at risk. Strengthened international cooperation and targeted policies are necessary to address disparities in wealth and access to social protection.

Objective 2: End Hunger and Achieve Food Security: The pandemic further strained global food systems, particularly affecting small-scale farmers and food producers in developing countries. Malnutrition rates worsened, with 21.3% of children under five suffering from stunting, and 6.9% facing wasting. Investment in sustainable agriculture is inadequate. Rising food prices, especially in regions like sub-Saharan Africa, threaten food security. Focused interventions are required to ensure equitable access to nutrition.

Objective 3: Ensure Healthy Lives and Promote Well-being: The COVID-19 pandemic severely disrupted global health systems, reversing gains in maternal and child health. Immunization programs were halted in 70 countries, increasing the risk of preventable diseases. Strengthening health systems, expanding Universal Health Coverage, and investing in pandemic preparedness are crucial for long-term resilience (Rosentreter, 2022).

Objective 4: Ensure Inclusive and Quality Education: The closure of schools affected 1.57 billion students, disrupting learning globally. Remote learning was inaccessible to many students, particularly in low-income regions where digital infrastructure was lacking. Urgent action is needed to address digital divides and ensure inclusive access to education. Investments in school infrastructure, including hygiene facilities, are vital to safely reopening schools (International Development Law Organization [IDLO], n.d.)

Objective 5: Achieve Gender Equality: The pandemic disproportionately affected women, increasing domestic violence cases by 30% in some regions. Women also faced greater economic insecurity due to unpaid caregiving responsibilities. Policies must focus on women's economic empowerment and gender parity in leadership roles. Addressing systemic gender inequalities is vital to ensuring balanced post-pandemic recovery.

Objective 6: Ensure Sustainable Water and Sanitation for All: The pandemic highlighted the importance of access to water and sanitation for public health, with 3 billion people lacking basic handwashing facilities. Water scarcity is also expected

Fig. 2 Key findings from the 2020 Sustainable Development Goals (SDGs) Report. The infographic outlines how the COVID-19 pandemic reversed or stalled progress across multiple goals, including poverty eradication, health systems, gender equality, economic growth, climate action, and global partnerships. Adapted from The Sustainable Development Goals Report 2020, (United Nations, 2020a, 2020b), © United Nations. Retrieved from https://www.un.org/sustainabledevelopment

to displace 700 million people by 2030. Investments in water management and sanitation infrastructure are essential for improving public health and ensuring sustainability (United Nations, 2015).

Objective 7: Ensuring Access to Affordable, Reliable, Sustainable, and Modern Energy for All: The international dependence on cheap and stable energy came starkly into focus with the COVID-19 pandemic. Medical facilities, central to managing the crisis, were severely strained by a lack of electricity. As of 2018, a quarter of such facilities worldwide did not have access to power, compounding healthcare provision in underserved areas. The shortfall highlighted the critical importance of energy access to support essential services.

Advancement toward energy efficiency has been slow, with improvement rates trailing behind the desired 3% per year. Capital flows to promote renewable energy in developing nations grew to $21.4 billion in 2017, but merely 12% found their way to Least Developed Countries (LDCs). This disparity shows that greater focus must be put on funding equity, particularly in areas such as sub-Saharan Africa, where deficits in energy persist.

The pandemic also hampered progress toward clean cooking technologies, endangering the health of virtually 3 billion people who use unsafe practices. Scaling up renewable energy programs is critical for achieving long-term climate targets. Strengthening energy efficiency policies, essential for curbing greenhouse gas emissions, should be expedited. While global financing for renewable energy is promising, its limited extension to the world's poorest countries raises the need for more universal approaches (United Nations Educational, Scientific and Cultural Organization [UNESCO], 2020).

Objective 8: Promoting Sustained, Inclusive, and Sustainable Economic Growth: The COVID-19 pandemic triggered the most severe global economic crisis since the Great Depression, with GDP per capita declining by an estimated 4.2% in 2020. The economic fallout disproportionately impacted workers in the informal sector, placing 1.6 billion individuals at risk of losing their livelihoods. Tourism, a key source of employment and economic growth, was historically challenged, with global tourist arrivals falling by as much as 850 to 1,140 million in 2020. Moreover, global unemployment rose, as many as 400 million jobs were lost in the second quarter of the year alone. Even before the pandemic, least-developed countries found it difficult to achieve a minimum annual economic growth rate of 7%, and productivity improvements were ever more under pressure (Global Witness, 2022).

To reduce the economic blow and build resilience, policies need to focus on revitalizing labour markets, occupational safety, and coverage for informal workers who suffered the most. Economies that are tourist-dependent, and especially small island developing states, have increased vulnerabilities that require bold and adaptive responses to recovery. Building climate resilience policies and adaptation can help strengthen their ability to absorb future shocks. Encouraging diversification in the economy will diminish dependence on foreign economies, promoting more sustainable and autonomous growth. Furthermore, augmenting international assistance in the form of specific funding and capacity-building interventions will be vital for restoring economies and yielding a more equitable and resilient recovery.

Objective 9: Building Resilient Infrastructure, Promoting Inclusive Industrialization, and Fostering Innovation: COVID-19 has significantly disrupted global infrastructure and industrialization. The aviation industry, a major economic contributor, suffered a 51% decline in passenger numbers between January and May 2020 compared to the same period in 2019. Similarly, manufacturing, already experiencing sluggish growth, plummeted further during the pandemic. Access to technology and financial services became crucial for resilience. While mobile connectivity is nearly universal, half of the world's population, especially in LDCs, remains offline. Small-scale industries face critical financial gaps, hindering their recovery. Investments in research and development are more essential than ever, not only for industrial recovery but also for addressing pandemic-driven challenges (United Nations, 2019).

Objective 10: Reducing Inequality Within and Among Countries: The pandemic exacerbated existing inequalities, disproportionately affecting vulnerable groups, including the elderly, disabled individuals, migrants, and refugees. Women with disabilities faced compounded discrimination, highlighting the intersectionality of inequalities. Income disparities widened as the wealthiest segments recovered faster than others, while workers saw their share of output diminish. Development assistance to poorer nations is at risk of compression due to global economic strain, potentially reducing the resources available for inequality-reducing measures. Migration policies, although increasingly comprehensive, need to address the vulnerabilities of migrant populations during crises. Policies fostering inclusive economic recovery and addressing systemic disparities are vital (United Nations, 2018).

Objective 11: Making Cities Inclusive, Safe, Resilient, and Sustainable: Urban areas, home to over 90% of COVID-19 cases, faced severe health and socioeconomic challenges during the pandemic. Public transport systems and open spaces became focal points for resilience. However, the proportion of urban populations living near accessible public spaces remains limited. The pandemic reversed progress in reducing slum dwellers, exacerbating their vulnerabilities. Yet, it also offered an opportunity for cities to rethink infrastructure, emphasizing sustainability and inclusivity. Cleaner air during lockdowns illustrated the potential of reduced emissions, inspiring cities to adopt long-term sustainable practices.

Objective 12: Ensuring Sustainable Consumption and Production Patterns: The pandemic underscored the urgency of transitioning to sustainable consumption. Unsustainable resource use continued, with e-waste generation increasing by 38% from 2010 to 2019, of which only 20% was recycled. Food supply chains also remain inefficient, with 13% of food lost during harvesting, transportation, and processing. Fossil fuel subsidies rose from $318 billion in 2015 to $427 billion in 2018, contradicting global climate goals. Recovery plans should pivot toward sustainable economic practices, ensuring companies improve the quality of sustainability reporting. Governments must also realign subsidies to support renewable energy.

Objective 13: Combating Climate Change and Its Impacts: COVID-19 highlighted the fragility of global systems in the face of disasters. While funding for climate action has increased, fossil fuel investments continue to outpace it. Only 85 countries have aligned disaster reduction strategies with the Sendai Framework, showing slow progress in resilience planning. Natural disasters, exacerbated by climate change, affected over 39 million people in 2018. The pandemic serves as a reminder of the interconnectedness of health, climate, and development. Urgent measures are needed to meet the Paris Agreement targets and strengthen climate adaptation in vulnerable regions (World Economic Forum [WEF], 2020).

Objective 14: Conserving and Using Oceans Sustainably: The temporary slowdown in human activity during the pandemic presented an opportunity for ocean recovery. Marine biodiversity protection expanded, with key areas covered rising from 30.5% in 2000 to 46.0% in 2019. However, ocean acidification continues to threaten ecosystems and livelihoods, particularly in artisanal fisheries critical for developing economies. The Agreement on Port State Measures has restricted illegal fishing, but greater efforts are required to address overfishing and ensure sustainable practices.

Objective 15: Protecting Terrestrial Ecosystems and Biodiversity: The pandemic revealed the links between biodiversity loss and human health, including the risk of zoonotic diseases. Forest loss persists at an alarming rate, with 10 million hectares destroyed annually between 2015 and 2020, primarily due to agricultural expansion. Wildlife trafficking remains a significant threat to ecosystems, while degraded land affects billions globally. Despite progress, less than half of key biodiversity areas are protected. Sustainable management of forests and stronger wildlife protection laws are essential for reversing biodiversity loss.

Objective 16: Promoting Peaceful and Inclusive Societies: The pandemic intensified challenges to peace and justice, with overcrowded prisons increasing the risk of COVID-19 transmission. Violence against civilians, including women and children, remains high, with human rights defenders and journalists often targeted. While more countries have enacted freedom of information laws, their implementation varies (Human Rights Watch, 2021).

Efforts to reduce homicide rates and ensure access to justice are progressing slowly. Addressing systemic violence and ensuring accountability at all levels is crucial for fostering inclusive societies.

Objective 17: Revitalizing Global Partnerships: COVID-19 disrupted global trade and financing, with remittances to low- and middle-income countries declining from $554 billion in 2019 to $445 billion in 2020. Foreign direct investment is projected to decrease by up to 40%, further challenging economic recovery in developing nations. The digital divide is a major obstacle and continues to separate the world as half of its population remains unconnected. Deepening global partnerships, guaranteeing credible data generation, and ranking investments in digital infrastructure are indispensable for an inclusive recovery.

The COVID-19 pandemic greatly impacted world development towards the SDGs, exposing and increasing long-standing challenges. Before the pandemic, significant progress had been achieved across sectors, but the crisis underscored health infrastructure weaknesses in the system, highlighting universal healthcare needs and enhancing pandemic preparedness. Marginalized populations were disproportionately impacted, highlighting the importance of tackling income inequality and widening social protection programs. While the pandemic's economic impact was central to the report, environmental goals, including climate action and biodiversity conservation, remained critical to the global recovery process (World Health Organization [WHO], 2019).

To speed up development towards the SDGs, considering these failures, some policy suggestions are a must. Raising social protection nets to encompass informal labour and vulnerable groups, especially the worst-affected recession victims, is crucial to achieving long-term resilience. Closing the digital gap entails huge investments in technology infrastructure among disadvantaged groups. Women's empowerment policies are a must, considering the uneven impact of the pandemic on women and girls, with efforts directed towards reducing gender-based violence. Health system strengthening through expanded financing, prioritizing universal health care access and pandemic readiness, is also key. Empowering small-scale farmers by training, subsidizing, and enhancing access to markets will boost food security, especially in the developing world. In addition, investment in water, sanitation, and energy infrastructure will be central to building resilience to future crises and enhancing public health outcomes.

The pandemic, though presenting enormous challenges, also presents a chance to rethink and reinforce policies that promote inclusive, sustainable development. Urgent action, coupled with continued investments, is required to reverse the setbacks and place the SDGs back on track by 2030. Targeted action on social protection, gender equality, health, education, and environmental sustainability is needed to create a resilient, equitable global community. These objectives can only be achieved with a collective and transformative strategy that incorporates sustainability in all facets of recovery and development. In the future, joint action, policy changes, and strategic investments will be essential in ensuring that the SDG goals are achieved by 2030 (Kumar et al., 2025).

6 Insulators Aspects not Considered as Agenda 2030 Items

The United Nations' Agenda 2030 for Sustainable Development provides a framework with 17 SDGs addressing pressing global challenges. However, significant issues remain overlooked or inadequately addressed. Below is a detailed exploration of 18 aspects that merit consideration within or alongside the agenda 2030 framework.

Mining activities are essential for industrial development but often result in severe environmental degradation. Issues such as groundwater contamination, air pollution, habitat destruction, and health hazards from toxic mining waste are frequently underregulated. Large-scale industrial mining contributes to water scarcity and soil degradation in vulnerable regions, while artisanal mining, though crucial for livelihoods, lacks safety protocols, further exacerbating environmental destruction. To address these challenges, international collaboration on green mining technologies and the development of stricter regulatory frameworks are critical. Agenda 2030 should incorporate mining-specific goals that focus on pollution mitigation and community rehabilitation (Miyah et al., 2022).

Smart cities, which use technology to improve urban living through energy efficiency, waste management, and citizen engagement, have been underexplored in the context of the SDGs, particularly under Goal 11 (Sustainable Cities and Communities). The potential of smart city technologies to enhance water management, promote renewable energy adoption, and bolster climate adaptation strategies has not been fully recognized. By explicitly referencing smart cities, governments would be encouraged to integrate digital solutions into urban planning, ensuring urban resilience to climate change and resource scarcity.

As environmental disputes become more complex and widespread, there is an increasing need for judicial intervention. However, Agenda 2030 does not prioritize the creation of specialized courts or alternative dispute resolution mechanisms for environmental matters. Empowering judicial bodies with resources, expertise, and international cooperation is essential for resolving these conflicts efficiently. Establishing specialized environmental tribunals and aligning judicial capacity-building with SDG 16 (Peace, Justice, and Strong Institutions) would improve the resolution of environmental disputes.

New environmental principles, such as climate justice, intergenerational equity, and the precautionary principle, are becoming increasingly vital in the face of evolving challenges. Although these concepts are recognized globally, they are not sufficiently integrated into Agenda 2030. The legal codification of emerging environmental principles would enhance accountability in international environmental governance. Institutions like the United Nations could lead efforts to embed these principles within global treaties and frameworks, ensuring they are more effectively applied (Deva et al., 2019).

AI offers transformative potential for addressing environmental challenges, including precision agriculture, biodiversity monitoring, and emission tracking.

However, the role of AI in environmental sustainability is not adequately covered in Agenda 2030. AI can optimize renewable energy grids, model climate scenarios, and predict disaster impacts. Yet, the high energy consumption of AI infrastructure and potential biases in decision-making may counteract sustainability efforts. Including AI as a cross-cutting enabler in the SDGs, with a focus on its responsible and sustainable deployment, is essential for maximizing its benefits.

Global environmental governance is hindered by fragmented legal regimes and inconsistent enforcement. Codifying environmental laws and principles is necessary to ensure uniform application across jurisdictions. While Agenda 2030 addresses environmental sustainability broadly, it lacks a focus on creating comprehensive, enforceable legal codes. Efforts should be made to develop international treaties that codify environmental obligations, with binding commitments for all signatories, ensuring a more coherent and effective global legal framework (Sharma et al., 2025).

Environmental governance, encompassing policies, regulations, and institutional frameworks, is crucial for the effective management of environmental issues. Despite its central role, governance remains implicit rather than explicit within Agenda 2030. Corruption, weak institutions, and a lack of stakeholder engagement often hinder effective governance. Establishing governance-specific targets within SDG 16 would promote transparent, participatory decision-making in environmental contexts, ensuring that environmental policies are implemented effectively and inclusively.

The recognition of the right to a healthy environment is gaining momentum, but it has not been universally adopted. Its explicit inclusion in Agenda 2030 would strengthen state obligations to safeguard environmental and public health. This right is particularly crucial for marginalized communities, who are often the most vulnerable to environmental degradation. Advocating for its universal recognition through international treaties and aligning it with existing human rights frameworks would enhance global environmental justice (World Resources Institute [WRI], 2021).

In addition to the right to a healthy environment, emerging rights such as access to digital infrastructure for sustainability and protection from environmental displacement should be included in global agendas. Expanding rights frameworks to reflect the evolving challenges of the Anthropocene era is essential. These rights should be enshrined in future updates to Agenda 2030 or supplemental frameworks, ensuring that they address the complex and multifaceted nature of contemporary environmental challenges.

Lack of environmental awareness remains a significant barrier to achieving sustainable development. While Agenda 2030 emphasizes education, it does not adequately focus on sustainability training. Incorporating mandatory sustainability education at all levels of schooling, as well as vocational training in green technologies, is crucial for equipping future generations to address environmental challenges. This aligns with SDG 4 (Quality Education) and could be strengthened through dedicated environmental literacy targets (Deva, 2021).

Climate-induced displacement, affecting millions globally, receives insufficient attention in Agenda 2030. The absence of legal recognition for climate refugees exacerbates their vulnerability and leaves them without protection. A global framework should be developed to recognize and protect environmentally displaced persons, linked to SDG 10 (Reduced Inequalities) and SDG 13 (Climate Action). This would ensure that those displaced by environmental disasters receive the support and protection they need.

Environmental justice, which ensures the equitable distribution of environmental benefits and burdens, is central to sustainable development but remains underexplored in Agenda 2030. Marginalized communities often bear the brunt of pollution and resource scarcity, exacerbating inequality. Institutionalizing environmental justice as a guiding principle for all SDGs would ensure fair resource allocation and legal redress mechanisms, promoting a more equitable approach to environmental governance.

Meeting sustainability challenges involves interdisciplinary solutions that consider natural sciences, social sciences, and technology. Agenda 2030 implementation, however, tends to be within siloed systems that limit cross-disciplinary specialist inputs in policy formulation and implementation. Such a siloed approach undermines the effectiveness of efforts aimed at addressing the complex and interrelated environmental issues in the world. To create comprehensive and effective solutions, it is important to promote collaborations among governments, the private sector, and civil society. These collaborations can tap into multiple viewpoints, resources, and expertise to propel sustainable development. It can also make decision-making better, resource management more efficient, and the achievement of the SDGs faster by incorporating technological innovation into policy systems. Building cross-sectoral strategies that harmonize economic, social, and environmental goals will make efforts at sustainability comprehensive and resilient. Adopting interdisciplinary collaboration is crucial to breaking through current barriers and making significant, long-term strides in the SDGs (Zhang and Han, 2025).

Effective environmental action requires effective coordination between global, national, and local institutions. Agenda 2030, however, does not have an adequate focus on interinstitutional mechanisms, with the consequence of inefficiency and duplication of resources. Building strong institutional frameworks for interinstitutional cooperation, particularly for coping with cross-frontier issues like pollution and conservation of biodiversity, would make global environmental governance more effective. Coordination ensures that environmental programs are more consistent, efficient, and effective. At the grassroots level, local communities and grassroots movements contribute significantly to fulfilling the Sustainable Development Goals (SDGs) through bottom-up strategies that involve people actively in sustainability efforts. Through local involvement in decision-making, policies are made sensitive to local needs and priorities and are hence implemented more effectively and inclusively. Participatory governance not only creates a sense of ownership and responsibility but also enhances the resilience of local systems. In addition, giving financial and institutional support to grassroots organizations enables them to mobilize resources, develop capacity, and scale their effects. Identifying and scaling up the efforts of local actors is critical to attaining the SDGs equitably and sustainably, building long-term progress from the bottom up.

While SDG 12 (Responsible Consumption and Production) addresses waste management, emerging contaminants such as microplastics and e-waste are not given adequate attention. Mismanaged waste poses significant environmental and public health risks, especially in low-income regions. Strengthening waste management targets and emphasizing circular economy models and innovation in recycling technologies would help mitigate the impacts of contamination and promote more sustainable consumption and production patterns.

The deforestation of the Amazon, a crisis with global ecological consequences, is glaringly absent from Agenda 2030. Deforestation leads to biodiversity loss, contributes to climate change, and displaces indigenous communities. Mobilizing international cooperation to fund conservation efforts, enforce deforestation moratoriums, and protect the Amazon is a global priority. Addressing this issue should be a key focus in the fight against climate change and biodiversity loss (Dehbi & Martin-Ortega, 2021).

Artisanal and small-scale mining is often excluded from sustainability dialogues, despite its significant environmental and socio-economic impacts. ASM contributes to mercury pollution, habitat destruction, and human rights violations. To mitigate these impacts, international standards should be established, and ASM communities should be supported in adopting sustainable mining practices. This would help balance the economic benefits of ASM with environmental and social sustainability.

While SDG 14 (Life Below Water) addresses marine conservation, it remains limited in scope. Issues such as illegal fishing, ocean pollution, and the loss of coral reefs require more robust attention. Unsustainable fishing practices and ocean acidification threaten marine biodiversity and livelihoods. Expanding marine conservation targets and incorporating stricter enforcement mechanisms and funding for vulnerable regions would help protect marine ecosystems and ensure their resilience in the face of climate change.

The agenda 2030 framework represents a monumental step toward sustainable development, but it has significant gaps that must be addressed. The issues highlighted above, including mining pollution, smart cities, environmental justice, and the need for specialized environmental courts, are essential for a more comprehensive, inclusive, and actionable global sustainability agenda. Integrating these overlooked dimensions will be crucial for advancing sustainability and equity in a rapidly changing world (Bexell et al., 2023).

7 A Fundamental Rights Perspective

The 2030 Agenda for Sustainable Development, which has been celebrated as a revolutionary master plan for global equality and sustainability, is globally acclaimed for its expansive scope. Yet, its execution has also been questioned about its capacity to sufficiently protect and advance core human rights. The ambiguous and passive actions under the agenda threaten to infringe on fundamental rights, such as the right to peace, life, free development of personality, a healthy environment, and the well-being of future generations. This article discusses these intrusions through a human rights lens, highlighting the necessity of remodulating the execution of the agenda to respect these essential rights.

7.1 The Right to Peace and Tranquillity

Peace and tranquillity are the building blocks of human well-being. While SDG 16 focuses on fostering peace, justice, and effective institutions, the approach of Agenda 2030 is ambiguous and inadequate in addressing the real sources of conflict and environmental degradation. Problems like environmental conflict and policy paralysis further weaken this inherent right. The displacement of indigenous peoples for resource extraction and the lack of effective international frameworks for resolving such conflicts aggravate this problem. Strengthening conflict resolution at the community level and mainstreaming peacebuilding into economic and environmental policy is a necessary step forward.

7.2 The Right to Life

The right to life is codified in international agreements, but Agenda 2030's inability to properly address environmental emergencies and unequal distribution of resources jeopardises this right. Climate change, weather-related disasters, and sea-level rise endanger vulnerable populations, while pollution and inadequate waste management lead to millions of avoidable deaths. A more focused strategy under SDG 3 (Good Health and Well-being) and SDG 13 (Climate Action) is needed, with a focus on life-saving interventions such as disaster preparedness and enhanced health infrastructure.

7.3 The Right to the Free Development of the Personality

The unfettered development of the personality involves autonomy, dignity, and personal decision-making. Agenda 2030 tends to neglect individual liberties in favour of collective advancement. Climate migration and excessive policy regulation, especially, hit marginalized groups in the Global South, inhibiting cultural practices and individual development. An inclusive approach with policies tied to the ambitions of impacted individuals and communities is paramount to protecting this right (Baziar and Parsa, 2024).

7.4 The Right to Free Development

Free development can be defined as the group's right of people to have self-guided progress. Agenda 2030's cookie-cutter formula tends to neglect the sovereignty of countries, especially those in the Global South. Structural disparities like resource exploitation by developed countries add fuel to the existing gaps. The focus on equal partnerships under SDG 10 (Less Inequalities) and SDG 17 (Partnerships for Goals) is vital to guarantee equal development for everyone.

7.5 The Right to a Healthy Environment

A healthy environment is at the heart of the enjoyment of all other human rights. Agenda 2030's failure to be implemented and its lack of adequate attention to environmental protection deny this right, however. Deforestation, loss of biodiversity, and industrial pollution go on unrelentingly, with poor communities bearing the greatest burden. Protecting this right requires immediate actions like giving greater priority to the restoration of ecosystems and stricter pollution controls (Ali and Yeon, 2019).

7.6 Ensuring the Well-Being and Security of Future Generations

The welfare of future generations, an integral part of sustainable development, is put at risk by the depletion of natural resources and neglect of climate debt. The policies of Agenda 2030 rarely take into consideration the long-term security of future generations, as which does not incorporate their interests into decision-making. Development projects need to have long-term impact assessments that consider the needs of future generations. The weaknesses of Agenda 2030 disproportionately affect vulnerable groups like indigenous peoples, women, and children. Land grabbing, forest destruction, and the absence of gender-sensitive policies widen these gaps. In addition, the voluntary character of the agenda restricts its enforceability, enabling states and corporations to give precedence to economic interests over human rights. Establishing an international legal system that combines sustainability objectives with human rights safeguards is critical to making accountability possible (Adamatti and Ferreira, 2023).

To position Agenda 2030 in tune with human rights, there must be a paradigm shift. This involves including human rights as part of each SDG target, embracing models of inclusive governance that empower people, turning the agenda into an enforceable format, and mainstreaming the areas of intersection among social, economic, and environmental rights. The ambiguous words and passive approach to implementing Agenda 2030 show important areas where it falters in promoting core human rights. Its rights-based design, addressing abuses pointed out by it, will be essential for making sure sustainable development favours all of humanity.

Revamping the Agenda in terms that respect peace, life, freedom, and ecological health will set the pace towards a future that is both equitable and more just.

8 Conclusions and Suggestions

The 2030 Agenda for Sustainable Development, as ambitious as its vision, has faced tremendous challenges in both conceptualization and implementation, revealing a wide gap between its intentions and the actuality of progress. Major issues like the poor choice and prioritization of agenda items, the lack of specialized knowledge and dedication among stakeholders, and the lack of binding accountability mechanisms have eroded its efficacy. In addition, excessive dependence on the COVID-19 pandemic as an excuse for underperformance has masked deeper, more structural issues. As of 2024, SDG progress is still limited, indicating the necessity for a thorough review of strategies, resources, and political will. Despite these setbacks, the agenda remains promising. With refreshed political commitment, increased stakeholder capacity, and reinforced international coordination, hope for reviving the agenda and reaching its goals by 2030 or beyond remains. For it to be a success, however, the international community must cure these structural deficits and operational inefficiencies and guarantee that the parameters of equity, justice, and human rights orient the way ahead. To save the purpose and goals of the 2030 Agenda, a strategic rethinking is necessary. The following recommendations provide a roadmap for reinvigorating global efforts and ensuring meaningful progress.

- Reassess and refine the SDGs: Conduct a thorough review to identify gaps and redundancies, and prioritize issues based on global urgency and feasibility. This should include extending the timeline for achieving these goals to ensure realistic, phased implementation.
- Strengthen Political Will and Accountability: Transition from voluntary participation to mandatory compliance mechanisms, with legally binding agreements. Independent monitoring bodies should be established to assess progress and enforce accountability.
- Enhance Stakeholder Capacity and Awareness: Invest in targeted training for policymakers and public officials to build expertise in sustainable development. Launch awareness campaigns to foster grassroots involvement and promote partnerships with academic institutions for deeper insights.
- Promote Interdisciplinary and Collaborative Approaches: Encourage cross-sectoral collaboration to address the interconnected nature of the SDGs, leveraging partnerships between governments, private entities, and international organizations for resource pooling and knowledge exchange.

- Allocate Adequate Resources and Financing: Mobilize innovative financial mechanisms, such as green bonds and carbon credits, ensuring equitable resource distribution. Special emphasis should be placed on supporting low-income nations to meet their SDG targets and addressing disparities in resource allocation.

Global case studies highlight the importance of effective governance, creative financing, and people-centred solutions to deliver the SDGs. Best practices typically include the smooth incorporation of SDGs within national policy and the continuity of policy between successive governments. International knowledge hubs are also important for the diffusion of best practices to other parts of the world, building cooperation, and building momentum.

While the 2030 Agenda was conceived as a roadmap to a more sustainable and equitable future, its implementation has shown a disconcerting gap between action and ambition. Closing this gap calls for a new focus on commitment, accountability, and partnership. By addressing the structural weaknesses and operational inefficiencies that impede progress, stakeholders can revitalize global action towards sustainable development. Realizing the SDGs requires commitment, creativity, and a steadfast commitment to the values of equity, justice, and human rights. With sincere engagement by member states and stakeholders, there remains a realistic way to achieve a sustainable and equitable world, either by 2030 or in the years thereafter.

References

Adamatti, B., & Ferreira, E. A. (2023). Enhancing corporate responsibility: The importance of human rights compliance systems and the shortcomings of Brazilian Decree N. 9.571/2018. *ESG Studies Review, 6*(1), e01579. https://doi.org/10.37497/esg.v6i1.1579

Ali, H. M., & Yeon, A. L. (2019). Integrating sustainable development goals within the corporate governance framework. *European Proceedings of Social and Behavioural Sciences, 70*, 1–10. https://doi.org/10.15405/epsbs.2019.10.1

Baziar, A., & Parsa, N. (2024). Legal challenges in renewable energy development: A comparative study of China and selected countries. arXiv preprint. https://arxiv.org/abs/2412.10203

Bexell, M., Hickmann, T., & Schapper, A. (2023). Strengthening the sustainable development goals through integration with human rights. *International Environmental Agreements: Politics, Law and Economics, 23*, 133–139. https://doi.org/10.1007/s10784-023-09605-x

Dehbi, F., & Martin-Ortega, O. (2021). Integrating human rights and the environment in supply chain regulations. *Sustainability, 13*(17), 9666. https://doi.org/10.3390/su13179666

Deva, S. (2021). Business and human rights: Alternative approaches to transnational regulation. *Annual Review of Law and Social Science, 17*, 139–158. https://doi.org/10.1146/annurev-law socsci-113020-074527

Deva, S., Ramasastry, A., Wettstein, F., & Santoro, M. (2019). Editorial: Business and human rights scholarship: Past trends and future directions. *Business and Human Rights Journal, 4*(2), 201–212. https://doi.org/10.1017/bhj.2019.14

Economic Commission for Latin America and the Caribbean (ECLAC). (2023). *Regional Progress and Challenges in the Implementation of the 2030 Agenda for Sustainable Development*. United Nations Publications.

Feiring, B., & König-Reis, R. (2020). Strengthening the sustainable development goals through integration with human rights. *International Environmental Agreements: Politics, Law and Economics, 23*, 133–139. https://doi.org/10.1007/s10784-023-09605-x

Food and Agriculture Organization of the United Nations (FAO). (2020). *The State of Food Security and Nutrition in the World 2020*. FAO.

Witness, G. (2022). *Defending tomorrow: the climate crisis and threats to environmental defenders*. Global Witness Publications.

Human Rights Watch. (2021). *World Report 2021: Events of 2020*. Human Rights Watch.

International Development Law Organization. (n.d.). Transforming legal frameworks: Urgent action to achieving Beijing+30 and the 2030 Agenda. https://www.idlo.int/news/events/transforming-legal-frameworks-urgent-action-achieving-beijing30-and-2030-agenda

International Energy Agency (IEA). (2022). *Renewables 2022: Analysis and forecast to 2027*. IEA Publications.

International Labour Organization (ILO). (2022). *World employment and social outlook: Trends 2022*. ILO Publications.

International Monetary Fund (IMF). (2021). *Fiscal monitor: Strengthening the credibility of public finances in the post-pandemic era*. IMF Publications.

Inter-American Development Bank (IDB). (2019). *Promoting economic growth and decent work in Latin America and the Caribbean*. IDB Publications.

Intergovernmental Panel on Climate Change (IPCC). (2022). *Sixth Assessment Report: Climate Change 2022—Impacts, Adaptation, and Vulnerability*. IPCC.

Kumar, R., Khanna, R., Singh, T., & Goel, R. (2025). Role of information and communication technology in poverty elimination with respect to SDG-2030 agenda. In *sustainability and sustainable development* (pp. 1–19). Apple Academic Press.

Lemos, M. C. (2024). Legal aspects of sustainable development: Governance mechanisms to balance growth and environmental protection. *Mayo RC Journal of Communication for Sustainable World, 1*(2). https://www.researchcorridor.org/index.php/mrcjcsw/article/view/161

Mantelero, A., & Esposito, M. S. (2024). An evidence-based methodology for human rights impact assessment (HRIA) in the development of AI data-intensive systems. arXiv preprint. https://arxiv.org/abs/2407.20951

Miyah, Y., Benjelloun, M., Lairini, S., & Lahrichi, A. (2022). COVID-19 impact on public health, environment, human psychology, global Socioeconomy, and education. *The Scientific World Journal, 2022*(1), 5578284.

Nagda, P., & Sharma, H. (2024). Towards environmental justice: Analyzing international legal frameworks for sustainable development. *Educational Administration: Theory and Practice, 30*(1), 651–655. https://doi.org/10.53555/kucy.v30i1.4244

Organisation for Economic Co-operation and Development (OECD). (2021). *The Inequality Paradox in Latin America: Challenges in the Implementation of SDGs*. OECD Publishing.

Rosentreter, H. (2022). Leveraging the linkages: How human rights data can advance SDG monitoring. International Institute for Sustainable Development. https://www.iisd.org/articles/insight/leveraging-linkages-how-human-rights-data-can-advance-sdg-monitoring

Sharma, C. P., Malhotra, R. K., & Kumar, R. (2025). The role of industrial policy in achieving sustainable development goals: A case of Uttarakhand. In *Sustainability and Sustainable Development* (pp. 67–82). Apple Academic Press.

United Nations. (2015). Transforming our world: The 2030 Agenda for Sustainable Development (A/RES/70/1). https://www.un.org/sustainabledevelopment/why-the-sdgs-matter/

United Nations. (2018). *The sustainable development goals report 2018*. United Nations Department of Economic and Social Affairs.

United Nations. (2019). *Quadrennial progress report on sustainable development goals in Latin America and the Caribbean*. United Nations Economic Commission for Latin America and the Caribbean.

United Nations. (2020a). *The sustainable development goals report 2020*. United Nations Department of Economic and Social Affairs. https://unstats.un.org/sdgs/report/2020/

United Nations. (2020b). *Sustainable development goals*. United Nations. https://www.un.org/sustainabledevelopment/
United Nations Development Programme (UNDP). (2021). *Human development report 2021: Inequality and sustainable development*. UNDP.
United Nations Educational, Scientific and Cultural Organization (UNESCO). (2020). *Global education monitoring report: Inclusion and education*. Paris: UNESCO Publishing.
United Nations Environment Programme (UNEP). (2021). Making peace with nature: A scientific blueprint to tackle the climate, biodiversity, and pollution emergencies. Nairobi: UNEP.
United Nations Human Rights Council. (2018). The need for an integrated approach to the implementation of the 2030 Agenda for Sustainable Development for the full realization of human rights, focusing holistically on the means of implementation (A/HRC/RES/37/25). https://digitallibrary.un.org/record/1485564
United Nations Development Programme. (n.d.). *Progress across selected SDGs: A regional analysis*. United Nations. Retrieved May 28, 2025, from https://www.undp.org
United Nations. (2020). *The Sustainable Development Goals Report 2020*. https://www.un.org/sustainabledevelopment/progress-report/
United Nations. (2024). *Quadrennial report on regional progress and challenges in relation to the 2030 Agenda for Sustainable Development in Latin America and the Caribbean. Economic Commission for Latin America and the Caribbean (ECLAC)*. https://www.cepal.org/en/publications/48976
Universal Rights Group. (2018). Towards a stronger connection between human rights and the 2030 Agenda for Sustainable Development: The role of the Human Rights Council. https://www.universal-rights.org/towards-stronger-connection-human-rights-2030-agenda-sustainable-development-role-human-rights-council/
World Bank. (2018). *Poverty and shared prosperity 2018: Piecing together the poverty puzzle*. The World Bank Group.
World Economic Forum (WEF). (2020). *The future of work report 2020: Reskilling for a Post-COVID World*. WEF Publications.
World Health Organization (WHO). (2019). *Universal health coverage: Progress report 2019*. WHO Publications.
World Resources Institute (WRI). (2021). *Tracking progress on the SDGs: A sustainable future for all*. WRI Publications.
Zhang, Y., & Han, H. (2025). Legal routes for accomplishing corporate environmental compliance against the 'carbon peaking and carbon neutrality' goals. arXiv preprint. https://arxiv.org/abs/2502.13543

Justice for All in the Digital Age: Evaluating India's Tech-Driven Legal Reforms and SDG 16

Shradha Baranwal

Abstract

Peaceful and inclusive societies are one of the core goals under the Sustainable Development Scheme of the United Nations. The progress of other goals under Sustainable Development Goals (SDGs *hereinafter* referred to as 'SDGs') majorly depends on peace, stability and commitment to the inclusion of any country. Ensuring peace, justice and maintaining inclusive societies is one of the guiding forces of the judicial wing of any state. Considering a highly multicultural and multi-dimensional country like India which has multiple languages and different belief systems it is important that judicial services are reachable to all. Of late there have been several technology-based innovations making access to justice easier reflecting on the language barrier, accessibility to the Courts, legal learning opportunities and access to legal documents in their preferred language. This paper investigates the use and impact of technology in ensuring access to justice for all. The paper attempts to find out the feasibility and success of such innovations in justice delivery system. The methodology used for the research is doctrinal and comparative. The researcher has analyzed the phase wise planning and budgetary allocation for implementation of Information and Communications Technologies (*hereinafter* referred to as 'ICT') tools in judicial system. The research suggests a steep rise in innovations and transition from electronic age to the age of Artificial Intelligence. The report reveals the efforts of the Supreme Court and other courts below in moving towards complete digitization of data. The Supreme Court, through eCourts Mission Mode Project (*hereinafter* referred to as 'e-Courts project'), Electronic Supreme Court Reports (*hereinafter* referred to as e-SCR), Supreme Court Vidhik Anuvaad Software (*hereinafter* referred to as 'SUVAS') and Supreme Court Portal

S. Baranwal (✉)
UPES, Dehradun, Uttarakhand, India
e-mail: sbaranwal@ddn.upes.ac.in

T. Choudhury et al. (eds.), *Open Access as a Knowledge Platform for Implementing the United Nations Sustainable Development Goals*, World Sustainability Series,
https://doi.org/10.1007/978-3-031-99605-4_15

for Assistance in Court's Efficiency (*hereinafter* referred to as 'SUPACE') portal is consistently making information available to all irrespective of physical boundaries. The study also reflects upon the innovations made by private practitioners. Towards the conclusion it is seen that though there are multiple efforts, the ultimate success is to facilitate utilization of the same through awareness and education so that access to justice is available to all.

Keywords

SGDs • Goal 16 • Judiciary • Technology • Access to justice • Artificial intelligence • Digitalization

1 Introduction

> *The right of effective access to justice has emerged with the new social rights. Indeed, it is of paramount importance among these new rights since, clearly, the enjoyment of traditional as well as new social rights presupposes mechanisms for their effective protection. Such protection, moreover, is best assured by a workable remedy within the framework of the judicial system. Effective access to justice can thus be seen as the most basic requirement—the most 'basic human right'—of a system which purports to guarantee legal right* (Cappelletti and Garth, 1978).

Access to Justice is one of the Goals under the Sustainable Development Goal 16. The United States Institute of Peace defined access to justice as the ability of people to seek and obtain a remedy through formal or informal institutions of justice for grievances in compliance with human rights standards. Access to Justice gained prominence in the background of tremendous changes and transformations of the world from the beginning of the twenty-first century, leaving both positive and negative lasting impacts. Need brought countries together and ideology separated them.[1] The world became bipolar, leading to conflicts, intimidation, and was tottering between impossible peace and improbable war as termed by Raymond Aron in 1948 (Bourekba, 2021). From 1945 to 1989 the cold war overshadowed the progress of every nation which came to an end with the fall of Berlin wall and disintegration of USSR. Post 1990 the world faced new challenges which reached to its peak in 2010s decade. The decade started with Russian control over Chechnya, withdrawal of Israel from southern Lebanon, terrorist attack on Unites States of America by AL-Qaeda, invasion of Afghanistan by North Atlantic Treaty Organization (*hereinafter* referred to as 'NATO'), invasion of Iraq, few ex-USSR

[1] The Nazi aggression towards USSR made USSR an ally of the allied powers in the World War 2. USSR was among the big 4 s having China, USA and UK as other prominent members. Multiple other nations also joining them. But post-World War 2 USA and USSR differed on ideological basis and that completely divided the world into two groups leading to unprecedented development of weapons, aggression, non-cooperation among countries etc.

countries joining NATO etc. Subsequently, the decade of 2010s witnessed Syrian civil war, 9/11 attack on United States of America, defeat of Muammar Gaddafi in Libya, end of United States occupation of Iraq, upgradation of status of Palestine from entity to non-member observer state by the United Nations (*hereinafter* referred to as 'UN'), Global financial crisis of 2007 which shattered the great economies of the world, followed by the exponential rise of social media, significant drop in birth rate, rise of Arab although the struggles never left them, acute migration crisis, climate change and consistent rise in the same, rise of nationalist politics, Brexit[2] etc., to name a few. The events listed aptly reflected the wisdom of Raymond Aron, who saw peace impossible and war improbable. Apart from Geo-Political conditions the decade of 2010s also witnessed the crisis world was going through due to continued struggle of nations on various fronts which also made it certain that Millenium Development Goals (*hereinafter* referred to as 'MDGs') are not going to succeed as imagined. This is when the Sustainable Development Goals were conceived and developed into a vision for coming decades. The Sustainable Development Goals (SDGs) were conceived during the UN Conference on Sustainable development in Rio de Janeiro in 2012. The 'Agenda 2030' hit off with 17 goals, 169 targets and 232 indicators having 5 dimensions in focus—*People, Planet, Prosperity, Peace and Partnership.* The objective was to develop some common goals for the world to achieve by replacing the earlier existing 8 MDGs.[3] MDGs had left a crucial impact on the world poverty and health index by effectively addressing HIV AIDS, increasing child mortality, reducing school dropout rates etc., but this was not enough considering increasing complexity and need of the world. MDGs were superseded with SDGs with lot of experience of success and relative failures. The vision under SDGs was very wide which not only covered the basics of human life but also took care of the governance side of it and duly recognized technical advancement the decade of 2010s made. A pictorial image of the SDGs in Fig. 1 will reflect the diverse development index fixed to be achieved by 2030.

What initially worked in favour of SDGs was the parallel international conferences and international frameworks on some of the key focus areas of SDGs which included Conference of Parties (*hereinafter* referred to as 'COP 21') Paris on Climate Change (2015), Sendai Framework on Disaster Risk Reduction (2015), the United Nations Conference on Sustainable Development (*hereinafter* referred to as 'UNCSD, 2012'), and the Global Climate Action Summit (2019). Notably most of the areas covered under various international instruments pertained to the climate protection call. With respect to Goal 16 of SDGs it is primarily seen as the municipal obligation of nations to make peaceful and resilient societies including

[2] Brexit—Brexit is an abbreviation of two English words: 'Britain' and 'exit' and refers to the withdrawal process of the United Kingdom (UK) from the European Union (EU).

[3] In the year 2000, 189 countries under UN came out with Millenium Development goals which set the multiple progress indexes to be achieved by the year 2015. This included poverty, hunger, education, gender parity, women empowerment, child mortality, maternal health, HIV/ AIDS, Malaria and other diseases, environmental sustainability and global partnership in development.

Fig. 1 SDGs by the United Nations. *Source*—United Nations—Department of economic and social affairs social inclusion. Available at https://social.desa.un.org/issues/disability/sustainable-development-goals-sdgs-and-disability

an important function within the governance that is adjudication and access to legal justice. The background and aspirations of the UN under goal 16 can be traced in the report of the High-Level Political Forum on Sustainable Development of UN which identifies sustainable development, peace and security as indispensable for each other. The report says that SDG 16 is both the outcome and enabler of other SDGs. Having peace, justice and strong institutions are preconditions for enabling and achieving other SDGs (High Level Political Forum on Sustainable Development, Review of SDG Implementation and Interrelations among goals, 2019). How peace, justice and strong institution affect the growth and overall justice in a society can be seen from the below mentioned Table1 pertaining to India in its judicial appointments which is created based on date collected from multiple sources.

The information above explains why it took so many years after independence to recognize a girl child as coparcener in a Hindu joint family. The extreme under-representation also explains why some communities saw social transformation in personal laws at a highly belated stage (Shah Bano Case, 1985) and (Mohd Abdul Samad Case, 2024). Also, disproportionate representation from states affected address to issues pertaining to peculiar state conditions. This also remains the fact that when the judiciary worked in a free environment it passed historic judgments but under pressure and with constant interference the impact was seen on the judgments as well. Thus, strong and free institutions are *sine quo non* of any democratic society. It would be further interesting to look at the aspirations India has with goal 16 and how much India has progressed under this goal in specific with its judicial system. The subsequent investigation would throw a light upon

Table 1 Data pertaining to the judicial appointments in India

Representation in Supreme Court on multiple parameters through number of years	1950–1990 (Era of Executive Supremacy)	1991–2010 (Era of Collegium)	2010–2024 National Judicial Appointments Commission (NJAC) and subsequent restoration of Collegium)
Gender parity no of Chief Justices of India (CJIs) versus female CJI	20:1	17:0	14:1 Expected by 2027
Number of women judges in the Supreme Court	1	3	7
Overall representation of women, backward community and minorities	Out of total 664 only 75 are women which constitutes only 11% of the total appointment. Similarly, there have been unequal and disproportionate representation from the minority community and backward community (Dev, 2022). With respect to CJI position till date only two backward community judges could make it namely, Justice K.G. Balakrishnan and Justice BR Gavai		
State wise representation	During the initial few decades only three High Court (HC) dominated in terms of appointments, Bombay, Madras and Calcutta. But with time the proportion got better and at till date almost 17 HCs have got representation in the Supreme Court (Padmanabhan and Devulapalli, 2018). In recent times Patna HC, Bombay HC, Delhi HC, Allahabad HC, Madras HC, Karnataka HC had been amongst the biggest spot gainers. While the impact of Calcutta HC has substantially reduced (Mihir, 2021). Since 1950 the far northeast has stayed unrepresented except for Tripura HC in Assam (Das, 2023)		

India's efforts towards goal 16 in its judicial set up. In specific with respect to assimilation of technological advancement in justice delivery and legal awareness for the vast and diverse area of India. But before one advance towards knowing technological advancements of the system, it is important to know the basic structure of that system which aspires to excel under Goal 16 of the SDGs.

2 Justice Administration in India *Institutional and Procedural Positioning of Courts*

India has an interesting model of justice administration which is not only institutionally unique but is unique in its approach as well. India has three tier structure of adjudication alongside parallelly running expert tribunals. The inspiration for Indian judicial structure comes from two sources one is its common law tradition of United Kingdom, and another is the desire of having free and fair judiciary like United States of America which granted the highest court of India the power of 'Judicial Review' and precedent. Structurally, the judicial system in India is robust

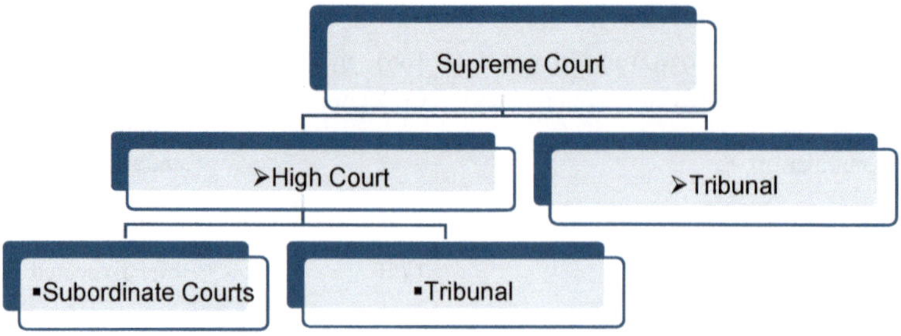

Fig. 2 Depicting hierarchical structure of the Judiciary in India

as it offers stability, uniformity and certainty in application of laws. This can be seen from the working of all the courts from top to bottom in India. This branch of Government is separate from the executive and legislative body of the Government of India and has strictly maintained the 'separation of powers' by including it in the basic structure of the Constitution.[4] Under the Indian Constitution a three-tier adjudicatory system is created which is bound by the hierarchy both in terms of structure and application of orders. On top of the hierarchy is the Supreme Court which is considered the protector and keeper of the constitutional mandate and values. Dr. Ambedkar considered the Supreme Court as 'one single integrated judiciary, having jurisdiction and remedial power in all cases under the Constitutional, Civil and Criminal Law and further considered this task as paramount in maintaining the unity of the country. The Supreme Court is followed by the High Courts in every state and below them are the district courts. With the passing of time since specialized bodies were needed due to the diverse areas of growth the 'Tribunals' were created which provided an alternate to the HC but never at par with the High Court being the quasi-judicial body. Some Tribunals are created in such a manner that they do not supplant the High Court but rather work under them such as Appellate Board under the Copyright Act, 1957. The hierarchy of institutional structure of adjudication in India can be seen in Fig. 2.

The jurisdiction at the lowest level is decided on 3 basis, pecuniary, territorial and subject matter. From the High Court onwards there is no such restriction as above. Apart from hierarchical structure the law also creates such system that no discrepancy can arise in the justice administration from top to bottom layer. The Constitution gives the power of *stare decisis* in favour of the Supreme Court which makes it a Court of precedent meaning thereby that all the Courts below including High Courts would be bound by the judgments of the Supreme Court (Article 141 of the Indian Constitution). This requirement of precedent helps in maintaining uniformity at all levels of adjudication in India which strengthens the system. The

[4] *Refer* Indira Nehru Gandhi v. Shri Raj Narain & Anr, AIR 1975 SC 2299. Available at https://indiankanoon.org/doc/936707/

district courts are first trial courts in India both under territorial as well as pecuniary jurisdiction up to a statutory prescribed limit. Appointments at this level are made by state judicial services exams. The State High Courts exercise administrative and supervisory control over the multiple district courts in a State. Moving to the functions of the Supreme Court it is called the Court of Justice for multiple reasons. Unlike trial courts, the Supreme Courts is not bound by the rigid bounds of law and in the capacity of protector of the Constitution it has expanded the ambit of constitutional protection to multiple rights beyond statute as well. (Right from the Kesavananda Bharti judgement, 1973) when the Supreme Court imposed the restriction of Basic structure on the Parliament, the Court has propounded the theory of Constitutional Morality (Navtej Singh Johar case 2018), accorded protection against Arbitrariness under Article 14 (Maneka Gandhi Case, 1978), and expanded the ambit of Article 21 to grant right to livelihood (Olga Tellis case 1986), right to remain silent (Bijoe Emmanuel case, 1986), right to safe environment (Virendra Gaur case, 1995), right to education (Mohini Jain case, 1992) and many more rights in its capacity of keeper and protector of the Constitution. This role of the Supreme Court also became relevant when technology made a way into access to justice. The detailed discussion on the Supreme Court's technological initiatives in access to justice is discussed later.

3 Access to Justice Under SDG16: *Technology and State Action in India*

Using twitter, WhatsApp, email, Facebook and many other technology-based communication modes and portals have become part of our day-to-day life which also includes the involvement of these in justice delivery process. This may be done from the perspective of evidence or depositions or processes or for more information purposes. The transition is also recognised under Article 13 of the United Nations Convention on the Rights of Persons with Disabilities (*hereinafter* referred to as 'CRPD') 2006 to which India is a party. The CRPD requires that state parties need to ensure effective access to justice for people with disabilities on an equal basis with others and for this the conventions look up to the digital transformation. Article 9 of the CRPD insists on having an infusion of technology in tapping modes of communication (The Global Initiative for Inclusive Information and Communication Technologies (popularly known as G3ict (2006) by UN Global Alliance for Information and Communications Technology (ICT) and Development). It is interesting to note that although Goal 16 has always been around since 2015 yet the development and dependency on technology can be traced to the aftermaths of corona virus disease (*hereinafter* referred to as 'COVID 19') pandemic which made the advantages of technology so evident. Electronic filing, attending court procedures online, getting information online, e-justice platforms all these contributed to reducing human contact and digitizing the process. This also had an impact on geographical limitations. Information in paper still has a smaller reach compared to online presence. Apart from addressing the boundary specific limitations technology can also introduce fairness, uniformity, and expedite the procedure and efficiency which is completely consistent with the mandate of the European Convention of Human Rights (*hereinafter* referred to as ECHR) for facilitating an independent and impartial judiciary (Article 6 of ECHR).

Talking about the transformation of justice by adoption of technology, Llagami Naureda (2024) proposes three stages of integration of technology in the justice delivery system. *First* is the electronic phase being the integration of electronic equipment into the working of the courts and advocacy. The *second* phase deals with the discovery of software leading to data processing and storage which substantially reduces the reliance on physical maintenance of data. The *final* stage is the stage of Artificial Intelligence (*hereinafter* referred to as 'AI') which is enabled to provide auto generated responses based on consistent patterns and hence the intervention of technology is not only limited to data storage and processing but is able to give and generate new information and conclusion based on pattern reading (Llagami Naureda, 2024). The technological interface is also useful in tracing the patterns of judgment to assess the probability and future predictions and this may also lead to aiding law professionals (Ruppert et al., 2018).

Apart from technology tools digital transformation has also affected the way institutions are created and function. In this regard, an assessment of the E-Courts (Electronic Courts) in India would be useful which defied the boundary norms and need for physical presence for availing justice. Responding to the Constitutional mandate of free and fair justice to all as part of fundamental rights, the Courts have consistently ruled in favour of strengthening both the procedural and institutional parts of the justice delivery system. Referring to United Nations Development Program (*hereinafter* referred to as UNDP), Access to Justice – Practice Note (2004) the Supreme Court has held that "access to justice in an egalitarian democracy must be understood to mean qualitative access to justice as well. Access to justice is, therefore, much more than improving an individual's access to courts, or guaranteeing representations. It must be defined in terms of ensuring that legal and judicial outcomes are just and equitable (Imtiyaz Ahmed case, 2012)." The Supreme Court identifying the foundation of access to justice in Anita Kushwaha case (2016) held 4 parameters as the essence of access to justice.

The state must provide an effective adjudicatory mechanism.

The mechanism provided must be reasonably accessible in terms of distance.

The process of adjudication must be speedy; and

The litigant's access to the adjudicatory process must be affordable.

All said and done while tracing the history of digitization of courts way back in 1990 the National Informatics Centre (*hereinafter* referred to as 'NIC') initiated computerization of India Courts alongside the Supreme Court. The Court Information System (*hereinafter* referred to as 'COURTIS') application enabled listing of routine activities in digital mode including filing of case, generation of cause list, issuance of order and judgment etc. Similarly, Judgment Information System

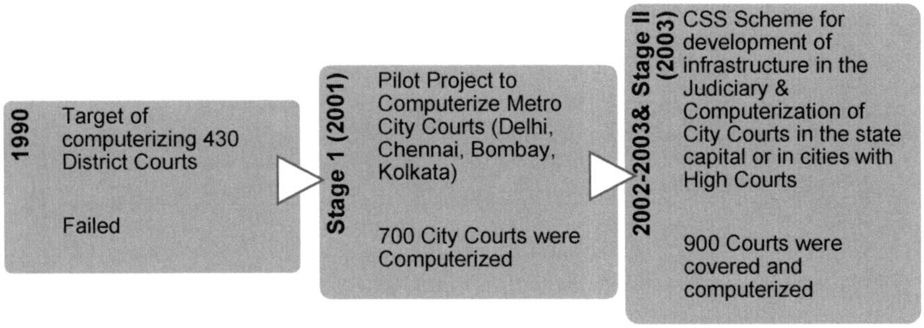

Fig. 3 Depicting digitization of courts in different phases. Graph generated based on data by National Judicial Academy. Available at (https://nja.gov.in/Concluded_Programes_2015-16/P-942%20PPT.pdf

(*hereinafter* referred to as 'JUDIS') provided entire text of all reported judgments. Some other applications such as National Informatics Centre providing information of pending case status (*hereinafter* referred to as 'COURTNIC') and List of Business Information System (*hereinafter* referred to as 'LOBIS') for generating daily and weekly cause list of High Courts along with providing information of fresh, disposed of and pending cases. These early efforts drastically reduced the burden of courts within a short span of time (Seetharam and Chandrashekaran, 2016). A gradual digitization of Indian Courts till phase II can be seen from the Fig. 3.

Till 2005 the focus was on digitization of data and making the courts computer and system enabled. But during all these periods the Supreme Court specifically recognised the value of accessibility in terms of distance and language. In this regard the progress of India is bit recent. The 2005 marked the eCourt policy of India with the Chief Justice Conference-2005 held in September 2004. This was followed by setting up of an 'E-Committee for Monitoring Use of Information Technology and Administrative Reforms in the Indian Judiciary'. The Committee submitted its report 'National Policy and Action Plan for Implementation of Information and Communication Technology (ICT *hereinafter referred as* ICT) in the Indian Judiciary—2005'. The project focuses on efficiency, transparency, productivity, accessibility, cost effectiveness, time bound services and affordability of justice delivery system. The implementation was conceived of being done in three phases as shown in Fig. 4.

As per the phases given above, the Committee, in its initial two phases which begin from 2007 with the extended deadline of 2015, focused on computerization of District Courts by incorporating system of Case information (*hereinafter* referred to as 'CIS'), training of judicial officers with technology and digitization of data available. Phase II of the project begins in 2014 with the cooperation between eCommittee, Department of Justice, National Informatics Centre (*hereinafter* referred to as NIC), Department of Electronics and Information Technology (*hereinafter* referred to as 'DietY') and the Ministry of Finance. This phase

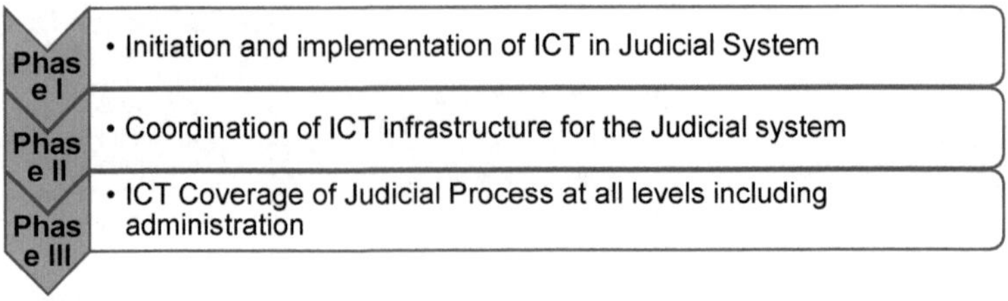

Fig. 4 Phase wise implementation plan of ICT in judiciary (Seetharam and Chandrashekaran, 2016)

focused on increasing the availability of the computer systems at various district courts and designated High Courts as implementing authority for the same. For facilitating the implementation process the government created multiple levels of expertise through various committees including High Court Computer Committee (*hereinafter* referred to as 'HCCC'), Central Project Coordinator (*hereinafter* referred to as 'CPC'), District Court Computer Committee (*hereinafter* referred to as 'DCCC') and a Nodal Officer for every Court Complex who is well conversant with the ICT concept (ecommitteesci.gov.in). Further Phase II was committed towards Free and Open-Source Solutions (*hereinafter* referred to as 'FOSS') making way for Judicial Knowledge Management System including Integrated Library Management system and creation of Digital Libraries. Some of the services accessible using mobile included eCourts Service App including information under Case Number Record (*hereinafter* referred to as 'CNR') a unique number assigned to each case filed in District and Taluka Courts, though case information system. This unique number can provide the entire case information under various heads such as Case Detail, Case Status, History of Case Hearing, Judgment and Order, Advocate under Case Status, My Cases etc.) The system also gives the option of exporting and importing data (sercices.ecourts.gov.in). Phase II also provided for National Judicial Data Grid (*hereinafter* referred to as NJDG) for qualitative information for courts, government and public.

Thus, phase II was able to accommodate multiple services in digital mode, and it was better streamlined compared to phase I. Now coming to Phase III, moving ahead from the Integrated Mission Mode Project covering Phase I and Phase II; the eCourt Phase III was approved by the Cabinet in 2023 spanning to four years with a financial coverage of Rs 7210 crore. Continuing the legacy of Phase I and Phase II the Phase III progresses towards digital, online and paperless courts which also includes e-filing/e-payments with e-Sewa Kendra in all Court complexes. The focus of Phase III is to achieve uniformity in all courts. The specific components of Phase III along with allocation of finance are mentioned in Fig. 5.

A close look at the chart shows the sensitivity of the government as it has considered the need for uneducated, and technology deprived population alongside taking specific consideration for the differently abled person. E-Sewa Kendra has been carefully planned for the people who are not technologically advanced. E-payments have enabled payment of court fees, fines and penalties from anywhere.

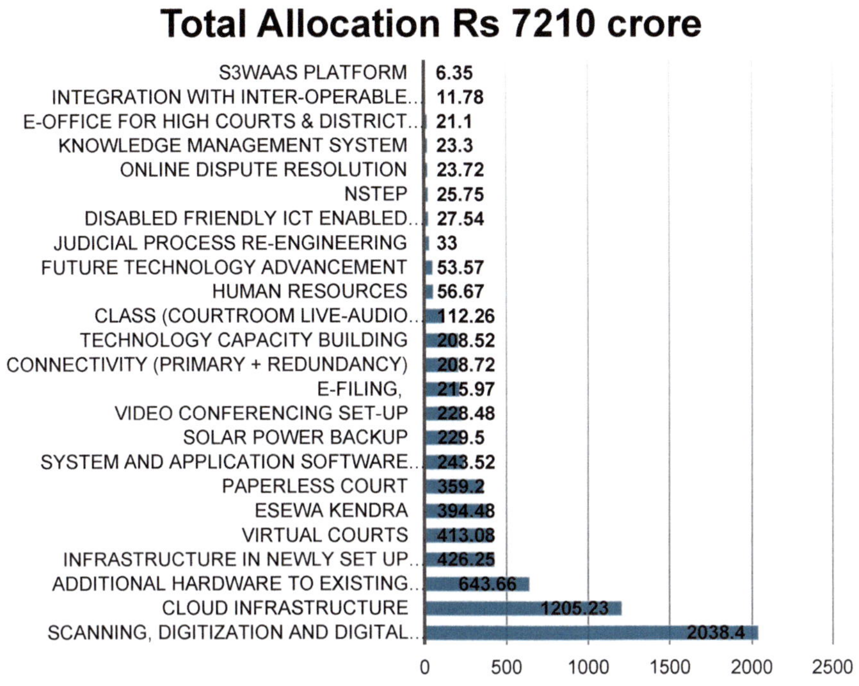

Fig. 5 Allocation of money towards digitalization of Indian Courts. All figures in Crore; Source—Department of Justice, Ministry of Law and Justice, Government of India. Available at https://doj.gov.in/phase-iii/

The phase III system also enabled automated delivery of court summons by further expanding the National Serving and Tracking of Electronic Processes (*hereinafter* referred to as 'NSTEP') which will reduce the delay in trials. The success of all three phases in terms of digitization of Courts can be traced from the report of Ministry of Law and Justice titled 'Digitization of Courts' (2023) available at Press Information Bureau of Government of India (pib.gov.in).

4 Access to Justice Under SDG16: *Technology, Supreme Court and Non-State Actors*

The Supreme Court of India has immense contributions towards technology integration into justice delivery process. In Swapnil Tripathi case (2018) a nine-judge bench recognised the need for access to public information, open justice and transparency in the justice delivery system. Answering the issue of viability of live streaming of court proceedings, the Court held that live streaming is manifested in public interest (Kumari and Shukla, 2024). Tracing the contribution of the Supreme Court right from 2005 when the eCourt Committee was formed it has been a keen observer of technological advancements in the country and has consistently tried to incorporate the same in making the justice delivery process more accessible, easier and transparent. Post e-Court Committee report the National e-Governance Plan (*hereinafter* referred to as 'NeGP') 2006 facilitated establishment of eCourts throughout India. The efforts of the Supreme Court of India have had a lasting impact on interconnecting SDG9 and SDG16. Technologies which help in influencing Court administration include Cloud Computing, AI, data bank, blockchain, metaverse and internet of things. As of now experimenting with Robots seems a distant goal for Indian judicial system. Technology intervention in justice administration picked pace due to COVID 19 which exposed the need to have contactless justice for all by breaking multiple barriers such as geographical distance, language barrier, scarcity of time, lack of resources being just few. Post COVID 19 the Government as well as the Judiciary adopted various measures to incorporate technology in justice administration this included video conferencing, creation of multiple portals and apps for providing pro-bono services etc. In fact, the Apex Court had upheld the benefits of video conferencing in multiple cases (Kalyan Chandra Sarkar v. Rajesh Ranjan@Pappu Yadav (2005), State of Maharashtra v. Praful B Desai (2003). To accelerate the integration of technology in justice system the Department of Justice adopted Designing Innovative Solutions for Holistic Access to Justice program (hereinafter referred to as 'DISHA program') for five years commencing from 2021–2026. Further, to increase legal literacy other apps were launched including Tele-Law and Nyaya Bandhu. Talking about Tele-Law app, it provides pre-litigation guidance which functions through lawyers, over phone call and through video conferencing at Common Service Centre. The services can be available on mobile as well. Through Tele-Law a needy and disadvantaged person can speak to panel lawyers of Tele-Law and seek legal guidance over digital platform with there being any need to meet physically. Nyaya Bandhu mobile app also helps in receiving pro-bono legal services. Also, the App has enabled justice in regional languages alongside Hindi and English.

In this regard, the advancement of the Supreme Court towards technology can be traced under 4 heads—e-Courts project, Electronic Supreme Court Records (e-SCR), Supreme Court Vidhik Anuvaad Software (SUVAS) and Supreme Court Portal for Assistance in Court's Efficiency (SUPACE). Multiple initiatives under e-Courts project have already been discussed under the head '*Technology and State Action in India'*. As discussed, the development in this regard spans around 3 phases. Currently, India is in its 3rd phase for which budgetary allocations and specific heads for expenditure of that budget has already been discussed. The implementation of e-Courts is entrusted with the Supreme Court which keeps track of implementation of targeted goals. Coming to e-SCR, this deals with maintaining the digital record of all Supreme Court Reports, judgment and orders. The information is made available on the Supreme Court portal and the same can also be accessed through mobile phones. This initiative reduces the burden of storing multiple hard copies of reports and judgments which was highly cost-inefficient and unsustainable. Also, at present, because of this initiative information can be accessed from any part of the world irrespective of distant geographical placement.

As a common challenge to educate and make aware a country having multiple languages and ethnicity with judicial judgments which are only produced in English, the Supreme Court thought about launching SUVAS which provides the translated version of the Supreme Court orders, judicial documents, judgments in nine vernacular language scripts and vice versa (vernacular languages include—Assamese, Hindi, Bengali, Marathi, Odiya, Kannada, Telugu, Tamil and Urdu). This is done through Artificial Intelligence, which also may be considered as the first experiment of AI with judicial services. In the course of time the information would be available in other regional languages as well. At present the process of translating some selective subject cases is in progress. Another portal which is going to change judicial functioning is SUPACE. SUPACE is a 100% AI-enabled portal which provides relevant information from bulky case files which otherwise is a gigantic and time taking task. This will increase the efficiency of the judges and researchers as they would be able to gather relevant information within a few seconds. Apart from Judiciary individual law practitioners are also experimenting with Artificial intelligence and other technical tools and techniques. Post COVID 19 platforms like Practice League and Law Gro offered cloud-based legal solutions which did not require the physical presence of the lawyers at the place of service. Similarly, online dispute resolution (ODR) is gaining popularity with an accelerated speed with the help of platforms like Presolv360 (platform for Online Dispute Resolution), CORD (Centre for Online Resolution of Disputes) and SAMA (an online dispute resolution platform which is also recognised by the Ministry of Law and Justice, Government of India). Talking about private sector and infusion of technology with law, the technology is experienced in two ways one is in house development of apps and portals and second is hiring the services from tech startups. India is experimenting with both. A renowned law firm, PSL Advocates and Solicitors, has in-house developed PSL Assist (web and mobile application) which helps in client management and knowledge management. The progress of cases in the firm can be traced in real time. Similarly, DDfact (indigenous tool for Royzz & Co—a Mumbai based law firm) has performed well in maintaining the record of deadlines and sends reminders to the clients. These tools have helped in reducing the burden of documentation and manual intervention. In this regard, the efforts of India's leading law firm Cyril Amarchand Mangaldas cannot go unmentioned. This firm led the way for law firms way back in 2017 by signing an agreement with Canadian AI software provider Kira to handle its contract lifecycle management on transactions. The firm was also the first to hire a C-suite level technology executive for the integration of technology into law. As a result, the firm launched 'Prarambh' the first Legaltech incubator in February 2019. This was developed aiming to nurture young entrepreneurs and to develop bespoke legal technology solutions for the Indian legal market (cyrilshroff.com). By the end of 2020 the LegalTech Market of India was valued at USD 385 million with prediction of huge growth. Some of the key areas which are covered under technology-based services include contract management and compliance, Legal marketplace, online dispute resolution, Legal research and Analysis, Document automation, Legal Assistance (Sarvam.AI for legal workflow and Jhana.AI for AI enabled paralegal services in

India), and vernacular solutions. Some of the tech giants in the legal field are managing over USD 1 trillion in contract management and the list is growing day by day (Vajpei and Dhillon, 2024).

Thus, as far as private players are concerned, they are making complete use of creativity and innovation for revolutionizing the legal field, yet everything has a dark side to it and so is technology. The rapid growth of technology raises ethical and legal concerns towards data safety and cyber-attacks. Also, there are chances that big players with resource abundance may take advantage of technology to create a wider divide between big and small legal practitioners and hence there is a need to involve the technology in the legal curriculum itself for maintaining quality and competition. A proper regulatory framework needs to be devised for making legal innovations data protection norms compliant both nationally and internationally. The challenges of funding and training are one of the limitations in digitizing access to justice. AI has its own limitations as it may lead to system generated bias which is highly difficult to detect and hence dataset selection, programming must be top notch. Ethical considerations in AI have been a part of constant debates since the time AI has been discovered and implemented. Apart from technical limitations, the hybrid model has not always ensured integrity.

Another difficulty with increased digitization of justice process includes concerns of intellectual property rights and liability. Post experiments with AI, most European countries are concerned with liability issues and are looking for ways to prioritize human interest over technology. For instance, France has balanced the infusion of AI in justice process by imposing the liability on a natural or legal person for any damage. By using the predication tools of AI, it has not considered AI as a replacement but only as a facilitator in maintaining consistency, efficiency and predictability of a judicial system. Having clarity on the goals makes the task of infusing technology into the justice process much easier and more effective. The last but not the least difficulty with technology in the legal system is the unsurety of its public acceptance. Individual to individual contact always has an emotional touch which machine learning might not be able to provide.

5 Conclusion

As discussed above India is making its way towards access to justice through technological advancement and for that all the organs of the state and non-state actors are both toiling hard. But given the vast area of India and diverse needs across country the task is humongous and time consuming. Also, considering the current wave of cyber-crimes it is less likely for the common man to prefer technology over human touch. Having said so India has a commitment to fulfill under SDG16 to which technology is an integral part and hence awareness and training seems to be a big area which requires attention. Although, the growth of technology and its use in India is on rise but we need to remember that most of the LegalTech services are on paid basis and hence small players in legal field cannot make use of the same which would make the legal profession highly imbalanced. Hence, there is a need to also devise a well-thought regulation which would not only streamline the growth of technology but would also ensure that no one can take undue advantage of resources.

References

Aneja, D., & Mathew, D. (2023). Smart automation and AI in the Indian judicial system: A case of organised irresponsibility? *Digital Future Labs, Goa*. Retrieved from https://assets-global.website-files.com/60b22d40d184991372d8134d/646315ae7153859ff45652c0_DFL%20FINAL%20web.pdf

Anita Kushwaha vs Pushap Sudan (Supreme Court, 2016). Retrieved from https://indiankanoon.org/doc/147862660/

Bailey, J., Burkell, J., & Reynolds, G. (2013). Access to justice for all: Towards an "expansive vision" of justice and technology. *Windsor Yearbook of Access to Justice*, *31*(2), 181–207. Retrieved from https://ir.lib.uwo.ca/cgi/viewcontent.cgi?article=1055&context=fimspub

Bhupatiraju, S., Chen, D.L., & Joshi, S. (2021). The promise of machine leaning for the courts of India. *National Law School of India Review, 33*(2). Retrieved from https://repository.nls.ac.in/cgi/viewcontent.cgi?article=1439&context=nlsir

Bijoe Emmanuel v. State of Kerala, 1986(3) SCC 615. Retrieved from https://indiankanoon.org/doc/1508089/

Cabral, J. E., Chavan, A., et. Al. (2012). Using technology to enhance access to justice. *Harvard Journal of Law & Technology*, *26*(1), 241–324. Retrieved from https://jolt.law.harvard.edu/articles/pdf/v26/26HarvJLTech241.pdf

Cappelletti, M., & Garth, B. (1978). Access to Justice: The newest wave in the worldwide movement to make rights effective. *Buffalo Law Review, 27*, 181–218. Retrieved from https://digitalcommons.law.buffalo.edu/cgi/viewcontent.cgi?article=1927&context=buffalolawreview

Chakravarthy, G. A. (2023). An assessment of emerging forms of technological innovation in justice delivery mechanism. *Nalsar Law Review, 8*(2). Retrieved from https://nalsar.ac.in/images/NLR_Vol%208_No-2.pdf

Das, A. (2023). How regionally representative is Supreme Court? Delhi, UP leads the list, North-Eastern states barring Assam unrepresented. *LiveLaw*. Retrieved from https://www.livelaw.in/top-stories/supreme-court-regional-representation-states-representation-220996

Department of Justice. (n.d.). *eCourt*. Retrieved from https://doj.gov.in/division/ecourts/

Dev, V. (2022). Judicial diversity in India: Exploring the 'unwritten' criteria of gender, religion, caste and regionality in appointments to the Supreme Court. Retrieved from https://theleaflet.in/equality/judicial-diversity-in-india-exploring-the-unwritten-criteria-of-gender-religion-caste-and-regionality-in-appointments-to-the-supreme-court

Donoghue, J. (2017). The rise of digital justice: Courtroom technology, public participation and access to justice. *Modern Law Review*, *80*(6), 995–1025. Retrieved from https://csja.gov.in/images/p1195/s_3_lct_court_and_case_management/jane_donoghue_rise_of_digital_justice.pdf

E-Committee Supreme Court of India. (2014). *Policy and action plan document phase* II *of the eCourts Project*. Retrieved from https://ecourts.gov.in/ecourts_home/static/manuals/PolicyActionPlanDocument-PhaseII-approved-08012014-indexed_Sign.pdf

e-Committee, Supreme Court of India. (n.d.). Information and communication technology in Indian judiciary. Retrieved from https://ecommitteesci.gov.in/project/brief-overview-of-e-courts-project/

eCourts Services App: About Us (n.d.). Retrieved from https://services.ecourts.gov.in/App/appaboutus.html#:~:text=The%20eCourts%20Project%20was%20conceptualized,by%20ICT%20enablement%20of%20Courts.

European Convention on Human Rights. (1950). *Council of Europe*. Retrieved from https://www.echr.coe.int/documents/d/echr/convention_ENG

For the first time the right of a Muslim women towards maintenance under section 125 CrPC was upheld in the case of Shah Bano in 1985 even though section 125 exists separately from the Muslim personal laws granting equal protection to all the communities. Sadly, the ratio of Shah Bano's case was reduced to nullity by subsequent passing of Muslim Women (Protection of Rights on Divorce) Act was enacted in 1986. Finally, it was only in the year 2024 that in the case of Mohd Abdul Samad v. The State of Telangana the Supreme Court upheld the application of section 125 CrPC on Muslim women.

Gaur, V. (1995). v. State of Haryana, 1994 SUPPL. (6) SCR 78, 2 SCC 577. Retrieved from https://indiankanoon.org/doc/27930439/

Goswami, P., & Goswami, A. (2024). Virtual justice: The role of technology in transforming criminal administration. *Integral Law Review, 3*. Retrieved from https://www.iul.ac.in/LawJournal/Pdf/Volume_III_Manuscript_1.pdf

High Level Political Forum on Sustainable Development. (2019). Review of SDG implementation and interrelations among goals. Retrieved from https://sustainabledevelopment.un.org/content/documents/23621BN_SDG16.pdf

Imtiyaz Ahmed vs State of Uttar Pradesh & Ors. (2012). 2 SCC 688. Retrieved from https://indiankanoon.org/doc/50352079/

Indian Kanoon (2024). *Mohd Abdul Samad vs. The State of Telangana on 10 July, 2024.* Retrieved from https://indiankanoon.org/doc/7964121/#:~:text=Mohd%20Abdul%20Samad%20vs%20The%20State%20Of%20Telangana%20on%2010%20July%2C%202024

Kalyan Chandra Sarkar v. Rajesh Ranjan@Pappu Yadav (2005). AIR 2005 SC 972. Retrieved from https://indiankanoon.org/doc/932761/

Kaur, J. (2014). Development in law and use of technology: Relevance for judge's research. *Nirma University Law Journal, 4*(1). Retrieved from https://docs.manupatra.in/newsline/articles/Upload/AA576CE6-2AFA-4797-811F-26CE6F7B81F7.pdf

Kaur, N., & Kaur, M. (2024). Role of artificial intelligence in the Indian Courts. *International Journal of Law, Policy and Social Review, 6*(1). Retrieved from https://www.lawjournals.net/assets/archives/2024/vol6issue1/5124.pdf

Khandbahale, S. (2022). Can technology defuse the ticking pendency bomb for the Indian judiciary system. *Vedhas Law Journal,* Retrieved from https://vedhaslawjournal.com/2022/04/20/can-technology-defuse-the-ticking-pendency-bomb-for-the-indian-judiciary-system/

Kumar, A., & Batra, A. (2024). Access to justice during COVID-19 pandemic: Role of technological interventions, *International Journal of Intellectual Property Rights (IJIPR), 14*(1). Retrieved from https://iaeme.com/MasterAdmin/Journal_uploads/IJIPR/VOLUME_14_ISSUE_1/IJIPR_14_01_002.pdf

Kumari, S., Shukla, P. (2024). Access to justice through use of legal technology. *Quest Journal of Research in Humanities and Social Science, 12*(3), 242–246. Retrieved from https://www.questjournals.org/jrhss/papers/vol12-issue3/1203242246.pdf

Llagami Naureda, M. (2024). The Use of the technology in justice system, *Global Journal of Politics and Law Research, 12*(4), 66–82. Retrieved from https://eajournals.org/gjplr/wp-content/uploads/sites/36/2024/07/The-Use-of-the-Technology-in-Justice-System.pdf

Maneka Gandhi, v. UOI, 1978 SCR (2) 621. Retrieved from https://indiankanoon.org/doc/1766147/

Mihir, R. (2021). Regional Representation at the Supreme Court [1985–2021]. *Supreme Court Observer*. Retrieved from https://www.scobserver.in/journal/regional-representation-at-the-supreme-court-1985-2021/

Ministry of Law and Justice. (2023). *Digitization of courts,* Press Information Bureau. Retrieved from https://pib.gov.in/PressReleaseIframePage.aspx?PRID=1896034

Mishra, S. P., & Arde, S. (2024). Digitalization and judicial review exploring the new horizon of rule of law. *International Journal of Creative Research Thoughts (IJCRT), 12*(4). Retrieved from https://www.ijcrt.org/papers/IJCRT24A4503.pdf

Mohd Abdul Samad v. The State of Telangana, Criminal Appeal No. 2842 of 2024 (Supreme Court). Retrieved from https://api.sci.gov.in/supremecourt/2024/3533/3533_2024_11_1501_53688_Judgement_10-Jul-2024.pdf

Mohini Jain v. State of Karnataka, AIR 1992 SC1858. Retrieved from https://indiankanoon.org/doc/40715/

Mukherjee, M. J., & Kumar, V. (2023). The judicial digital empowerment in India. *International Journal of Novel Research and Development (IJNRD), 8*(4). Retrieved from https://www.ijnrd.org/papers/IJNRD2304093.pdf

Navtej Singh Johar, v. UOI, AIR 2018 SC 4321. Retrieved from https://indiankanoon.org/doc/168671544/

Olga Tellis v. Bombay Municipal Corporation, AIR 1986 SC 180. Retrieved from https://indiankanoon.org/doc/709776/

Padmanabhan, V., & Devulapalli, S. (2018). How the Supreme Court has evolved since 1950. Retrieved from https://www.livemint.com/Politics/iK7w9InnxqgsELzcf26HRN/How-the-Supreme-Court-has-evolved-since-1950.html

Patne S. D., & Pimpalshende, P. (2024). The significance of science and technology in legal proceeding. *International Journal of Research Publication and Review, 5*(8). Retrieved from https://ijrpr.com/uploads/V5ISSUE8/IJRPR32646.pdf

Rattan, J., & Rattan, V. (2022). Role of information and communication technologies in the metamorphosis of justice administration in India: A legal study. *Indian Journal of Public Administration. 69*(1). Retrieved from https://doi.org/10.1177/00195561221109048

Ruppert, E., Hartung, D., Sitting, P., Gschwander, T., Rönneburg, L., Killing, T., & Biemann, C. (2018). LawStats: Large scale German court decision evaluation using web service classifier. In International *Cross- Domain Conference for Machine Learning and Knowledge Extraction* (pp. 212–222). Springer.

Seetharam, S., & Chandrashekaran, S., (2016). eCourts in India: From policy formulation to implementation. Retrieved from https://vidhilegalpolicy.in/wp-content/uploads/2019/05/eCourtsinIndia_Vidhi.pdf

Sharma, J., & Imran, M. (2021). The significance of technology in the Indian legal system and legal education: A comprehensive examination. *Russian Law Journal, 9*(1).

Simshaw, D. (2022). Access to A.I. Justice: Avoiding an inequitable two-tiered system of legal services. *Yale Journal of Law & Technology, 24*, 150–226. Retrieved from https://yjolt.org/sites/default/files/simshaw_-_access_to_a.i._justice.pdf

Tejaswi, M. (2024). Legal fraternity may explore Gen AI to reduce errors, enhance efficiency. *The Hindu*. Retrieved from https://www.thehindu.com/business/legal-fraternity-may-explore-gen-ai-to-reduce-errors-enhance-efficiency/article68328911.ece

The Cold War (1945–1989). (n.d.). *CVCE.EU*. Retrieved March 17, 2025, from https://www.cvce.eu/content/publication/2011/11/21/6dfe06ed-4790-48a4-8968-855e90593185/publishable_en.pdf

The State of Maharashtra v. Dr. Praful B. Desai (2003), AIR 2003 SC 2053. Retrieved from https://indiankanoon.org/doc/560467/

Tripathi, S. Supreme Court of India, AIR 2018 SC 4806. Retrieved from https://indiankanoon.org/doc/43629806/

Vajpei, S., & Dhillon, N. (2024). LegalTech made in India: A global game-changer in legal innovation. Retrieved from https://www.iltanet.org/blogs/shreya-vajpei/2024/09/20/legaltech-made-in-india-a-global-game-changer-in-l

Verma, K. (2018). e-Courts Project: A Giant leap by Indian judiciary. Retrieved from https://core.ac.uk/download/pdf/268231072.pdf

Vivek, T., & Nilakshi. (2024). Artificial intelligence in the India judiciary: A systematic analysis of potential applications and challenges in addressing case backlogs. *Journal of Trends and Challenges in Artificial Intelligence, 1*(3). Retrieved from https://jai.aspur.rs/archive/v1/n3/3.pdf

Waseem, S., Sharma, A., & Kumar, A. (2023), *NUJS Journal of Regulatory Studies, 8*(2). Retrieved from https://www.nujs.edu/wp-content/uploads/2023/08/vol8iss2-3.pdf

Wolf, M. J. (2012). Collaborative technology improves access to justice, *Legislation and Public Policy, 15*(4), 759. Retrieved from https://nyujlpp.org/wp-content/uploads/2013/01/Wolf-Collaborative-Technology.pdf

Are Biofuels Environmentally Sustainable? Evidence from Augmented Panel ARDL Approach

Mohammad Younus Bhat and Syed Mohd Shahzeb

Abstract

The purpose of this study is to examine the impact of bioenergy on CO_2 emissions across the highly bioenergy-consuming countries during 2000–2020. While acknowledging the role of The study also brings out the role of bioenergy in realising sustainable development goals. bioenergy, we estimated the transmitted effects of bioenergy on CO_2 emissions through an augmented fixed effect panel data setting. The empirical model is enhanced to account for potential heterogeneity and endogeneity among the nations to get robust estimates. The findings indicate that bioenergy and renewable energy are effective in decreasing carbon emissions and may serve as an alternate strategy for alleviating overall environmental degradation. The correlation between energy use and economic growth is resulting in a substantial rise in carbon emissions. Nonetheless, we assert that the bioenergy industry must undergo certain essential technical and organisational reforms. The study enhances this field of research by examining the relationship between CO_2 emissions, economic growth, and bioenergy use in a through manner to elucidate the real nature of the nexus.

Keywords

Biofuel consumption • CO_2 • Emissions • Economic growth • Sustainabilty • ARDL

M. Y. Bhat (✉)
Global Economics and Finance Cluster, School of Business, University of Petroleum and Energy Studies, Dehradun, India
e-mail: younus.bhat@ddn.upes.ac.in

S. M. Shahzeb
School of Management, IILM University, Greater Noida, India

T. Choudhury et al. (eds.), *Open Access as a Knowledge Platform for Implementing the United Nations Sustainable Development Goals*, World Sustainability Series,
https://doi.org/10.1007/978-3-031-99605-4_16

Abbreviations

GHGs	Greenhouse Gasses
IEA	International Energy Agency
IRENA	International Renewable Energy Agency
CO_2	Carbon emission
EG	Economic Growth
NREC	Non-renewable Energy Consumption
BIOFC	Biofuel Consumption
VECM	Vector Error Correction Model
ADF	Augmented Dicky-Fuller Test
PP	Phillip Perron
ARDL	Autoregressive Distributed Lag
PMG	Pooled Mean group
REC	Renewable Energy Consumption

1 Introduction

Energy is a basic requirement for the growth and development of any country. It is equally important for satisfying crucial human ends and extending our future capabilities. Energy, in all its forms, is a vital aspect of human societies. Energy and human well-being are closely aligned to each other and there exists a strong correlation between energy use per capita and Human Development Index (Foust et al., 2015). The use of fossil fuels is dangerous as they have damaged the environment. Vehicles that use fossil fuels release carbon dioxide and other greenhouse gases, which significantly alters the climate. As a result, the economic growth driven by the use of energy derived from fossil fuels has been the main contributor to the edifice of greenhouse gases (GHGs), leading to the threat of global warming. Global greenhouse gas emissions, particularly CO_2 emissions, have constantly gone up by 2.1% in 2017 compared to 2016, whereas world CO_2 emissions from the combustion of burning fossil fuels have increased 2.1% per year from 2000 to 2017. Further, the estimates suggested that the global human population has reached the $7.6 billion$ mark in 2019 (IPCC, 2022). It means the world population has more than doubled in the past fifty years increasing by over 150 people per minute. This population growth is intensifying the persistent need for more and more energy in the future.

In the case of emerging economies, which are on the path towards industrialization are consuming fossil fuels and pumping up the Greenhouse Gasses (GHGs) into the atmosphere. Nevertheless, the most viable route to mitigate the impact of greenhouse gases, including CO_2 emissions, is by reducing consumption levels of fossil fuels by substituting conventional energy with efficient use of natural gas. Towards this, World Summit on Sustainable Development (2002) has reiterated and committed itself to encourage the enhancement of renewable energy resources to achieve sustainable consumption and production. This study aims to assess the effects of bioenergy use and economic expansion on emissions in countries with significant bioenergy usage from 2000 to 2020. The study brings out the role of biofuels in realizing sustainable development goals. Within this ambient, we estimated bioenergy effects on the CO_2 emissions through an augmented fixed effect panel data setting of whose results reveal that biofuels and renewable energy tend to reduce the carbon emission significantly. Further, it is observed that economic growth and non-renewable energy consumption are leading to a significant increase in carbon emissions.

The rest of the paper is structured in the following way. In the next section, we elaborate on the main contributors to greenhouse gases and elucidate the potential benefits of biofuels in avoiding these emissions. The data and technique used to analyse the relationships between the factors taken into consideration for this study are described in Sect. 3. Section 4 provides a summary of the findings from the econometric study, and Sect. 5 discusses the estimated findings. Section 6 wraps up the report by offering more comprehensive policy recommendations for embracing bioenergy as a safe and sustainable renewable energy source.

2 Bioenergy as Sustainable Innovation

Carbon dioxide (CO_2) emissions are critical in augmenting atmospheric CO_2 concentrations, which account for approximately two-thirds of radiative forcing contributing to drastic climate changes. The other industrial processes account for about 60% of total anthropogenic GHG emissions since preindustrial times. Unchecked, climate change will cause severe, pervasive, irreversible impacts on plants, animals, and people—think more intense and regular extreme weather events, ocean acidification, and, more visibly, the rising sea level. Carbon dioxide emissions have increased in the form of the combustion of more fossil fuels, making environmental resources more vulnerable. As reported in Fig. 1, we observe a steady growth of CO_2 emissions. The rising emissions are worrisome, and there is a need to have a systematic way for a sustainable resource base for a balanced climate. Therefore, reducing CO_2 emissions from using fossil fuels is essential to lessening the negative social effects of climate change. China is the top contributor to CO_2 emissions contributing 30% of total global emissions (Fig. 2). The other significant contributors include the USA, EU, Japan, India, and the Russian Federation. The two emerging economies, India and China, together accounted for 37% of the total global emissions. In order for all countries to achieve sustainable development, reducing the effects of climate change is essential and a key policy objective. In line with the Paris Agreement, all nations have pledged to reduce CO_2 emissions so that the increase in global average temperature can be brought down within limits. The goal to limit global temperature implicitly needs a shift towards a low-carbon energy sector, which currently accounts for two-thirds of global emissions. To provide reductions in CO_2 emissions by 2050, renewable energy sources accompanied by energy efficiency gains can provide nearly 90%

reductions. Therefore, renewable energy is considered a vital constituent that can contribute significantly by adhering to the Paris Agreement.

The primary policy objectives for creating a sustainable bio-economy are to diminish reliance on non-renewable resources and to address and alleviate global climate change. The objective of dependence on bioenergy as an alternate energy resource is to enhance further economic growth accompanied by job creation and improving the nation's balance of trade (Fig. 3). Other goals that are crucial for this transition include ensuring food security and managing resources from nature in a proper and sustainable approach. There has been a huge global support

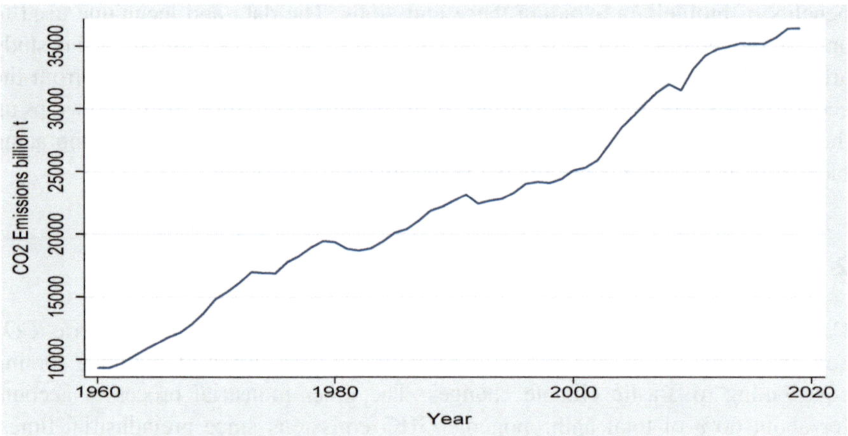

Fig. 1 Global CO_2 emissions (billion tons)

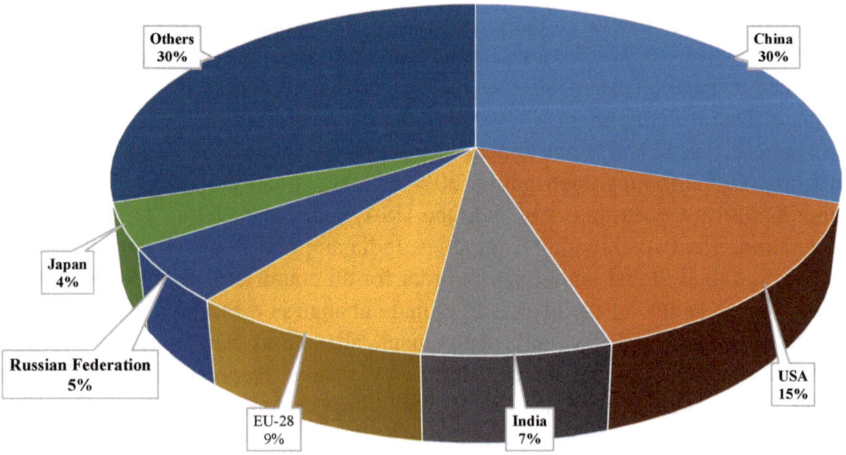

Fig. 2 Share of global carbon emissions by Country (billion tons)

through major policy initiatives for the production and use of renewable energy sources relating particularly to climate change and bioenergy (Elobeid et al., 2013). These initiatives towards a sustainable bio-economy include, among others, "the United States Environmental Protection Agency's Renewable Fuel Standard (RFS2)", the American Clean Energy and Security Act (2007), Brazil's National Climate Change Policy (2009), Renewable Fuels Regulations in Canada (2006), Energy and Climate Change Package, EU (2009), Biofuel Policy, India (2018). The fundamental idea of these policy measures include minimising reliance on fuel imports, reducing fuel prices, stimulating economic growth, enhancing the rural economy, and addressing climate change through adaptation and mitigation strategies. Bioenergy can help reduce the devastating effects of climate change while also reducing carbon emissions in the energy industry. Biofuels as a source of energy have a greater potential to reduce greenhouse gas emissions (see Fig. 4). To better understand the possible effect of the use of renewables like biomass energy, the International Renewable Energy Agency (IRENA) has developed a calculator estimating the greenhouse gas (GHG) emissions that are avoided each year as a consequence of renewable energy deployment in a country.

Biofuels, apart from being cost-competitive and environmentally sustainable, can be an alternate energy source. Their efficient consumption is warranted if it mitigates market failures resulting from adverse environmental impacts. Biofuels, as a clean energy source, provide various advantages for the environment contingent upon the specific feedstocks utilised. For example, feedstocks like switchgrass may provide a better habitat for wildlife, and others like miscanthus have greater potential to mitigate greenhouse gas emissions. Feedstocks that can enhance biodiversity and other ecological benefits are Indian grass and Big Bluestem (Khanna, 2008). With technological advancements, biofuels offer significant environmental benefits, helping to internalize externalities and support a sustainable mix of feedstocks. Aligning energy and climate policies—such as leveraging the inverse

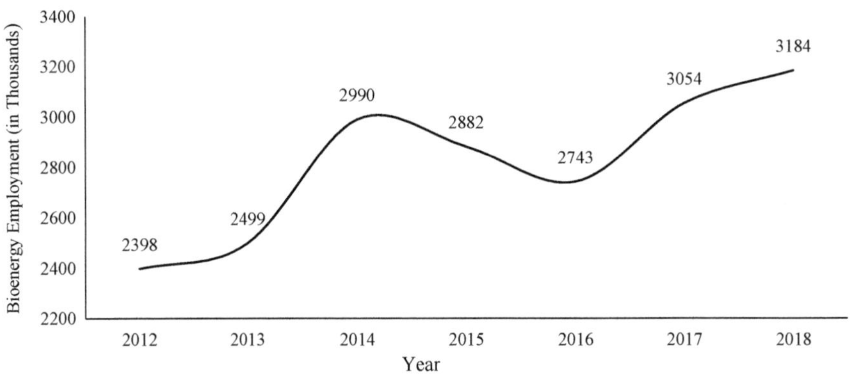

Fig. 3 Global bioenergy employment, 2012–2018 (Thousands)

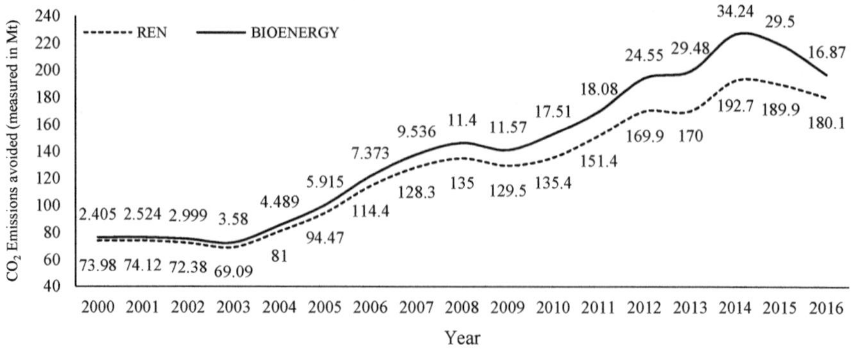

Fig. 4 CO_2 Emissions (Mt) Avoided during 2000–2016

relationship between tax credits and carbon footprints—can incentivize the production of low-carbon feedstocks. Additionally, policy incentives across countries can promote feedstocks that enhance biodiversity and improve ecosystem services. Harnessing energy from biomass can play a crucial role in mitigating the adverse effects of climate change while accelerating the decarbonization of the energy sector.

2.1 Bioenergy as a Driver of SDG's

Bioenergy Serves a vital function in the UN Sustainable Development Goals (SDGs) helping to support infrastructure development, Climate change resilience, Innovation-driven industrial growth and socio-economic integration. Bioenergy plays a significant role in advancing Sustainable Development Goal 9 (SDG 9), which focuses on developing robust infrastructure, promoting inclusive and sustainable industrialization, and fostering innovation. As a renewable energy source derived from organic materials such as crops, waste, and algae, bioenergy reduces reliance on fossil fuels, supports rural industrial development, and drives sustainable economic growth. By providing clean and consistent energy, bioenergy helps industries lower operational costs and emissions while creating local industries, particularly in rural areas. Innovations in bioenergy, including second-generation biofuels from non-food crops, algae-derived fuels, and waste-to-energy systems, illustrate the potential for a circular economy and technical progress. Countries like Brazil exemplify how bioenergy can transform economies and align with SDG 9. Brazil's sugarcane-based ethanol industry, one of the largest in the world, has reduced oil dependence while contributing to job creation and industrial growth. The government's Proálcool Program and investments in biogas projects, which convert agricultural residues into energy, have strengthened rural economies and reduced greenhouse gas emissions. Furthermore, Brazil's research initiatives in second-generation biofuels and a well-established ethanol refuelling infrastructure highlight the integration of innovation and infrastructure in achieving sustainable industrialization.

With its decentralized energy capacity, it is particularly crucial in enabling energy access in remote and developing regions. For instance, decentralized biogas systems in India, have fortified local energy resilience and decreased reliance on traditional, fossil fuel-sourced grids. Furthermore, bioenergy provides a firm and baseload power source complementing the non-continuous renewables, such as solar and wind. Beyond infrastructure, bioenergy's socio-economic impact extends into sustainable consumption, production and equitable growth. It also contributes to SDG 8 (Decent Work and Economic Growth) and SDG 12 (Responsible Consumption and Production) through the revalorization of agricultural and forestry residues. It minimizes industrial emissions and helps circular economic models. Brazil's use of sugarcane bagasse to produce ethanol demonstrates this integration. The biomass value-chain also provides rural employment and promotes equitable growth. With the implementation of appropriate regulatory mechanisms, unsustainable land-use change can be minimised, and food production is likely to be avoided. Advancement in clean technologies is imperative to scale bioenergy sustainably. Algae-based biofuels, biogenic feedstock gasification, and enzymatic hydrolysis for cellulosic ethanol are a few of the more innovative methods being developed that could make biofuels more efficient and diminish dependence for food crops. Sweden excels in energy recovery from waste, achieving 99% recycling, thus helping countries meet Sustainable Development Goal 7 (Affordable and Clean Energy) and SDG 12. Further, linking bioenergy with carbon capture can enhance its per-capita climate mitigation potential. To realize the full potential of bioenergy innovations, persistent barriers such as land-use conflicts, high technology costs, and limited research and development (R and D) capacity in the developing world demand substantive action. To address such barriers, policies like carbon pricing, sustainability mandates (EU Renewable Energy Directive II) and organizational facilitation signified by U.S. Bioenergy Technologies Office, which can help overcome these challenges. Lasting outcomes of bioenergy relies on robust environmental sustainability measures. Denmark, Brazil, and others could serve as examples, backed by certificates Roundtable on Sustainable Biomaterials (RSB) and Forest Stewardship Council (FSC) to meet ecological standards. National policy frameworks and international collaboration like GBEP, EU-Africa partnerships and Brazil–Mozambique knowledge transfer signify SDG 17 (Partnerships for the Goals). Equitable growth impacts covering employment generation in the U.S to healthier air environment and gender equity in Nepal which highlights bioenergy's potential in advancing robust, equitable, and sustainable development globally.

Despite its promise, bioenergy faces challenges such as feedstock sustainability, high initial investment costs, limited access to advanced technologies in some regions, and inconsistent policies. However, opportunities exist to overcome these barriers through international collaboration, technology transfer, and the development of decentralized bioenergy systems that enhance energy access and resilience in remote areas. By integrating bioenergy into waste management systems and establishing sustainable supply chains for feedstocks, countries can accelerate progress toward SDG 9.

In conclusion, bioenergy is a powerful driver of sustainable industrialization, innovation, and resilient infrastructure. Countries like Brazil showcase its transformative potential in achieving economic, environmental, and social benefits. To maximize its impact globally, sustained investment in technology, capacity building, and supportive policy frameworks is essential. Bioenergy's alignment with SDG 9 underscores its importance as a cornerstone of sustainable development in the energy and industrial sectors.

3 Data and Theoretical Foundation

This section outlines the data, describes the variables, and details the econometric methods employed to derive the results. Theoretically, the model specification aligns with the framework where carbon emissions are influenced by economic performance and energy consumption. However, this study additionally incorporates biofuel and renewable energy consumption as factors impacting carbon emissions. A panel of countries was examined, including the USA, Canada, India, China, the Korean Republic, Japan, and Brazil, all of which have established blending targets aimed at increasing the use of biofuels and renewables to mitigate carbon emissions. These countries were purposely chosen for analysis as these countries have designed biofuel policies to avoid greenhouse gas emissions. The study considers two-panel data models, fixed effects and random effects, as both of them present different modalities of panel data estimation. The intercept, slope coefficients, and error term of the model are subject to various assumptions. We prefer panel data models over other models because they offer a robust measure of the impact across different cross-sections over a given timeframe. Further, panel data models also succeed in capturing heterogeneity in both cross-sectional units and time dimensions. In the present endeavor, we intend to capture the influence of nonrenewable energy consumption (NREC), renewable energy consumption (REC), biofuel consumption (BIOFC), and gross domestic product per capita (GDPPC) on carbon dioxide emissions. A panel data framework can be used to estimate the relationship, with emissions as the dependent variable. The dependent variable is impacted by a number of independent variables, including GDP per capita (GDPPC), biofuel consumption (BIOFC), renewable energy consumption (REC), and non-renewable energy consumption (NREC). Consequently, the empirical approach employed to examine the causal relationship may be expressed as follows:

$$CO_{2it} = f(GDPPC_{it}, NRENC_{it}, REC_{it}, BIOFC_{it}) \tag{1}$$

After logarithmic transformation, Eq. (1) results in the following equation:

$$LnCO2_{it} = \beta_0 + \beta_1 LnGDPPC_{it} + \beta_2 LnNREC_{it} + \beta_3 LnREC_{it}$$

Table 1 Variable description and data source. *Source*: Author's compilation

Variable	Description	Measurement	Data sources
CO_2	Carbo emission per capita	Carbon emissions (kt)	World bank development India
GDPPC	Gross domestic product (per capita)	GDP (constant 2010 US$)	World bank development India
NREC	Non-renewable energy consumption per capita	Fossil fuel energy consumption (% of total)	World bank development India
REC	Renewable energy consumption per capita excluding modern biofuels	Renewable energy consumption per capita (in kilogram)	International energy agency
BIOFC	Consumption of modern biofuels	Biofuel consumption per capita (in kilogram)	International energy agency/ International renewable energy agency

$$+ \beta_4 LnBIOFC_{it} + \varepsilon_{it} \tag{2}$$

where; β_0 is the intercept; $\beta_1, \beta_2, \beta_3, \beta_4$ are the coefficients of the model and ε_{it} is the error term. CO_2 is Carbon Dioxide Emissions measured in per capita terms, GDPPC is Gross Domestic Product Per Capita $(ataconstantprice, 2010 = \$100US)$ used as a proxy for economic growth, NREC is the consumption of non-renewables measured in kg of oil equivalent per capita, REC is the per capita consumption of renewables (% of total final energy consumption) and BIOFC is the consumption of biofuels (% of total renewable energy consumption).

Data concerning all variables, such as CO_2, GDP per capita (GDPPC), non-renewable energy consumption (NREC), renewable energy consumption (REC), and biofuel consumption (BIOFC), was sourced from various entities, including development indicators compiled by the World Bank (WB), the U.S. Energy Information Administration (EIA), the International Energy Agency (IEA), and the International Renewable Energy Agency (IRENA), among others (refer to Table 1).

To address heteroscedasticity concerns and guarantee effective, reliable outcomes, the model's variables were converted into natural logarithms (ln) (Lau et al. 2014). Additionally, a series of tests, including Im, Pesaran, and Shin; ADF-Fisher; PP-Fisher; Levin, Lin and Chu; and Breitung, were conducted to assess the level of stationarity. These tests are essential for identifying unit roots and preventing spurious estimations. To determine the relationship between variables, estimates were obtained through a series of models such as Pooled OLS, Random Effects, and Fixed Effects on the stationary data, the results of which are presented in the next section. Finally, PMG estimator through the ARDL model was employed to estimate the short-run as well as long-run estimates besides examine the causal relationships (For a detailed discussion on the application of the ARDL technique see Mensah et al. (2019).

Table 2 Descriptive statistics of variables

	CO_2	GDPPC	NREC	REC	BIOFC
Mean	9.039	24,924.2	3908.519	377.873	20.265
Maximum	20.179	51,015.130	8441.185	2010.88	51.787
Minimum	0.967	829.359	417.382	0.690	0.040
Std.Dev	6.284	19,678.510	2824.775	590.638	17.430
Skewness	0.220	0.089	0.308	1.654	0.513
Kurtosis	1.712	1.243	1.606	4.482	1.741
Jarque–Bera	8.111	13.644	10.162	57.480	11.536
Probability	0.017	0.001	0.006	0.000	0.003
Observations	147	147	147	147	147

4 Results and Discussion

4.1 Descriptive Statistics of Variables

In this section, the descriptive statistics for the variables employed in the analysis are displayed. The overall average for all variables considered in the analysis is positive. The average magnitude of CO_2 emissions is $9.04 ktpercapita$ equivalent of oil, with a standard deviation of $6.29kt$. The average bioenergy in the panel of countries examined in this study is $377.87Mteo$, with a standard deviation of Table 2, indicating that the mean GDP per capita is 24924.2, while the mean renewable energy consumption per capita stands at 20.26, with standard deviations of 19678.51 and 17.43, respectively. Correlations between the parameters considered in the analysis are shown in Table 3 below. The correlation coefficients (r) are strong for the EG, NREC, and CO_2 variables ($allr > 0.80$) whereas the r are low for REC and BIOFC. The correlation matrix presented in Table 3 illustrates a positive correlation of GDPPC and NREC with CO_2 but a negative correlation of REC and BIOFC with CO_2 emissions. Column II focusses on CO_2 emissions, which have a significant positive association with both non-renewable energy consumption (NREC) and economic development (GDPPC).

4.2 Cross-Sectional Dependence and Panel Unit Root Test

Table 4 presents the outcomes of the cross-sectional dependence tests conducted by Friedman, Frees, and Pesaran. The findings demonstrate insufficient evidence to exclude the null hypothesis, indicating that the variables display cross-sectional independence. Multiple unit root tests were performed to assess the unit root qualities of the data, with the findings presented in Table 5. The results indicate that all variables are stationary in their first difference, hence allowing for the rejection of the null hypothesis of a unit root for all variables in their first difference. The

Table 3 Pair-wise correlations between the variables

Variables	CO_2	GDPPC	NREC	REC	BIOFC
CO_2	1.000				
GDPPC	0.868*** (0.000)	1.000			
NREC	0.979*** (0.000)	– 0.932*** (0.000)	1.000		
RENC	– 0.637*** (0.000)	– 0.518*** (0.000)	– 0.594*** (0.000)	1.000	
BIOFC	– 0.057*** (0.000)	– 0.131*** (0.000)	– 0.062*** (0.000)	0.493*** (0.000)	1.000

Note p-values are in parenthesis;*** indicates significance at 1% level

Table 4 Cross-sectional dependence test results

Test	Statistics
Pesaran's test of cross-sectional independence	– 1.163(pr=1.755)
Frees' test of cross-sectional independence	0.421
Friedman's test of cross-sectional independence	12.457(pr=0.053)

Note Average absolute values of the of-diagonal elements is (0.399). Critical values for Free's Q distribution for 1% is (0.335), for 5% is (0.226) and for 10% is (0.172)

Pedroni panel co-integration analytical approach is employed to examine the long-term relationship among the variables. The findings of the Pedroni study, presented in Table 6, indicate the potential for a prolonged link among GDP per capita, energy consumption, carbon emissions, alternative energy consumption, and biofuels utilisation. Kao suggested an alternate technique to ascertain the persistent relationship between carbon emissions, economic growth, the utilisation of renewable and non-renewable energy sources, and biofuel use. The rejection of the null hypothesis of no cointegration at the 1% significance level for the carbon emissions equation indicates, as presented in Table 7, a long-term relationship among the variables. The Kao test corroborates the findings of the Pedroni test, indicating long-term cointegration among the variables examined in this study.

4.3 Panel Regression Estimates

This section provides a summary of the panel regression results, which are presented in Table 8. Before analysing the formal nexus between CO_2 bioenergy and other macroeconomic indicators, a systematic model specification criterion has to be followed. To choose between the various specifications we adopted the standard

Table 5 Panel unit root test results

Test	Level					First Difference				
	CO_2	GDPPC	NREC	REC	BIOFC	ΔCO_2	$\Delta GDPPC$	$\Delta NREC$	ΔREC	$\Delta BIPOFC$
Im, Pesaran and shir	2.871 (0.998)	3.497 (0.999)	3.978 (1.000)	3.402 (0.999)	0.844 (0.800)	− 4.895*** (0.800)	− 4.261*** (0.000)	− 5.279*** (0.000)	− 3.157*** (0.000)	− 4.397*** (0.000)
ADF-Fisher	6.794 (0.942)	2.971 (0.999)	2.331 (0.999)	11.199 (0.670)	12.698 (0.550)	47.728*** (0.000)	42.664*** (0.000)	51.550*** (0.000)	42.953 * ** (0.000)	43.527*** (0.000)
PP-Fisher	4.332 (0.993)	4.425 (0.992)	2.206 (0.999)	8.220 (0.877)	8.365 (0.869)	48.571 * ** (0.000)	50.877*** (0.000)	60.732*** (0.000)	79.760*** (0.000)	60.089*** (0.000)
Levin, Lin and Chu	− 0.968 (0.166)	2.042 (0.979)	1.445 (0.925)	0.088 (0.535)	− 1.871 (0.030)	− 6.738*** (0.000)	− 5.136*** (0.000)	− 7.489*** (0.000)	− 6.410*** (0.000)	5.146*** (0.000)
Breitung	2.493 (0.993)	2.392 (0.991)	1.182 (0.88)	2.703 (0.996)	1.227 (0.890)	− 5.306*** (0.000)	-3.193*** (0.000)	− 4.746*** (0.000)	− 2.465*** (0.006)	10.023*** (0.000)

Note p–values in parenthesis *** indicates significance at 1%

Table 6 Pedroni Co-integration test results

Within-dimension (Panel statistics)			Between dimension (individual statistics)		
Test	Statistics	Prob.	Test	Statistics	Prob.
Pedroni (1999)					
υ Statistic	– 1.252	0.895	Group *ρ statistics*	2.948	0.998
ρ Statistic	2.542	0.995	Group PP statistics	– 5.568***	0.000
PP Statistic	– 0.678	0.249	Group *ADF* statistics	– 2.433***	0.008
ADF Statistic	– 2.018***	0.022			
Pedroni (2004) *weighted statistics*					
υ statistic	– 1.513	0.935			
ρ statistic	1.722	0.958			
PP Statistic	– 4.503***	0.000			
ADF Statistic	– 3.231***	0.001			

Note *** and ** indicates significance at 1 and 5% respectively.

Table 7 Kao panel cointegration test

	Statistic	Prob.
ADF	– 3.586***	0.000
Residual variance	9.09E-05	
HAC variance	9.92E-05	

Note *** represents significance at 1% level.

Hausman test. The test statistics suggested the fixed effect model be more appropriate. Therefore, the results are interpreted considering the fixed effects model presented in column IV of the Table 8.

The results suggest that gross domestic product and energy consumption contribute to an increase in carbon emissions, while the adoption of renewables and biofuels helps reduce them. We can interpret coefficients of each variable in semi-elasticity terms explaining the percentage impact explanatory variables have on carbon emissions. For example, a one% increase in GDPPC and NREC will increase CO_2 emissions further by 0.266 and 0.885% respectively. However, a one% increase in REC and BIOFC will avoid CO_2 emissions by 0.005 and 0.023% respectively. The results unequivocally demonstrate a positive relationship between energy consumption, gross domestic product per capita, and carbon emissions, suggesting that these factors play a role in global warming in these countries. Bioenergy and renewable energy use in these nations significantly reduce greenhouse gas emissions.

Table 8 Panel regression results

Dependent Variable: (ΔCO_2)					
Variables	Pooled OLS regression	Random effects GLS regressions		Fixed-effects regressions	
$\Delta GDPPC$	– 0.221*** (0.032/0.000)	0.192*** (0.063/0.002)		0.266*** (0.065/0.000)	
$\Delta NREC$	1.229*** (0.050/0.000)	0.932*** (0.076/0.000)***		0.885*** (0.076/0.000)	
ΔREC	– 0.062*** (0.018/0.001)	– 0.020 (0.027/0.447)		– 0.005 (0.027/0.859)	
$\Delta BIOFC$	0.006 (0.006/0.370)	– 0.017*** (0.006/0.004)		– 0.023*** (0.006/0.000)	
Cons.	– 2.450 (0.085/0.000)	– 3.148 (0.163/0.000)		– 3.296 (0.160/0.000)	
N	105	105		105	
F(4,100)	1040.11	*Waldchi*2(4)	2262.1	F(4,94)	588.33
Prob > F	0.000	*Prob > chi*2	0.000	Prob > F	0.000
R-squared	0.976	$R - squared$	0.930	$R - squared$	0.915

Note The numbers in parenthesis is represent the standard error. *** Significant at 1% level; ** significant at 5%; * significant at 10% level

4.4 Pooled Mean Group (PMG) Estimates

Table 9 shows the short- and long-term coefficients that were found from the elasticity of carbon emissions for the group of countries that this study looked at. It also shows the results of the PMG analysis. The variables GDPPC and NREC show a notably positive impact on carbon emissions, whereas REC and BIOFC reveal a negative impact on carbon emissions. Assuming all other factors remain constant, a one% increase in GDP per capita leads to a 0.16% t rise in carbon emissions, while a similar increase in non-renewable energy consumption results in a 1.22% increase in emissions. An increase of one percent in the use of REC and BIOENC results in a decrease in carbon emissions of 0.002 and 0.01%, respectively.

5 Discussion

This analysis employs a sophisticated econometric approach to investigate the relationships between carbon emissions, economic growth, renewable and non-renewable energy usage, and biofuel consumption from 2000 to 2020 across a panel of nations with significant biofuel intensity. Now, to summarise the empirical facts of data using these applied econometric tools; While passing through the unit root tests report, we can perceive their variations were not stationary at their levels, and it doesn't refer to both ways every time it attains stationarity at their first differences. This is inconsonant with the findings of Adewuyi

Table 9 PMG estimation results

Dependent var.	CO_2	GDPPC	NREC	REC	BIOFC
Long-run coefficients					
CO_2		0.156**	1.215***	– 0.002	– 0.011***
FDPC	– 0.948***		1.313***	– 0.077	– 0.014
NREC	0.702***	0.211***		0.001	0.005*
RENC	– 0.131	0.268***	– 0.612***		0.036***
BIOFC	– 5.067**	8.619**	– 5.779*	– 0.864	
ECT	– 0.83***	– 0.907***	-0.446***	– 0.647***	– 0.493***
Short-run coefficients					
ΔCO_2		0.152	0.015	– 0.056	– 0.021
$\Delta GDFFC$	0.174***		0.107	0.064	0.019
$\Delta NREC$	0.009	0.291***		0.011	– 0.017**
ΔREC	– 0.752*	– 0.344	1.082		0.035
$\Delta BIOFC$	6.086***	– 4.299	– 1.399	4.362*	

Note: ***Significant at 1% level; ** significant at 5% level; * significant at 10% level.

and Awodumi 2017. Secondly, the results of the Pedroni and Kao tests confirmed the presence of long-run co-integration among the variables. *Third*, panel regression analysis demonstrates that carbon emissions are positively linked to GDP per capita and non-renewable energy use, while renewable energy and biofuel consumption exhibit an inverse relationship with emissions. *Finally*, the Pooled Mean Group (PMG) panel ARDL method was employed to analyse both short-term fluctuations and long-term equilibrium relationships in the estimates. Results revealed that per capita GDP and non-renewable energy consumption impact carbon emissions positively; however, renewable energy and biofuel consumption impact carbon emissions negatively. Consistent with prior empirical work, the findings validate the economic theories justifying the utilisation of renewable energy, biofuels, and conventional energy sources (Mensah et al., 2019; Bilan et al., 2019; Farhani and Rejeb, 2012). The empirical estimations presented above yield important policy implications, aiding in the comprehension of the dynamic relationships among the analysed variables. This understanding can guide policymakers in crafting policies informed by these variables. This indicates the necessity for enhanced international cooperation to seek out cleaner and more sustainable energy options, with a focus on renewables and biofuels specifically. Additionally, this work verifies the predictions regarding the growing application of bioenergy as a reliable, sufficient, and sustainable energy source. Liquid biofuels as a form of bioenergy consumption can enhance the connection between energy use and environmental sustainability. The significant outcome from the data analysis indicates that bioenergy shows a higher potential compared to other renewables in mitigating carbon emissions. Less greenhouse gases, which cause climate change at a global level, are released in the countries promoting these types of energy (renewable energy and biofuels). Less greenhouse gases, which cause climate change at a global level, are released in the countries promoting these types of energy (renewable energy and biofuels). The research notes that though the continuation of economic growth with enjoyment of fossil fuel consumption in each country results in increased carbon emissions, renewables and biofuels in particular are critical to providing solutions to this global challenge. It is essential to implement energy efficiency measures and increase investments in renewable energy sources, including solar power, wind power, and natural gas. These actions would increase energy security, reduce pollution, and maintain economic well-being—all without an economic cost.

6 Conclusion and Policy Implications

This study aims to significantly advance and stimulate research on the interplay between carbon emissions, economic growth, and energy consumption, including renewable, non-renewable, and biofuel sources. Fossil fuel burning in automobiles contributes substantially to damaging climate change by releasing carbon dioxide and other greenhouse gases. Global warming is a real hazard because economic expansion fuelled by energy consumption from fossil fuels continues to be a significant source of greenhouse gas (GHG) emissions. Against this backdrop, bioenergy has been a major achievement in the sense that it has enabled nations to strengthen national energy security by substituting significant quantities of imported fossil fuel products with biofuels. The consumption of biofuels holds significant potential for promoting ecological sustainability and enhancing climate security within contemporary energy systems. It is true that there are studies that show the link between climate change, economic performance, and energy use. However, almost none of them have used a variety of modern econometric techniques to look at the effects of bioenergy use on lowering emissions. The present study examined a group of countries with significant bioenergy consumption, utilising data from 2000–2020 to explore this relationship. Estimated results indicate that there is a positive and significant relationship between economic growth and energy consumption with emissions, while renewable energy and bioenergy consumption show a negative relationship with emissions. The estimated results hold considerable practical significance for policymaking in these countries. As a policy recommendation, the governments in the sampled nations should strive to advance climate change mitigation strategies within both the industrial and energy sectors. Our analyses suggest that we can achieve this by integrating biofuels and other renewable sources into the energy portfolio. The findings suggest that nations worldwide should advocate for the adoption of biofuels and strengthen collaboration to foster the sustainability of bioenergy systems in light of increasing global temperatures. Biofuels, as a sustainable, clean, and renewable energy source, play a crucial role in addressing the potential devastating impacts on humanity in the future.

It is suggested that public and private players in these countries should encourage the consumption of bio and renewable energy sources and limit their dependence upon non-renewables. Moving forward, it is essential for policymakers globally to employ suitable assessment methods to influence the implementation of sustainable bioenergy. A full analysis of the effects on the environment must first look at the different features, both beneficial and bad, of the various bioenergy cropping systems. The idea is that for bioenergy production to be sustainable, there needs to be good governance, a lot of public investment, solid scientific research, voluntary certification, and well-thought-out policies that support the long-term use of biodiversity. Through these strategies, bioenergy can contribute to realising its potential to mitigate significant environmental and extreme climatic changes that pose a serious threat to the basic survival of humanity. The main point of our argument is that policymakers and governments around the world need to work together to support the long-term growth of bioenergy systems. This is especially important now that global temperatures are rising and could have terrible effects on people.

References

Alternative fuels data center: Energy independence and security act of 2007. (n.d.). https://afdc.energy.gov/laws/eisa.html

Bilan, Y., Streimikiene, D., Vasylieva, T., Lyulyov, O., Pimonenko, T., & Pavlyk, A. (2019). Linking between renewable energy, CO_2 emissions, and economic growth: Challenges for candidates and potential candidates for the EU membership. *Sustainability, 11*(6), 1528.https://doi.org/10.3390/su11061528

Climate and energy package (2009). (n.d.). European environment agency. https://www.eea.europa.eu/policy-documents/climate-and-energypackage-2009

Elobeid, A., Carriquiry, M., Secchi, S., & Yu, T. (2013). Economics of bioenergy. *Economics Research International, 1*, 1–3.

Farhani, S., & Rejeb, J. B. (2012). Energy consumption, economic growth and CO2 emissions: Evidence from panel data for MENA Region. *DOAJ (DOAJ: Directory of Open Access Journals)*. https://doaj.org/article/43bcc5b6373d4fa1b1e83656a9cbb0ff

Foust, T. D., dougArent, M. B. A., de CarvalhoMacedo, I., Goldembergc, J., Chanakya, H., MacielFilhob, R., Francisco, E. B., Nigro, A. M., Tom L. R., Saddlerf, J., Samsethg, J., & Somerville, C. R. (2015). SCOPE Bioenergy & Sustainability -Energy Security.

IPCC. (2022). Emissions trends and drivers. In: IPCC, 2022: Climate change 2022: Mitigation of climate change. Contribution of working group III to the sixth assessment report of the intergovernmental panel on climate change. Cambridge University Press, Cambridge, UK and New York, NY, USA. https://doi.org/10.1017/9781009157926.004

Khanna, M. (2008). Cellulosic biofuels: Are they economically viable and environmentally sustainable? *Choices, 23*(3), 16–21.

Lau, L., Choong, C., & Eng, Y. (2014). Investigation of the environmental Kuznets curve for carbon emissions in Malaysia: Do foreign direct investment and trade matter? *Energy Policy, 68*, 490–497. https://doi.org/10.1016/j.enpol.2014.01.002

Meijl, H. V., Smeetsa, E., & Zilberman, D. (2015). *SCOPE bioenergy and sustainability—bioenergy economics and policies.*

Mensah, I. A., Sun, M., Gao, C., Omari-Sasu, A. Y., Zhu, D., Ampimah, B. C., & Quarcoo, A. (2019). Analysis on the nexus of economic growth, fossil fuel energy consumption, CO2 emissions and oil price in Africa based on a PMG panel ARDL approach. *Journal of Cleaner Production, 228*, 161–174. https://doi.org/10.1016/j.jclepro.2019.04.281

Mohiuddin, O., Asumadu-Sarkodie, S., & Obaidullah, M. (2016). The relationship between carbon dioxide emissions, energy consumption, and GDP: A recent evidence from Pakistan. *Cogent Engineering, 3*(1), 1–16.

Msangi, S., Sulser, T., Rosegrant, M., Valmonte-Santos, R. (2007). Global scenarios for biofuels: Impacts and implications. *Farmacja Polska, 4*(2), 1–18.

Msangi, S., Ewing, M., & Rosegrant, M. (2010). Biofuels and agricultural growth: Challenges for developing agricultural economies and opportunities for investment, In Khanna et al. (Eds.). *Handbook of bioenergy economics and policy*. Springer New York: USA; 2010.

Muhammad, H., & Muhammad, A. (2015). Energy consumption, economic growth and emission nexus in Pakistan. *Asian Journal of Applied Sciences, 8*, 27–36.

Nogueira, L. A. H., Leal, M. R. L. V., Fernandes, E. C. M., Chum, H. L., Díaz-Chavez, R., Endres, J. M., Mahakhantg, A., Otto, M., Seebalucki, V., Wielen, L. (2015). *SCOPE bioenergy and sustainability—sustainable development and innovation.*

Ozturk, I., & Acaravci, A. (2013). The long-run and causal analysis of energy, growth, openness and financial development on carbon emissions in Turkey. *Energy Economics, 2012*(36), 262–267.

Pedroni, P. (2004). Panel cointegration: Asymptotic and finite sample properties of pooled time series tests with an application to the ppp hypothesis. *Econometric Theory, 20*(03). https://doi.org/10.1017/s0266466604203073

Pimentel, D. (2012). Biofuels cause malnutrition in the world'. In D. Pimental (Ed.), *Global economic and environmental aspects of biofuels*, CRC Press. USA.

REN21. (2013). Renewables 2013 global status report, Renewable energy policy network for the 21st century.

Salahuddin, M., Gow, J., & Ozturk, I. (2015). Is the long-run relationship between economic growth, electricity consumption, carbon dioxide emissions and financial development in Gulf Cooperation Council Countries robust? *Renewable and Sustainable Energy Reviews, 2015*(51), 317–326.

Samuel, A.-S., & Phebe, A. O. (2017). A multivariate analysis of carbon dioxide emissions, electricity consumption, economic growth, financial development, industrialization, and urbanization in Senegal. *Energy Sources, Part b: Economics, Planning, and Policy, 12*(1), 77–84.

Shahbaz, M., Mahalik, M. K., Shah, S. H., & Sato, J. R. (2016). Time-varying analysis of emissions, energy consumption, and economic growth nexus: Statistical experience in next 11 countries. *Energy Policy*, 33–43.

Shahbaz, M. (2012). Lean HH Does financial development increase energy consumption? The role of industrialization and urbanization in Tunisia. *Energy Policy, 40*, 473–479.

Soytas, U., & Sari, R. (2009). Energy consumption, economic growth, and carbon emissions: Challenges faced by an EU candidate member. *Ecological Economics, 68*, 1667–1675.

Taheripour, F., & Tyner, E. W. (2008). Ethanol policy analysis-what have we learned so far? *Choices, 23*(3), 6–11.

Tang, C. F. (2008). A re-examination of the relationship between electricity consumption and economic growth in Malaysia. *Energy Policy, 36*, 3077–3085.

UNEP. (2009). *Towards sustainable production and use of resources: Assessing Biofuels*, UNEP, Paris.

UNEP GRID Arendal. (2011). Accessed March 2019. http://www.grida.no/publications/vg/biofuels/

United Nations. (n.d.). World summit on sustainable development, Johannesburg (2002) | United Nations. https://www.un.org/en/conferences/environment/johannesburg2002

Williams GL, Dahiya A, Porter P. Introduction to Bioenergy, in Dahiya, A. (ed.), Academic Press, London: UK;2015.

Open Access as a Strategic Tool for Stakeholder Engagement in SDGs: A Case Study Approach

Nzeyimana Bahati Shabani, Tiwari Shuchi, G. N. Sai Shankar, and Mariaselvam Sheela Mary

Abstract

Open Access (OA) is presented as a transformative strategy to accelerate the achievement of the United Nations Sustainable Development Goals (SDGs) by democratizing knowledge access and bridging critical information gaps among governments, Higher Educational Institutions (HEIs), Non-Governmental Organizations (NGOs), and the private sector. This chapter investigates the nuanced ways these stakeholders leverage OA initiatives to address specific SDG challenges, emphasizing the strategic importance of OA in fostering collaboration and equitable development. The study reveals that governments utilize open data to enhance evidence-based policymaking, particularly in climate action, public health, and economic development. HEIs promote open research and disseminate findings that drive innovation and community engagement. NGOs employ OA tools to implement data-driven projects in underserved sectors and benefit from OA by accessing collaborative research to develop sustainable practices and eco-friendly technologies, contributing to a circular economy. However, the study also identifies significant challenges to widespread OA adoption, including funding constraints that disproportionately affect researchers in developing countries, legal and intellectual property complexities, and the influences of

N. B. Shabani (✉) · M. S. Mary
PG& Research Department of Environmental Science, Bishop Heber College (Autonomous) Affiliated to Bhrathidasan University, Tiruchirappalli 620017, India
e-mail: bahatishabani.res.es@bhc.edu.in

T. Shuchi
Department of Chemistry, VNR Vignana Jyothi Institute of Engineering and Technology (VNRVJIET), Hyderabad, Telangana, India

G. N. S. Shankar
Department of Ecology with Ecological Informatics, School of Informatics, Kerala University of Digital Science, Thiruvananthapuram, Kerala, India

T. Choudhury et al. (eds.), *Open Access as a Knowledge Platform for Implementing the United Nations Sustainable Development Goals*, World Sustainability Series,
https://doi.org/10.1007/978-3-031-99605-4_17

cultural perceptions on OA adoption. To address these challenges, the chapter underscores the necessity of implementing sustainable funding models, developing standardized OA policies and licensing frameworks, and fostering a culture of open knowledge sharing. The findings emphasize OA's pivotal role in achieving the SDGs by promoting equitable access to information, fostering innovation, and enabling effective collaborations. It concludes by advocating for policy changes, collaborative frameworks, and innovative funding mechanisms to support OA initiatives, ensuring a more inclusive and sustainable future.

Keywords

Open access (OA) • Sustainable development goals (SDGs) • Stakeholders • Higher educations institution (HEIs) • Non-governmental organizations (NGOs) • Private sectors • Knowledge gap

1 Introduction

The United Nations Sustainable Development Goals (UNSDGs) is an international effort to tackle significant challenges such as poverty, hunger, education, climate change, inequality and more (Gil et al., 2019). The SDGs were implemented in 2015, replacing Millennium Development Goals (MDGs), with an excellent framework of 17 SDGs, 169 targets, and 232 indicators (Table 1). However, as the world approaches the 2030 deadlines for the SDGs agenda, the imperative for action is inefficient, as shown in the UN 2024 report that only 17% of all SDGs are on track globally. The potential to achieve these objectives remains unfulfilled due to huge barriers, including a continued absence of access to knowledge and resources, especially in developing countries (Rigg, 2024). At the heart of this challenge lies a critical question: How do we ensure equal access to all key stakeholders-governments, HEIs, NGOs, and the private sector to the information needed to drive meaningful progress of SDGs? In this scenario, OA becomes a panacea and forms a transformative mechanism for breaking barriers to traditional and modern information sharing to scientific research, data, and educational material. OA access provides unrestricted access to research, data, and educational materials, removing barriers to a knowledge gap and sharing. It guarantees academic findings and scientific discoveries, and everyone has free access to technological advancements, encouraging equity, cooperation, and inclusivity (Cudennec et al., 2022).

OA fills gaps that encourage cooperation while minimizing inequalities through the equal contribution of areas and sectors for SDG advancements. For example, free access to peer-reviewed research through repositories like DOAJ and UNESCO's OA platform (United Nations Department of Economic & Social Affairs-Statistics Division, 2021); it is crucial to realize its relevance to different stakeholders to evolve from ad-hoc projects to ecosystem-wide transformation

Table 1 United nations sustainable development goals (UNSDGs) with targets and indicators (UN, 2015)

Goal	Title	Targets	Indicators
1	No poverty	7	13
2	Zero hunger	8	13
3	Good health and well-being	13	28
4	Quality education	10	12
5	Gender equality	9	14
6	Clean water and sanitation	8	11
7	Affordable and clean energy	5	6
8	Decent work and economic growth	12	16
9	Industry, innovation, and infrastructure	8	12
10	Reducing inequalities	10	14
11	Sustainable cities and communities	10	15
12	Responsible consumption and production	11	13
13	Climate action	5	8
14	Life below water	10	10
15	Life on land	12	14
16	Peace, justice, and strong institutions	12	14
17	Partnerships for the goals	19	25

engines and moving beyond traditional models of technology transfer toward co-creation paradigms that value diverse knowledge system equally. Every group plays a unique role in promoting the SDGs, and OA is the glue that connects their work; the main questions our study wants to address are: How do diverse stakeholders (government, HEIs NGOs, and the private sector) utilize OA to address specific challenges in achieving SDGs and what are resultant impacts on SDG progress across different context? What innovative solutions and policy frameworks can be implemented to overcome the key challenges hindering widespread and equitable OA adoption, particularly in resource-constrained settings, and maximize its contribution to sustainable development goals.

Governments make policies based more on evidence when using open data; accurate and timely information is crucial for fighting climate change, improving public health, and maintaining food security. For instance, open repositories of climate data allow policymakers to develop custom adaptation and mitigation measures. Higher Education Institutions (HEIs) (DESA, 2019): HEIs are essential for generating knowledge supporting sustainable development as research and innovation centers. OA increases the impact of academic research by making it accessible to a global audience, encouraging cross-border collaboration, and enabling.

1.1 Non-Governmental Organizations (NGOs)

NGOs frequently work at the grassroots level to address specific SDG goals like education, health, and gender equality. OA gives these organizations the data and research they need to develop effective programs and advocate for systemic change (Borelli et al., 2018). It also ensures that localized knowledge informs global practices.

1.2 Private Sector

Industrial and commercial sectors contribute to realizing the SDGs by applying sustainability innovations and conducting Corporate Social Responsibilities (CSR) initiatives (Sharma et al., 2025). OA would enable businesses to access innovative research. Thus, development in eco-friendly technologies, resource efficiency practices, and circular economy models can be implemented. Funding constraints, intellectual property issues, and infrastructure limits have been addressed as actionable strategies for stakeholders overcoming challenges.

Through OA, disparities brought about by institutional, financial, and geographic barriers are addressed, allowing a wide range of stakeholders to participate in and contribute to the global SDG agenda (Hunger et al., 2022). To create successful interventions, governments rely on evidence-based policymaking backed by publicly available data (WHO, 2018). By expanding the reach of their research, HEIs use OA to foster international collaborations, creative thinking, and incentive innovation. While businesses can access state-of-the-art research promoting eco-friendly technologies and sustainable practices, NGOs utilize OA resources to create focused, data-driven programs (Maphosa, 2021).

This chapter combines theory, practices, and policy recommendations to demonstrate what OA needs for an equitable, sustainable, and collaborative future. It is designed to inspire and energize scholars, policymakers, practitioners, and industry leaders as a core constituent of the world effort toward achieving the SDGs.

2 The Role of Open Access in Advancing Sustainable Development

OA is transformative in supporting global sustainable development by providing equitable and free access to research, knowledge, and data. This work will discuss how OA can achieve the UNSDGs by bridging knowledge through various gaps, aiding innovation, and bringing together through forces-like collaboration on a world scale (Wright et al., 2022). This includes Accessibility, where outputs should be accessible to everyone, cutting through the cost, geography, or institutional privilege walls. Moreover, highly compatible scholarly works, data, and resources should be shared and integrated smoothly across platforms to foster worldwide

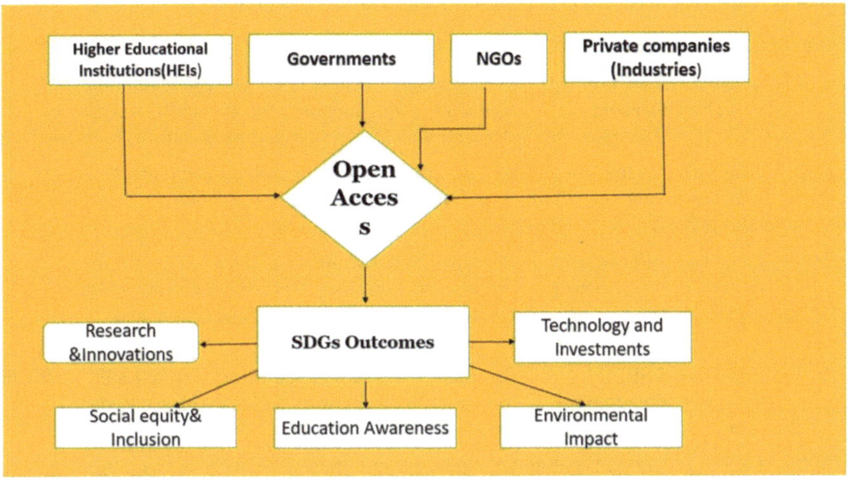

Fig. 1 Open access, a powerful tool to connect key stakeholders to successful SDG outcomes

collaboration and efficiency**.** OA including research output, both peer-reviewed papers, books and reports, and data educational resources with full reuse rights (Fig. 1), illustrates how OA acts as a central mechanism to connect key stakeholders, including Government, HEIs, NGOs and Private companies facilitates the achievements of sustainable development Goals (SDGs) outcomes such as Innovations and research development, Technology and investment, social equity and more, It highlights the crucial of working together as stakeholders from different sectors for advancing the sustainability.

OA enables the reusability of knowledge, which can be reused, republished, and adapted when appropriate crediting is made. The Budapest OA Initiative in 2002 created the basis for a global movement of OA, while the Berlin Declaration on OA in 2003 further solidified the promise of making scientific and cultural heritage available. Such initiatives highlight that open knowledge empowers communities, gives equal learning opportunities, and fosters innovation (Nakicenovic, 2000).

2.1 Timelines in Open Access Publishing and Contributions to SDGs

Making data freely accessible to the general public began in the fifteenth century and continued until the nineteenth century. The German inventor and craftsman Johannes Gutenberg's invention of the printing machine and the establishment of publishing in 1454 transformed the way knowledge was shared. They established the foundation for universal education in line with SDG 4: Quality Education. The condensed time series that highlights significant turning points in the development

Table 2 Timelines in open access publishing and contributions to SDGs

Year	Event / Initiative	Impact / Relevance to SDGs
1454	Invention of the printing press by Gutenberg	Revolutionized access to knowledge, laying the foundation for equitable education (SDG 4: Quality Education)
1665	Publication of the first scientific journal /(Journal des sçavans)	Created the formalized scholarly communication system, supporting SDG 4 (Quality Education)
1991	Launch of ArXiv.org (Open-access repository for preprints)	Provided free, unrestricted access to physics and mathematics research, fostering open knowledge exchange. (SDG 9: Industry, Innovation, and Infrastructure)
1997	Launch of SciELO (Scientific Electronic Library Online) in Brazil	Boosted access to regional research from Latin American countries, reducing inequalities (SDG 10: Reduced Inequalities)
2000	Launch of BioMed central (OA scientific journal publisher)	A significant milestone in biomedical OA publishing, supporting healthcare and public well-being (SDG 3: Good Health and Well-Being)
2001	Budapest OA initiative (BOAI)	Formalized the principles of OA, calling for unrestricted access to scientific research (SDG 4: Quality Education, SDG 9: Industry, Innovation, and Infrastructure)
2003	Berlin declaration on OA	Strengthened global commitment to making scientific knowledge accessible, influencing global SDG efforts

of OA and its relationship to sustainable development objectives is shown below (Table 2).

OA Affecting Social Life Open Access (OA) (https://open-access.network/en/information/open-access-primers/history-of-the-open-access-movement) has far-reaching effects beyond academia, allowing research to be accessed equitably by various societal groups, such as citizen scientists, medical patients, health advocates, NGOs as indicated in (Table 2) and those requiring adapted formats for accessibility. Removing paywalls supports continued lifelong learning by democratizing knowledge, as in Article 27 of the UN Declaration of Human Rights (Asher et al., 2022). This is how an approach to facilitating knowledge use creatively and innovatively can improve public involvement in science. Citizen science initiatives like Zoon Verse and Galaxy Zoo exemplify public contributions to research, emphasizing the need for accessible literature to avoid redundant efforts.

3 The Impact of Open Access on Sustainable Development Goals (SDGs)

See (Table 3).

3.1 Key SDGs Influenced Through Open Access

OA contributed a significant improvement to the advancing the various SDGs by promoting access to information and knowledge that is essential for addressing specific challenges and promoting sustainable development, such as:

3.1.1 Good Health and Well-being (SDG 3)

OA is crucial in improving global outcomes by providing access to the latest medical research, clinical guidelines, and public health data. This enables healthcare professionals to stay informed about best practices, researchers to collaborate on solutions to health challenges, and the public to reliable information correlating with clean water access (Muzigaba et al., 2022). For example, during COVID-19, Technology (OA) contributed to the rapid sharing of research findings, accelerating the development of vaccines and treatments.

3.1.2 SDG 4: Quality Education

OA is instrumental in achieving SDG 4 by democratizing access to educational resources and fostering inclusive learning environments. Openly licensed educational materials, such as Open Educational Resources (OER), provide affordable and accessible learning opportunities, particularly in underserved communities and developing countries where traditional educational resources may be limited. OER Commons site hosts openly licensed educational content in over 50 countries. A UNESCO **report from 2023 showed that 70% of developing countries rely on OA sources** for academic quality content, providing equal access to knowledge and enhancing learning outcomes worldwide. OA enables educators to strengthen their pedagogical approaches and improve learning outcomes. Furthermore, OA facilitates lifelong learning by providing individuals with access to knowledge and learning materials beyond formal education settings. By removing barriers in its policy implementation to educational resources, OA can empower learners of all ages and backgrounds to acquire the knowledge and skills necessary to thrive in a rapidly changing world. Makes education accessible to high-quality learning resources in underserved areas—for instance.

3.1.3 SDG 13: Climate Action

OA encourages climate action through easy access to the critical environmental data required. For example, Copernicus OA Hub offers free access to satellite imagery, which is essential for monitoring the effects of climate change. Climate Watch, 2022 claims, “80% of climate policies in low-income regions were using OA resources to achieve accurate modeling and effective adaptation strategies.”

Table 3 Recent open access publications included SDGs by Stakeholders (2018–2024)

Stakeholder	SDG focus	Work title/case study	References and year	Published In
UN	UNSDGs	SDG good practices	(UN DESA, 2019)	Report
World bank, UNESCO	SDG 1: No poverty	Ensures marginalized communities access poverty-alleviation research for targeted policy-making	(Ajay bang et al. 2021)	Open data for inclusive development
Academia	SDG 2: Zero hunger	Sustainable development goal 2: Improved targets and indicators for agriculture and food security	(Gil et al., 2019)	Springer, Ambio
Government &HEI	SDG 2: Zero Hunger	Impact pathways from climate services to SDG2 ("zero hunger"): A synthesis of evidence	(Hansen et al., 2022)	Elsevier, climate risk management
PLOS	SDG 3: Good health and healthy being	Open-access medical journals ensure equitable access to health research, fostering better outcomes	(Plos, 2019)	Plos for medicine
Government & HEIs	SDG 4: Quality Education	Enhancing educational access through open repositories	(York University, 2022)	York University studies
UN Women and Coar	SDG 5: Gender equality	Are we getting there?	(Barros & Leite, 2023)	Health (San Francisco)
UNESCO, HEI	SDG 6: Clean water and sanitation	Informs sustainable water management and sanitation systems in water-scarce regions	(Nzeyimana & Mary, 2024)	OA to water data
PLOS	SDG 7: Affordable and clean energy	Share research on affordable and renewable energy technologies globally	(Plos, 2023)	PLOS Energy: Renewable Solutions

(continued)

Table 3 (continued)

Stakeholder	SDG focus	Work title/case study	References and year	Published In
DOAJ	SDG 8: Decent work and Economic growth	Supports economic research for policymakers and businesses to promote inclusive growth	(Doaj, 2019)	DOAJ: Economic Growth through Sustainable Development
Industry (Tech Sector)	SDG 9: innovation	OA Data for sustainable tech development	(Doaj, 2020)	DOAJ
World Bank Group	SDG 10: reducing equalities	Supports research addressing social inequality and systemic disparities globally	(World Bank Group, 2019)	World bank
FAO	SDG 11: sustainable cities	Provides open resources to foster sustainable urban planning initiatives	(Sessa et al., 2022)	FAO
Plos, Industry expert	SDG 12: responsible consumption and production	Guides industries in adopting sustainable production models through shared data	(Plos, 2023)	Sustainable supply chains: OA innovations
Climate scientist	SDG 13: climate action	Equips stakeholders with climate science for actionable mitigation and adaptation strategies	(DOAJ, 2023)	Climate focused journal
UNESCO	SDG14:Life below water	Science for the sustainable development goals SDG 14: Life below water	(Zealand, 2023)	UNESCO Science report, IGI global
FAO, NGOs	SDG 15: Life on land	Open data supports the conservation of biodiversity and sustainable land-use policies	(Valters, 2017)	FAO's Forest Data for Biodiversity Preservation
Harvard University	SDG 16: Peace and justice	Public access to policy research for transparent governance	(Harvard University Library 2021)	OA Portal

(continued)

Table 3 (continued)

Stakeholder	SDG focus	Work title/case study	References and year	Published In
UN, PLOS, COAR	SDG 17: Partnership for the Goal	Partnerships for Sustainable Development Goals: A Legacy Review Towards Realizing the 2030 Agenda	(UNEP, 2023)	United Nations Department of Economic and Social Affairs
UNESCO	SDGs in HEI	Assessment of SDGs in HEI	(Caeiro et al., 2020)	UNESCO
Academia	HEI	ESG reporting in HEI China	(Mo & Wang, 2023)	Heliyon

3.1.4 SDG 17: Partnerships for Goals

Shared access has opened up research and data to the public and enhanced partnerships. The African Open Science Platform has bridged knowledge gaps between African countries and supported collaborative research ventures. By 2023, the platform had shown 45% growth in joint projects between African and European institutions. It means that the potential and valuable OA holds the potential to produce fair and great public partnerships. All the shown data point towards a crucial role that OA could play in hastening the progress of many SDGs. Removing barriers to information also promotes innovation and makes knowledge a tool for sustainable development accessible to everyone (Kumar et al., 2025).

The SDGs embraced by the United Nations in 2015 aimed to combat global problems through inclusive strategies, such as fighting against poverty, inequality, and degradation of the environment. It created these goals to be "mutually reinforcing," which means their holistic strategy makes progress in one dimension to benefit the others. Illustrations and statistics in different dimensions amply depict these intertwining effects. The Ways to Achieve **No Poverty (SDG 1)** and **Zero Hunger (SDG 2)** have seen success through sustainable and effective agricultural practices. A clear example of Rwanda's investments in smallholder farming using innovative irrigation techniques and facilities have helped reduce poverty rates by 10% between 2016 and 2020 while boosting food security. Initiatives are tied to **Quality Education (SDG 4)**, as literacy and vocational training empower farmers to adopt modern methods, improving productivity and income.

The WHO reported in 2022 that countries improving **Clean Water and Sanitation (SDG 6)** experienced a 30% reduction in waterborne diseases like cholera and dysentery. These outcomes underscore how environmental goals, such as **Life below Water (SDG 14)** and **Life on Land (SDG 15)**, complement health and economic objectives by preserving ecosystems that communities depend on for clean water and sustainable livelihoods. In energy, the move to Affordable and Clean Energy (SDG 7) has directly enabled Climate Action (SDG 13), as seen in India, which saw renewable energy capacity rise to 174 GW in 2023. This means the country has cut its annual greenhouse gas emission rate by 21% since 2005, which aligns with its Paris Agreement commitments. Such a transition fights climate change and drives Industry, Innovation, and Infrastructure (SDG 9) by creating green jobs and increasing energy security.

Data also illustrate the impact of Gender Equality (SDG 5) on more comprehensive ends. McKinsey Global Institute calculates that increasing gender parity could contribute $12 trillion to global GDP by 2025. Empowerment programs for women, such as microfinance schemes in Bangladesh, have raised millions out of poverty and improved health and education results. Global partnerships (SDG 17) will drive these transformations. The COVAX initiative has delivered over 2 billion COVID-19 vaccine doses to 92 low- and middle-income countries, illustrating the power of international cooperation in winning **Peace, Justice, and Strong Institutions (SDG 16) that** are being promoted and are equipped with equal access to healthcare and other facilities. (Fig. 2) shows how OA impacts SDG improvement and serves as a Central hub, drastically contributing to progress in Quality education SDG 4, Climate Action SDG 13, and Good Health well-being SDG 3, which are essential for advancing these goals.

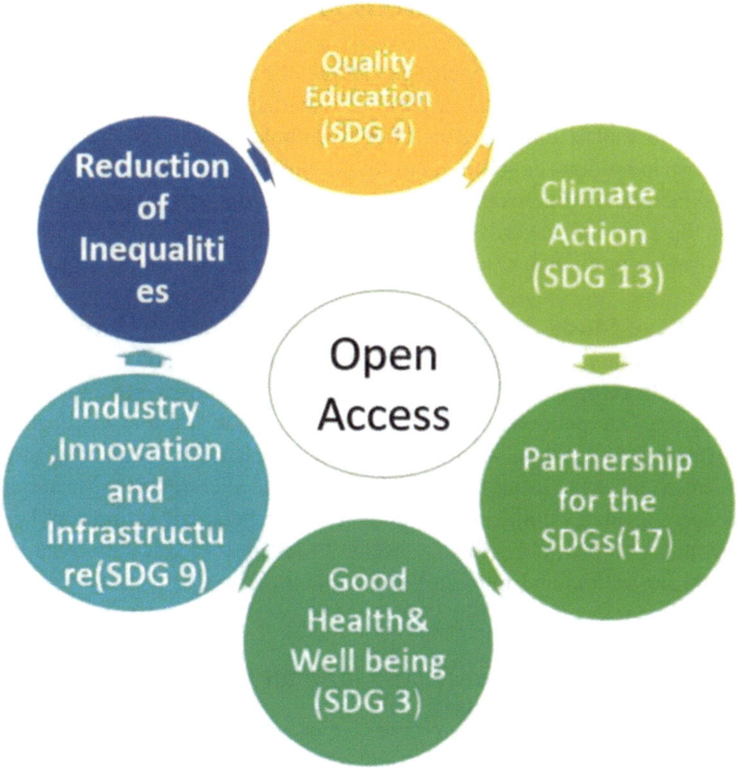

Fig. 2 Open access as a source of sharing data for impacting the advancement of SDGs

3.2 Successfully Global Case Studies of Sustainable Development Advanced

3.2.1 Case Study 1: Global Open Data for Agricultural Development (GODAN)

The Global Open Data for Agriculture and Nutrition (GODAN) initiative exemplifies the renewed power of OA in promoting sustainable agriculture and food security. it contributes primarily to SDG 2: Zero Hunger) GODAN provides agricultural and nutritional data access, enabling researchers, farmers, and policymakers to access and utilize data to advance agrarian practices, increase food production, and solve malnutrition problems.

Through the facilitation of the sharing of data on crop varieties, soil conditions, weather patterns, and market information, GODAN empowers stakeholders to make data-driven decisions that promote productivity increase and improve food security, particularly in developing countries. This case study plays a crucial role in demonstrating how OA contributes to SDG advancement and has reported that GODAN contributed to 15% of agricultural productivity in countries that use its resources, showing the impact of OA on SDG 2. (https://www.godan.info/about).

3.2.2 Case Study 2: Open Access to Climate Data (Open ClimateData.Org)

OpenClimateData.org is a compelling example of how OA could accelerate climate action (SDG 13) by promoting access to essential climate data and information. This platform makes climate data freely available to researchers, policymakers, and other stakeholders, enabling them to use and analyze trends and determine the influence factors. This helps them to develop mitigation and adoption strategies and monitor the progress towards climate goals initiatives. This makes collaboration and innovation in climate research and policy implementation possible.

3.2.3 Case Study 3: Open Access in Health Research (PLOS)

The Public Library of Science (PLOS) demonstrates the power of OA in transforming health research and contributing to SDG 3(Good Health and Well-being). It fosters the OA publication by providing access to various medical and scientific research, including studies on different diseases, treatments, and public health interventions. Through its policy, PLOS accelerates the dissemination of knowledge, facilitates collaboration among researchers, and empowers healthcare professionals and the public to have access to recent research data. This improves the healthcare sectors and mitigates global health issues.

3.3 Benefits of Open Access in Meeting the Sustainable Development Goals

OA is one transformative tool for achieving the SDGs of democratizing access to knowledge, encouraging innovation, and solid global cooperation. The OA tool helps solve issues concerning policymaking, research work, and communities with well-founded decisions and provides justice-based resources. These OA remove paywalls, thus making all research and data accessible and peer-reviewed, as in the case of GODAN *initiatives,* where access to agricultural data enhances food security. According to the World Bank report, countries' use of GODAN resources improved agricultural productivity rate in 2023 by **15%** compared to the non-usage by those countries.

3.3.1 Innovation Through Knowledge Sharing

We all love free and open-access materials; science says we are more attached to free material; these help researchers, students, and old age ones to find free available material to gather knowledge on Unrestricted access to OA data fuels innovation in addressing climate change, public health, and sustainability issues. OpenClimateData.org has empowered climate scientists to develop more precise mitigation and adaptation strategies. In 2022, 40% of the adaptation projects funded by the UNFCCC relied on research supported by OpenClimate-Data.org. Similar platforms highlight how OA accelerates global progress through collaborative problem-solving and resource optimization.

3.3.2 Support Global Cooperation for Sustainable Development

Encourages equal partnerships by filling knowledge gaps between regions, allowing collaboration in solving global problems. *DOAJ* indexes over 19,000 peer-reviewed journals and has significantly improved international cooperation among scholars. A study carried out in 2023 showed that researchers from low- and middle-income countries published 30% more cross-border articles after access to DOAJ, which underscores the role of OA in promoting inclusive global cooperation.

4 Case Studies of Key Stakeholders Using Open Access for SDG Implementation

This chapter highlights the key case studies influenced by OA implementation in a government initiative, national open data policy, and OA, the platform for public policy research to support SDG-informed decision-making. The Private sector, where we found the advancement of industries and corporate companies, is scaling up their production due to this target of SDGs with enormous data available in both traditional and advanced technology (Biswas et al., 2024). However, we cannot forget the contribution of HEI in shaping the workforce through critical thinking. How OA impacts the environment, health, or education data has helped shape evidence-based policies, especially for SDG targets. It will also demonstrate the Challenges and Solutions related to data standardization, funding for OA platforms, and recommendations for improving government-led OA initiatives.

4.1 The Indian Government Initiative Towards Open Access

The Indian government's National Data Sharing and Accessibility Policy (NDSAP), launched in 2012, promotes the proactive release of non-sensitive data collected by various government departments. Using the Open Government Data (OGD) Platform India, this initiative offers free access to essential datasets influencing policymaking in health, education, environment, and urban development (USAID Global Waters, 2020). These enable data-driven research and decision-making that goes a long way in helping towards the Sustainable Development Goals (SDGs). For example, for **SDG 3** (Good Health and Well-being), the monitoring of disease outbreaks and the assessment of healthcare policies are done through Ministry of Health data. **SDG 4 (Quality Education)**: Datasets on school enrollment rates help policymakers identify gaps in education access. **SDG 13 (Climate Action)**: Environmental datasets, such as sustainable water management, guide climate change adaptation strategies. This has shown the significant impact of shaping evidence-based Policies on Health Sector policies where real-time health data were available during the COVID-19 pandemic, allowing for swift policy actions, such as region-specific lockdowns and vaccination drives. It also empowered researchers and NGOs to create predictive models for healthcare resource allocation.

4.2 Role of HEIs as Knowledge Hubs in Both Developing and Developed Countries

Open education data has improved initiatives like **Samagra Shiksha Abhiyan** by identifying underperforming districts and allocating targeted resources to improve outcomes. Some case studies are examples of a university-led OA project on climate action or sustainable agriculture research (Aravind, 2024). How have the university's open research findings supported community efforts in sustainability or enabled partnerships with local governments and NGOs? (Table 4).

Table 4 indicates that the above SDGs have been impacted positively by OA globally and in the Indian people context: Several IITs (Indian Institutes of Technology) are actively focusing on implementing Sustainable Development Goals (SDGs) and promoting OA research. With its Department of "Energy Science and Engineering" and its open-access publications, IIT Bombay underlines SDG no. 7 (Affordable and Clean Energy), SDG no.9 (Industry, Innovation and Infrastructure), and SDG no.13 (Climate Action). One of the projects," Open Database on Smart Cities to Create a Better Urban Ecosystem," for SDG 11(Sustainable Cities and Communities). This and some outreach projects funded by the institute are why we are still open to accessing our research and the central SDGs IIT Delhi. IIT Kanpur focuses on SDG 6 (Clean Water and Sanitation), SDG 9, and SDG 13 by making its water management and environmental sustainability research OA. IIT Madras focuses on SDG 2 (Zero Hunger), SDG 7, and SDG 13 by disseminating OA to renewable energy and sustainable agriculture. IIT Kharagpur focuses on

Table 4 Key metrics and indicators for the SDGs 4, 6, 7, 9, 12, and 13 have significant influence by open access (Aayog, 2024)

SDG Goals	Metric	Global value (2022–23)	India value (2022–23)
SDG 4	Literacy rate	86% (Global)	74% (India)
SDG 6	Households with clean water access	2 billion lack access	60% of rural households covered under Jal Jeevan Mission
SDG 7	Renewable energy share	~30% of global energy mix	114 GW capacity, 60 GW from solar
SDG 9	FDI in the industrial sector	$1.5 trillion globally	$81 billion inflows in 2022
SDG 12	Plastic recycling rate	9% globally	30% of urban areas recycled
SDG 13	CO_2 emissions	36.6 billion tons (Global)	2.88 billion tons in 2022 (India)

SDG 4, SDG 9, and SDG 11 by considering open-access research areas, including infrastructure, water management, and education. IIT Guwahati focuses on SDG 6, SDG 13, and SDG 15, known as Life on Land, emphasizing environmental sustainability and open-access publications. IIT Roorkee focuses on SDG 9 and also on SDG 11 and SDG 13, hence providing support for open-access research in urban sustainability, climate action, and infrastructure. Lastly, IIT Ropar focuses on SDG 7, SDG 9, and SDG 13, advancing renewable energy research and encouraging open-access dissemination. These IITs are making significant contributions to SDG achievement through research, innovation, and the promotion of OA to knowledge, advancing sustainability in India and globally-the role of HEIs in bridging the gap between academic research and real-world applications for sustainable development (https://www.niti.gov.in/reports-sdg), (Table 5).

4.3 Non-Governmental Organizations (NGOs) and Civil Society Organizations

Various Non-governmental organizations (NGOs) and civil society organizations have contributed significantly to India's healthcare, education, environmental protection, and social welfare. A case study of an NGO utilizing open-access resources shows how organizations can take data-driven projects to tackle environmental, healthcare, or educational issues. By using OA resources, these organizations can access critical research, cooperate internationally, and monitor progress toward specific SDG indicators, making more informed decisions and contributing better to sustainable development. However, challenges exist in making OA resources accessible to everyone in remote areas (Hansen et al., 2022). To address this, collaboration and initiatives that make OA resources accessible to marginalized communities must be undertaken to ensure everyone benefits from

Table 5 Indian institutes projects and their alignment with sustainable development goals (SDGs)

S. No.	Institute	Project	SDGs	References
1	Indian institute of science (IISc), Bengaluru	Initiatives on renewable energy technologies and climate-resilient agriculture	**SDG 7: Affordable and clean energy** Advances in renewable energy ensure access to sustainable, clean energy **SDG 13: Climate action** Climate-resilient agriculture aids in mitigating and adapting to climate change **SDG 2: Zero Hunger** Promotes sustainable agriculture to enhance food security	http://dccc.iisc.ac.in
2	Jawaharlal Nehru University (JNU), Delhi	Focuses on sustainable water management and agriculture	SDG 6 (Clean Water and Sanitation), SDG 2 (Zero Hunger) and SDG 12 (Responsible Consumption and Production) Both work synergistically to ensure water-efficient and resilient agricultural practices	https://www.jnu.ac.in/ses-area-iii-research
3	Indian Agricultural Research Institute (IARI), New Delhi	Research on crop genetics and sustainable farming practices	**SDG 2: Zero hunger** (end hunger, achieve food security and improved nutrition, and promote sustainable agriculture) and **SDG 12: Responsible consumption and production** (ensure sustainable production patterns)	https://krishikosh.egranth.ac.in/
4	University of Hyderabad, Hyderabad	Emphasis on biodiversity and climate adaptation, available in its institutional repository	**SDG 13: Climate action** and **SDG 15: Life on land**. These goals focus on addressing climate change, protecting biodiversity, and promoting sustainable land use	http://www.uohyd.ac.in/

(continued)

Table 5 (continued)

S. No.	Institute	Project	SDGs	References
5	Banaras Hindu University (BHU), Varanasi	Research on organic farming and water conservation	**SDG 2 (Zero hunger)** and **SDG 6 (Clean water and sanitation)** focus on sustainable agricultural practices, improving food security, and enhancing water management for long-term environmental health	http://bhu.ac.in/
6	Anna University, Chennai	Dedicated to clean water technologies and sustainable urban planning	**SDG 6: Clean water and sanitation**, which works mainly on ensuring the availability and sustainable management of water and sanitation for all; **SDG 11: Sustainable cities and communities**, which aims to make cities and human settlements inclusive, safe, resilient, and sustainable	https://www.annauniv.edu
7	Amity University, Noida	Study on climate change mitigation strategies	**SDG 13: Climate action**, which focuses on taking urgent action to combat climate change and its impacts	https://www.amity.edu/
8	Aligarh Muslim University (AMU), Aligarh	Studying the socio-economic impacts of climate change on agriculture	**SDG 2: Zero hunger**, which aims to end hunger, achieve food security, improve nutrition, and promote sustainable agriculture	https://www.amu.ac.in/

(continued)

Table 5 (continued)

S. No.	Institute	Project	SDGs	References
9	Kerala agricultural University (KAU), Thrissur	Concentration on agroforestry and climate-resilient farming practices	**SDG 13: Climate action** and **SDG 15: Life on land**. These practices help mitigate climate change, enhance biodiversity, and promote sustainable land management	https://kau.in/
10	University of Jammu	Supports sustainable practices in Kesar (saffron) and honey production,	**SDG 2: Zero hunger** and **SDG 12: Responsible consumption and production**. Production, fostering agricultural innovation, improving local livelihoods, and encouraging eco-friendly production methods	https://www.jammuuniversity.ac.in/

it. The work of NGOs and CSOs would continue to contribute significantly to achieving the SDGs. Here are some examples of leading NGOs and CSOs in India (Table 6):

4.4 Private Sector and Academia-Industry

In response, the private sector and industry increasingly adopted OA to promote sustainable innovation and push toward the Sustainable Development Goals. For instance, an example is a corporate initiative or an industry-funded open-access project aimed at developing low-carbon technologies with the help of research institutions. With such collaborations, companies can avail themselves of advanced research, expedite the development of eco-friendly technologies, and contribute to the sustainability practices within an industry.

In such a way, OA has enormously helped these organizations by bringing together R and D activities aligned with SDG's goal, improving collaboration, and promoting sharing of data and solutions that contribute to environmental sustainability. However, private sector OA can be slowed due to intellectual property concerns and competition dynamics. Balancing open sharing with proprietary interests remains a key challenge for which innovative solutions are needed- for example, adaptive licensing models and collaborative frameworks that protect business interests while facilitating transparency and shared progress toward sustainable outcomes. (Fig. 3) provide the key similarities in the publication in open access, particularly in SDG 7. Both statistics show affordable and clean energy and

Table 6 Notable NGOs and CSOs in India driving progress towards achieving the SDGs

S. No	NGOs/CSOs	Project	SDGs addressed	Resources
1	Kailash Satyarthi Children's Foundation New Delhi, India	Works towards eradicating child labor and promoting education for every child	**SDG 4:** Quality education—ensuring equitable education and promoting lifelong learning opportunities for all **SDG 8:** Decent work and economic growth sustainable economic growth, employment, and decent work for everyone, focusing on eliminating the curse of child labor **SDG 16:** Peace, Justice, and Strong Institutions Protecting children from exploitation	https://satyarthi.org.in/
2	SEEWA (Social services organization in Hyderabad, Telangana	A women-led organization empowering underprivileged communities with a pan-India presence through a vast network of volunteers	**SDG 5: Gender equality**: Empowering women and promoting gender equality in leadership and decision-making **SDG 1: No poverty**—addressing the needs of underprivileged communities and working towards poverty alleviation **SDG 10: Reduced inequality**: focusing on reducing inequality and promoting social inclusion for marginalized groups **SDG 4: Quality education**: improving access to education for underprivileged sections, particularly women and girls **SDG 8: Decent work and economic growth:** Providing economic opportunities and improving livelihoods for the underprivileged	https://seewa.org/

(continued)

Table 6 (continued)

S. No	NGOs/CSOs	Project	SDGs addressed	Resources
3	The Akshaya Patra foundation	Provides nutritious meals to schoolchildren in India, focusing on improving educational outcomes and combating malnutrition	**SDG 2: Zero hunger:** tackling malnutrition and ensuring access to nutritious food **SDG 4: Quality education:** improving educational outcomes by better health and healthy nutrition for students **SDG 3: Good Health and Well-Being:** Promoting and awarding children's overall health is crucial for their growth and learning	https://akshayapatra.org/
4	Art of living foundation	Offers wellness programs and conducts humanitarian initiatives to uplift rural communities	**SDG 1: No poverty:** working to improve the livelihoods and well-being of rural communities, these initiatives help reduce poverty **SDG 3: Good health and well-being**: offering wellness programs directly contributes to better health outcomes **SDG 10: Reduced inequality:** these programs help bridge the gap between rural and urban communities, addressing social and economic inequalities **SDG 11: Sustainable cities and communities**– promoting sustainable development in semi-rural and rural areas supports the creation of resilient, better, and inclusive communities	https://www.artofliving.org/

(continued)

Table 6 (continued)

S. No	NGOs/CSOs	Project	SDGs addressed	Resources
5	Sadguru foundation	Support the foundation's primary mission of promoting a more sustainable, health-conscious, and environmentally aware society	**SDG 3: Good health and well-being**—through initiatives like Isha's health programs and campaigns for mental and physical well-being **SDG 6: Clean water and sanitation**—the foundation promotes water conservation through the **Rally for Rivers** initiative, aimed at rejuvenating rivers in India **SDG 13: Climate action**—the foundation actively works on environmental conservation through programs such as tree planting and sustainable agriculture **SDG 15: Life on land**—through efforts like the **Cauvery Calling** project, aimed at promoting soil regeneration and tree plantation **SDG 4: Quality education**—the foundation's educational initiatives, like the Isha Vidhya schools, focus on providing quality education to underprivileged children	https://sadgurufoundation.com/

(continued)

Table 6 (continued)

S. No	NGOs/CSOs	Project	SDGs addressed	Resources
6	SEEDS India (Sustainable Environment and Ecological Development Society)	**Focus** on Disaster resilience and sustainable development, training local communities in disaster risk reduction and recovery, particularly in flood-prone and disaster-stricken areas	**SDG 1: No poverty.** by improving disaster resilience and supporting recovery efforts, SEEDS helps reduce the vulnerability of impoverished communities **SDG 9: Industry, innovation, and infrastructure**: through the construction of climate-resilient infrastructure, SEEDS contributes to sustainable and innovative solutions for vulnerable communities **SDG 11: Sustainable cities and communities:** SEEDS focuses on disaster risk reduction and building resilient communities, ensuring they can hold or withstand future disasters **SDG 13: Climate action**: promoting climate-resilient practices and training communities in disaster preparedness, SEEDS works towards mitigating and adapting to climate change	https://www.seedsindia.org/

(continued)

Table 6 (continued)

S. No	NGOs/CSOs	Project	SDGs addressed	Resources
7	Smile foundation	**Focus on good** education and accessible healthcare for underprivileged children and families	**SDG 4: Quality education**—through programs like "Mission Education," which focuses on educating underprivileged children **SDG 3: Good health and well-being**—through "Smile on Wheels," a mobile healthcare initiative that brings medical services to rural and underserved populations **SDG 1: No poverty**—by providing livelihood opportunities for marginalized communities **SDG 10: Reduced inequalities:** by supporting marginalized communities and working to reduce disparities in access to good education, accessible healthcare, and economic opportunities	https://www.smilefoundationindia.org/
8	Pratham education foundation	Pratham focuses on improving the quality of education in underprivileged areas across India. They use innovative learning models such as the Annual Status of Education Report (ASER) and the "Read India" program to bridge learning gaps, improve literacy, and increase school enrollment	**SDG 4: Quality education**—by improving the quality of education, bridging learning gaps, enhancing literacy, and increasing school enrollment rates **SDG 10: Reduced inequality**—focusing on education for underprivileged areas, helping reduce educational disparities **SDG 1: No poverty**—empowering children through education reduces poverty in marginalized communities	https://www.pratham.org/

(continued)

Table 6 (continued)

S. No	NGOs/CSOs	Project	SDGs addressed	Resources
9	Goonj	Disaster relief and community development through resource recycling	**SDG 1: No poverty** providing underserved rural areas with necessities like clothing, sanitation, and infrastructure **SDG 3: Good health and well-being**—improving sanitation and hygiene directly impacts community health **SDG 8: Decent work and economic growth** creating rural employment through its "Cloth for Work" program **SDG 12: Responsible consumption and production:** promoting recycling and reducing waste through reusing urban cloth materials **SDG 13: Climate action**—contributing to sustainable practices by reducing textile waste and utilizing recycled materials **SDG 11: Sustainable cities and communities**: building resilient communities through disaster relief and infrastructure development	https://goonj.org/
10	HelpAge India	HelpAge-India mainly focuses on improving the lives of senior citizens by advocating for their rights and laws for them	**SDG 3: Good health and well-being:** providing healthcare services, organizing health camps, and promoting elderly welfare programs **SDG 10: Reduced inequalities:** advocating for the rights of senior citizens and working to create a more inclusive society **SDG 16: Peace, justice, and strong institutions**—ensuring the protection and rights of senior and older people through advocacy and policy engagement	https://www.helpageindia.org/

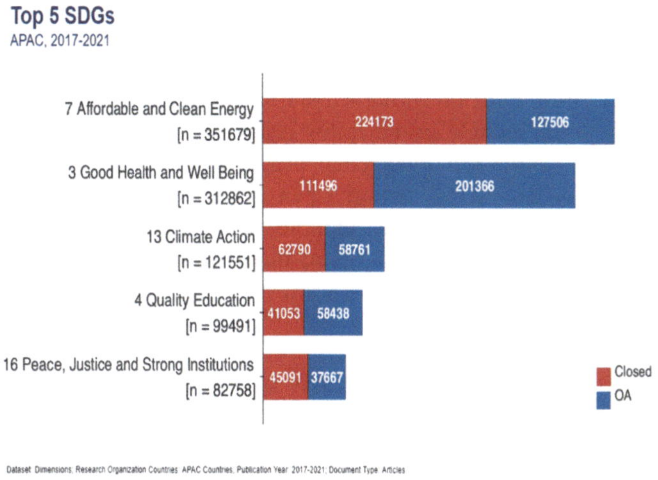

Fig. 3 Top 5 SDGs published in Asia–Pacific (APAC) and Nanyang technological university NTU from 2017–2021 (https://www.dimensions.ai/blog/moving-towards-an-open-access-future/)

good health and well-being. This reveals that the global and local areas cooperate towards the 2030 agenda.

4.5 Industries as the Catalysts for a Better and Sustainable Future

Industries have the immense potential to raise global sustainability by serving as agents of change development and platforms for innovation. Those roles extend beyond economic development to encompass environmental stewardship and societal well-being. By integrating green technologies, fostering partnerships with Higher Education Institutions (HEIs), and adopting robust Corporate Social Responsibility (CSR) initiatives, industries can realign their operations with the more significant aims of the United Nations Sustainable Development Goals (SDGs). Innovations from **Tesla's electric vehicles to intelligent factories by Siemens** demonstrate how technology can decouple economic growth from environmental degradation. The adoption of circular economy principles further underlines that industries can close their loops on waste and, through by-products, make a valuable resource. Models of facilitating these partnerships are also found in many harmonized regulatory frameworks, such as the European Union's Horizon Europe program. OA platforms also democratize knowledge, enabling researchers and industries worldwide to share insights and aiding progress. The way to industrial sustainability will call for a multi-stakeholder approach. Leading a culture of innovation and collaboration, industries could lead the way in addressing joining these technologies, HEI partnerships, and CSR, which is about more than just mitigating lousy impact. Impacts on reimagining the role of industries in society. By leveraging their immense capacity for innovation and influence, the industry and the private sector progressively appreciate the value of OA in advancing sustainable innovation and moving toward the SDGs. These partnerships facilitate businesses to bank on cutting-edge research, expedite the introduction of green technology, and include sustainability principles into business operations.

However, competition and intellectual property issues can make OA adoption problematic in the private sector. Open sharing must be weighed against proprietary interests, and innovative solutions in the form of adaptive licensing arrangements and collaborative mechanisms that advance business interests while supporting openness and joint progress toward sustainable objectives are necessary.

5 Challenges and Limitations in Expanding OA for SDGs and It is Solutions

While OA holds immense potential to facilitate the achievement of SDGs, its equitable and mass implementation is hindered by several challenges. The challenges must be overcome by solutions specifically tailored to the unique conditions of developed and developing countries.

5.1 Challenges Confronting Developing Countries on Open Access

Developing nations tend to experience specific and heightened challenges in embracing OA, further marginalizing their research communities and constraining their input to the global knowledge economy.

5.1.1 Funding Drawbacks and Infrastructure Limitations

Developing countries often face inadequate research and development budgets, which equates to limited funds for creating and maintaining OA infrastructure. This includes having insufficient funding for digital repositories, reliable internet connection, and technological knowledge to enable OA activities. Most research institutions and universities in these nations may lack the financial capacity to pay OA publication fees (APCs) or subscribe to OA journals, significantly hindering researchers from publishing and accessing OA material. Sub-Saharan Africa severely lacks finance for digital libraries and sustainable server infrastructure, hindering OA initiatives' growth.

5.1.2 Digital Divide and Technological Disparities

The disproportionate application of information and communication technologies via the digital divide is a foundational challenge to OA adoption in the global south. Socioeconomic characteristics like distant internet connectivity, particularly in rural towns, deter the use of OA materials and researchers who are part of the

OA movement. Furthermore, unequal access to technology and digital literacy may constrain researchers from creating and disseminating OA content.

5.1.3 The Legal and Policy Structures

Most developing nations do not have open and conducive legal and policy frameworks for OA, causing uncertainty and discouraging its adoption. Copyright legislation and intellectual property rights cannot be very supportive of OA, and there could be a lack of policies that require or encourage OA publishing. This can discourage institutions and researchers from adopting OA and restrict the development of OA initiatives.

5.1.4 Brain Drain and Lack of Recognition

Developing country researchers can also be deprived of the prestige and recognition of publishing in OA journals, which are sometimes viewed as lower quality than their traditional subscription-based counterparts. This can lead to the "brain drain," in which able researchers migrate to developed nations for better opportunities to work where OA is more established and recognized.

5.1.5 Language and Cultural Barriers

Language barriers also hinder access to and involvement in the worldwide OA community since most of the available OA content is in English. Cultural differences in publishing traditions and etiquette can also slow the uptake of OA in certain developing nations.

5.2 Limitations in Expanding Open Access

5.2.1 Funding and Sustainability

The creation and maintenance of OA resources require significant funding. For example, digitizing, hosting, and updating repositories require robust infrastructure, technical expertise, and consistent funding. This is highly challenging in developing regions, where resources are already thin. There are some case points in sub-Saharan Africa where many universities and research institutions do not have funds to finance digital libraries and sustain the servers required. There are initiatives such as African Journals Online to fill the void; however, they are heavily donor-funded and, therefore, not a long-term sustainable solution. Some OA journal models charge APCs, dissuading researchers in low-income areas from producing and publishing their output. Some platforms now have fee waiver provisions for PLOS ONE, but the policies are inconsistent; many researchers cannot pay the publishing fee.

Potential Impact: Without sustainable funding mechanisms, OA will be an inequitable model where better-resourced regions can overrule other areas, thus violating the ethos of inclusivity ingrained in SDGs.

5.2.2 Legal and Intellectual Property (IP) Issues

There is no doubt that OA can transform the way knowledge is shared, but the legal complexity surrounding it deters stakeholders from embracing it in its entirety. The IP laws and copyright policies vary widely across countries, creating a patchwork and uncertain environment for researchers, publishers, and institutions. This makes the OA movement challenging and slows its progress toward gaining the Sustainable Development Goals (SDGs).

5.2.3 The Complexity of Copyright and Licensing

At the heart of OA is the principle that knowledge should be freely available to everyone. However, the reality is far more complicated. Copyright laws, designed to protect the creators of intellectual property, often work against the open dissemination of knowledge. Researchers, particularly those in developing countries, usually face a dilemma: Should they publish their work in traditional subscription-based journals that are more prestigious but restrict access, or opt for OA platforms that make their research freely available but come with risks.

5.2.4 Misunderstandings and Misuse of Licenses

Even when researchers decide to go the OA route, the challenge does not end there. Licensing tools, such as Creative Commons (CC) licenses, ensure that OA content is used appropriately. However, these licenses can be misunderstood or misapplied. For example, a CC by license allows anyone to use, distribute, or modify a work as long as the original author is credited. While this may sound simple, not all stakeholders understand the nuances. Others will misapply the freely available resources for commercial purposes and fail to apply the licensing requirement, while others will over-constrain access inappropriately due to poor license choice.

A Common Misuse Example: Some non-profit-making institutions and training centers download an OA article on different subjects to evolve a training manual for local communities. However, they do not assign the original authors, which breaches the CC license terms of copyright. Although their intention was noble, their failure to attribute authors invoked a breach of copyright law. This may eventually lead to legal suits that deter some researchers from publishing their papers as OA.

Without a well-defined and implementable framework globally, the vagueness associated with legal jurisdiction discourages the adoption of OA. Researchers fear losing control over their work, publishers worry about loss of revenue, and institutions have no frameworks for the guidance of their stakeholders through copyright and licensing complexity.

5.3 Wider Impact of Open Access in Speeding up Sharing Data

Research Hoarding: Scientists may want to lock their results behind paywalls, which can deny critical access to knowledge for developing SDGs like Good Health and Well-Being (SDG 3) or Climate Action (SDG 13). Reduced Collaboration: The threat of litigation acts as a disincentive for transborder and interdisciplinary collaboration, thereby delaying the work of global research endeavors. Inequity: The complexity of the law will fall more on the shoulders of developing country researchers and institutions, worsening the digital divide.

5.3.1 Towards a Brighter Way Forward

A way forward requires cooperation at the global level, yet action needs to be taken locally. The creation of standardized policies to ease the landscape of OA should come from concerted efforts between governments, international organizations, and advocacy groups. Models for harmonization in OA policy have been found in initiatives like the Budapest OA Initiative and Plan S; their scope has to be broader. Education is just as vital. This will allow researchers like Dr. Arjun to share their work freely without fearing exploitation in the name of copyright or licensing. Workshops, clear guidelines, and institutional support systems would achieve this. The result is not about removing barriers alone but creating a culture of trust and collaboration, whereby knowledge is free to flow for all benefits.

5.3.2 Cultural and Institutional Resistance

Another obstacle to OA is institutional and cultural inertia. Most research and academic institutions are still bound by the traditional publishing paradigm, which equates high-impact subscription journals with the only measure of quality and prestige. In some regions of Europe, researchers are evaluated chiefly based on what they have published in high-impact journals like Nature or Science. This creates a conflict of interest, as these journals are often behind paywalls, discouraging researchers from publishing on OA platforms. Moreover, specific revenue-dependent sectors, such as pharmaceutical industries, perceive OA as threatening their proprietary research and development models. This resistance slows down the pace at which critical information, such as data on drug development, is made publicly available—a serious impediment to achieving SDG 3 (Good Health and Well-Being). The reluctance to embrace OA perpetuates inequities in knowledge dissemination, with well-funded institutions continuing to dominate global research output (Fig. 4).

Graphically illustrated above, the Scholastica survey conducted in the United Kingdom(UK) highlights the prominence of fully OA models, with 80% of publishers adopting fully OA publishing. Hybrid and Green OA models are less commonly used, with only 30% of respondents implementing these approaches. This fits within the larger pattern of promoting entirely OA journals as the future

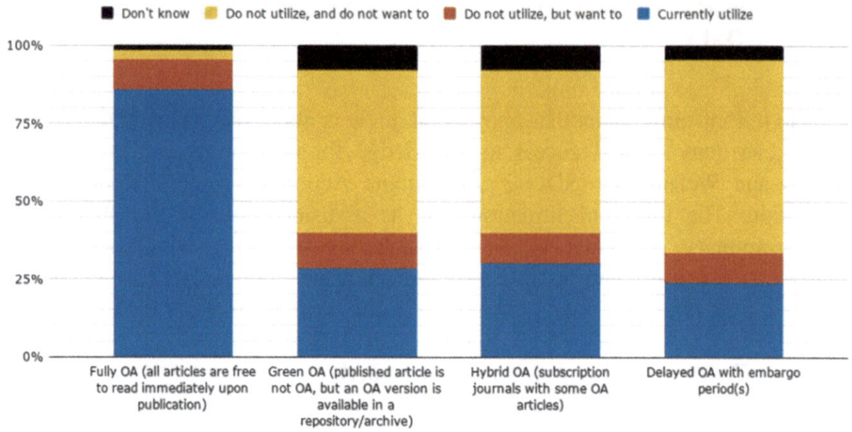

Fig. 4 How the scholar and scientist use journal options to publish their work. (https://blogs.lse.ac.uk/impactofsocialsciences/2020/10/21/open-access-is-here-to-stay-but-who-will-pay/)

gold standard for learned societies and universities (.https://blogs.lse.ac.uk/impactofsocialsciences/2020/10/21/open-access-is-here-to-stay-but-who-will-pay/). Stats are key to monitoring progress towards sustainable development goals. Statistics on SDGs, Climate and development assistance report data from the World Bank, and other leading indicators of the Progress of SDGs are tracked country-wise on the UN SDG Tracker.

International organizations such as the Global Health Data Exchange release health data to the public so that people become aware and follow up on the health-related SDGs. These are low-cost, high-efficiency portals where policy-makers and researchers can get an understanding of (follow and track) the multi-country efforts that the world has put in. Data may still be used to guide the decision-making necessary to ensure that our progress is back on track to reach the SDGs.

6 Strategies and Recommendations for Enhancing Open Access for SDG Achievement

Expanding OA to support SDGs requires a multi-pronged approach involving policy changes, collaborative models, and innovative funding mechanisms. Below are actionable recommendations for fostering a culture of OA.

6.1 Policy Recommendations for Stakeholders

Global Policy Harmonization: Policy efforts should focus on harmonizing OA policies at the global and regional levels. This will involve international collaboration to unify standards, guidelines, and laws that facilitate seamless OA implementation and interoperability. Incentivizing OA Adoption: Governments and institutions must enact policies encouraging OA adoption by researchers, including OA publishing in research assessment measures and funding OA projects.

Closing Equity Gaps: All new policies must aim to bridge equity gaps in OA, particularly for researchers and institutions from developing countries. This involves providing special aid for infrastructure development, capacity development, and economic assistance to enable them to engage fully in the OA environment.

Governments, Higher Education Institutions (HEIs), NGOs, and industries must champion policies prioritizing OA.

6.1.1 Governments

Governments can establish mandates requiring publicly funded research to be openly accessible. For example, the **European Commission's Horizon 2020** program mandates that all research outputs be OA. Similar policies in developing countries could accelerate knowledge dissemination.

6.1.2 HEIs

Universities can include OA publications as valid faculty evaluations and promotion metrics. For instance, **Harvard University** encourages its researchers to publish in OA journals through institutional repositories.

6.1.3 NGOs

Organizations like **UNESCO** should launch advocacy programs for OA in low-income regions, providing technical support and training for researchers.

6.1.4 Industries

Corporates can incorporate OA into their Corporate Social Responsibility (CSR) strategies, contributing data or funding OA platforms as part of their commitment to SDG 17 (Partnerships for the Goals). By embedding OA into institutional frameworks, stakeholders can collectively shift norms and practices toward greater inclusivity and collaboration.

6.2 Collaborative Frameworks and Partnerships

Effective OA expansion requires partnerships between public and private sectors and international collaborations.

6.2.1 Public–Private Partnerships (PPPs)

Can bridge resource gaps by pooling expertise and funding. For example, the HINARI Access to Research in Health Programme, led by WHO, collaborates with publishers to provide low-cost or free access to scientific journals in low-income countries. International Collaborations: Cross-border initiatives like Open AI models can connect European research-based outputs to global audiences. International donors and organizations support such models that developing regions can replicate.

This makes local communities participate in the OA process to make the information relevant and accessible. First, research findings must be translated into local languages to engage and be helpful to the local communities.

The **SCIELO Project**, which started in Brazil, showcases how regional collaboration can promote OA. It has expanded across Latin America, Spain, and Portugal, enabling access to thousands of research articles. These partnerships ensure that OA resources are created, sustained, and utilized effectively.

6.3 Innovative Funding Models

Addressing financial barriers to OA requires innovative funding models that distribute costs equitably, such as Government Grants: Governments can appropriate some of the funds allocated to their research budget to support the OA infrastructure. In the U.S., for instance, the National Institutes of Health (NIH) requires that all the research it sponsors be deposited in the free-access PubMed Central repository. *Non-Profit Support*: Philanthropic organizations like the Bill and Melinda Gates Foundation can be pivotal in funding OA initiatives. It has also sponsored COKA (Collaborative Knowledge Amplification), which develops OA resources in the health sector. Industry Sponsorships: Companies may sponsor OA journals or repositories as part of their sustainability activities. Elsevier, for example, is working with institutions to offer free access to critical, valued assets in areas such as renewable energy and climate change. To ensure that trust contributes, exploratory models for other blockchain-based platforms for OA-related micro-funding projects would be established to create accountability and transparency in funding distribution; it will be revolutionary.

7 Conclusion and Future Study Avenue

OA is a critical milestone for enhancing and speeding up worldwide movement toward the Sustainable Development Goals (SDGs), which this paper depicts. OA is a tool that can be used to democratize the distribution and consumption of research and information, thereby allowing their use in decision-making and the pursuit of sustainable development by people, communities, and nations alike. Moreover, although the potential of OA is excellent, the realization of its full potential is not without its challenges. Obstacles such as funding constraints, legal complexities, and cultural resistance pose significant barriers. However, this chapter has demonstrated the strategies that can be overcome through the collaboration of stakeholders these issues through targeted policies, fostering collaborations across sectors, and exploring innovative funding models; stakeholders can create a more inclusive and equitable knowledge-sharing ecosystem that benefits everyone, particularly those in under-resourced regions. Knowledge is the backbone of development. Whether it is climate-resilient agricultural practice development (SDG 13: Climate Action), creating equitable health systems (SDG 3: Good Health and Well-being), or simply pushing for gender equality (SDG 5), access to correct and on-time research plays a critical role. OA helps to break these barriers of old-fashioned publishing by allowing knowledge that is otherwise restricted to costly subscriptions or institutionally well-endowed institutions. For instance, a small healthcare NGO in sub-Saharan Africa may not be able to afford subscription-based medical journals. However, with OA, they may get access to state-of-the-art research on the prevention and treatment of diseases, thus better equipping them to service communities. Similarly, local researchers find their contributions ignored under the traditional system. However, they are given room to become part of global knowledge bases through OA, thus offering solutions to global problems with roots in specific regions.

Funding opportunities, of course, are one of the most pressing challenges facing OA expansion. However, implementing OA resources requires a significant financial investment, including creating, maintaining, and upscaling such resources. That task is even more complex in developing areas, where research agencies operate on highly charged budgets.

Public–private partnerships are the new financing methods that allow the government to collaborate with businesses to jointly support open-access programs, preventing the funding burden from falling entirely on one institution.

For instance, pharmaceutical companies could support OA journals focused on medical research, benefiting public health and industry innovation. Secondly, non-profit support exists: Foundations like the Bill and Melinda Gates Foundation have already demonstrated the impact of funding OA initiatives. Increasing such sponsorship can fill gaps for under-endowed institutions. Research funding and collaborative platforms: Any platform encouraging the public to invest small amounts may tap into their collective power so that communities interested in knowledge sharing can support various OA projects.

- Simplifying the legal Complexity landscape around OA requires sorting out twisted copyright and licensing issues. Uncertainties in these matters deter the majority from fully adopting OA for fear of losing rights to the intellectual property of their work. International efforts must, then, be in the direction of:
- Standardizing OA policies: International treaties could standardize copyright laws, norms, and licensing practices, reducing country-based inconsistencies in significant numbers.

- Educational Initiatives: Several webinars, conferences, and straightforward standards will be offered to make using license tools such as Creative Commons easy for academics and their institutions to share their work.
- Enforcement Mechanisms: Clear policies and enforcement actions help to foster system confidence by preventing OA content from being misused.

7.1 Fostering Cultural and Institutional Change

Cultural resistance to OA, often based on traditional publishing and revenue concerns norms, is another significant barrier. Some institutions equate OA with lower-quality research, while others fear losing financial gains from subscription-based models. Overcoming this requires a paradigm shift, such as Promoting Success Stories by highlighting case studies where OA has driven tangible development outcomes that can help change perceptions. For example, this is witnessed with the HINARI Access to Research in Health program, providing complete and unrestricted access to health journals in low-income countries. Such success stories have proved the vast effects of OA. Governments and funding agencies can publish all the publicly funded research on OA platforms, encouraging wider adoption. Building Institutional Capacity: Universities and research organizations must invest in building internal OA policies and infrastructure, making it easier for their members to participate in the movement. The success stories have proved the large-scale implications of OA. Governments and funding agencies can publish all their publicly funded research on OA platforms and encourage its wider adoption.

7.2 Collaborations and Partnerships

OA thrives on collaboration. No single entity, whether a government, institution, or NGO, can overcome the challenges of OA alone. The pooling of resources, sharing best practices, and creating synergies through partnerships across sectors and borders can be made. For example, the technological and publishing sectors significantly build and manage open-access platforms. OpenAIRE's international model shows the setting up of a coming border crossing project, OpenAIRE, which is also running across Europe. This broader incorporation and implementation of more types would engage Global South actors and accelerate this path. OA: How local communities matter to policy and infrastructure The path to a robust OA ecosystem requires concerted effort and commitment from all stakeholders, including governments, researchers, publishers, industries, and communities. Governments must lead the way by enacting supportive policies and funding mechanisms. Researchers and institutions must champion the cause by prioritizing OA in their work. Publishers need to innovate their business models to align with the principles of openness and equity.

References

Aravind, S. (2024). *Insurance Repositories in India: A Digital Path to Simplified*. https://doi.org/10.38193/IJRCMS.2024.6603

Asher, S., Bhowmick, A., Campion, A., Lunt, T., & Novosad, P. (2022). Big, open data for development: A vision for India. *India Policy Forum, 18*, 1–41.

Barros, P. P., & Leite, J. P. (n.d.). Are we getting there ? *Health (San Francisco)*.

Biswas, S., Adhikary, M., Alam, A., Islam, N., & Roy, R. (2024). Disparities in access to water, sanitation, and hygiene (WASH) services and the status of SDG-6 implementation across districts and states in India. *Heliyon, 10*(18), Article e37646. https://doi.org/10.1016/j.heliyon.2024.e37646

Borelli, S., Conigliaro, M., & Pineda, F. (2018). Urban forests in the global context. In *Unasylva. 69*(250)

Cudennec, C., Sud, M., & Boulton, G. (2022). Governing open science. *Hydrological Sciences Journal, 67*(16), 2359–2362. https://doi.org/10.1080/02626667.2022.2086462

Demeter, M., Jele, A., & Major, Z. B. (2021). The international development of open access publishing: A comparative empirical analysis over seven world regions and nine academic disciplines. *Publishing Research Quarterly, 37*(3), 364–383. https://doi.org/10.1007/s12109-021-09814-9

Gil, J. D. B., Reidsma, P., Giller, K., Todman, L., Whitmore, A., & van Ittersum, M. (2019). Sustainable development goal 2: Improved targets and indicators for agriculture and food security. *Ambio, 48*(7), 685–698. https://doi.org/10.1007/s13280-018-1101-4

Hansen, J., List, G., Downs, S., Carr, E. R., Diro, R., Baethgen, W., Kruczkiewicz, A., Braun, M., Furlow, J., Walsh, K., & Magima, N. (2022). Impact pathways from climate services to SDG2 ("zero hunger"): A evidence synthesis. *Climate Risk Management, 35*(July 2021), 100399. https://doi.org/10.1016/j.crm.2022.100399

Hunger, Z., Data, G. O., Conference, I., Data, O., States, U., Data, O., Charter, A., Nations, U., Bank, W., Summit, F. G., City, N. Y., Nations, U., Assembly, G., Revolution, S. D., Godan, T., York, N., Conference, G. N., Collaboration, S., Ministers, S., … Science, F. (2022). *GODAN Journey*. 7–9.

Kumar, R., Khanna, R., Singh, T., & Goel, R. (2025). Role of information and communication technology in poverty elimination with respect to SDG-2030 agenda. In *Sustainability and sustainable development* (pp. 1–19). Apple Academic Press.

Liang, Q., Tian, K., Li, L., He, Y., Zhao, T., Liu, B., Wu, Q., Huang, B., Zhao, L., Teng, Y., Dwivedi, Y. K., Hughes, L., Kar, A. K., Baabdullah, A. M., Grover, P., Abbas, R., Andreini, D., Abumoghli, I., Barlette, Y., Suzuki, G. (2022). Life cycle assessment on recycled e-waste concrete. *Science of the Total Environment, 926*(June), 170853. https://doi.org/10.1016/j.cscm.2022.e01412

Maphosa, V. (2021). Students' awareness and attitudinal dispositions to E-Waste management practices at a Zimbabwean University. *Journal of Information Policy, 11*, 562–581. https://doi.org/10.5325/jinfopoli.11.2021.0562

Mo, F., & Wang, D. D. (n.d.). Emerging ESG reporting of higher education institutions in China. *Heliyon, 9*(11), e22527. https://doi.org/10.1016/j.heliyon.2023.e22527

Muzigaba, M., Chitashvili, T., Choudhury, A., Were, W. M., Diaz, T., Strong, K. L., Jackson, D., Requejo, J., Detjen, A., & Sacks, E. (2022). Global core indicators for measuring WHO's paediatric quality-of-care standards in health facilities: Development and expert consensus. *BMC Health Services Research, 22*(1), 1–19. https://doi.org/10.1186/s12913-022-08234-5

Nakicenovic, N. (2000). Global greenhouse gas emissions scenarios. *Technological Forecasting and Social Change, 63*(2–3), 105–109. https://doi.org/10.1016/s0040-1625(99)00117-1

Nzeyimana, B. S., & Mary, A. D. C. (2024). Sustainable sewage water treatment based on natural plant coagulant: Moringa oleifera. *Discover Water, 4*(1). https://doi.org/10.1007/s43832-024-00069-x

Plos, T., & Editors, M. (2019). Health and medical research for all: The challenge remains open. *PLoS Medicine, 16*(10), 2017–2019. https://doi.org/10.1371/journal.pmed.1002970

Rigg, J. (2024). The sustainable development goals (SDGS). *The Companion to Development Studies*, 253–257. https://doi.org/10.4324/9780429282348-52

Sessa, R., Ni, J., Martin, G., Lévite, H., Ferrari, C. Z., & Marocchino, C. (n.d.). A Share your comments on this report's objectives and proposed content as outlined above . Do you find the proposed scope comprehensive to analyze and discuss the key issues concerning the role of urban and peri-urban food systems in achieving food.

Nzeyimana, B. S, Mary, & M. S. (2025). Micro / Nano-plastic pollution in aquarium systems : https://doi.org/10.4018/979-8-3693-3447-8.ch010

Sharma, C. P., Malhotra, R. K., & Kumar, R. (2025). The role of industrial policy in achieving sustainable development goals: A case of Uttarakhand. In *Sustainability and sustainable development* (pp. 67–82). Apple Academic Press.

Siviwe, B. (n.d.). *Open access and The Sustainable Development Goals : The Nexus.*By Siviwe International Open Access Week Presentation at UNISA / UP.

UN DESA. (2019). Good practices, success stories and lessons learned in SDG implementation. Sustainable Development Knowledge Platform, March 2020, 1–3. https://sustainabledevelopment.un.org/sdgs/goodpractices

UNEP. (2015). Partnerships for Sustainable Development Goals: A legacy review towards realizing the 2030 Agenda (18 p). United Nations Department of Economic and Social Affairs. https://sustainabledevelopment.un.org/sdinaction/publication/partnerships-a-legacy-review

United Nations Department of Economical and Social Affairs—Statistics Division. (2021). SDG indicator metadata. https://ourworldindata.org/natural-disasters, December 20. https://unstats.un.org/sdgs/metadata/files/Metadata-01-04-01.pdf

USAID Global Waters. (2020). *Water for the World Country Plan: D. R. Congo.* August 2020, 1–6. https://www.globalwaters.org/sites/default/files/wfw_drc_country-plan.pdf

Valters, P. (2017). *Biodiversity for food and agriculture biodiversity for food and agriculture.*

WHO. (2018). *WHO global water, sanitation and hygiene.* 52. https://www.who.int/water_sanitation_health/publications/global-water-sanitation-and-hygiene-annual-report-2018/en/

World Bank Group. (2019). *Sustainable Development Goal 10—Reduced Inequalities: Progress and Prospects. 2018*, 1–15.

Wright, C., Ritter, L. J., & Gonzales, C. W. (2022). Cultivating a collaborative culture for ensuring sustainable development goals in higher education: An integrative case study. *Sustainability (Switzerland), 14*(3). https://doi.org/10.3390/su14031273

Zealand, N. (2023). *Science for the Sustainable Development Goals SDG 14: Life Below Water.* 1–9.

GPSR Compliance

The European Union's (EU) General Product Safety Regulation (GPSR) is a set of rules that requires consumer products to be safe and our obligations to ensure this.

If you have any concerns about our products, you can contact us on ProductSafety@springernature.com

In case Publisher is established outside the EU, the EU authorized representative is:

Springer Nature Customer Service Center GmbH
Europaplatz 3
69115 Heidelberg, Germany

Batch number: 10755329

Printed by Printforce, the Netherlands